THE EVOLUTION OF AIRPORT DESIGN

This is the first book to comprehensively cover the evolution of airport design, from the start of commercial aviation in 1919 to the present day. Many books have been written about airport design at a particular moment in history, but none have rigorously considered why, where, when and how the ideas we now take for granted originated.

This book traces the history of airport design considering the philosophies adopted by designers, the functional layouts they have developed and the resultant form of the airport through a series of 40 case studies divided into 7 eras of approximately 20 years each. The themes include:

- The philosophies underpinning airport design.
- The evolution of design responses.
- How airports have avoided obsolescence.
- Identification of the key turning points.
- The evolution of master plans and terminal concepts in response to increasing traffic volumes.
- The future of airports in terms of environmental sustainability and the Covid-19 hiatus.

The case studies are international, covering the USA, Germany, the UK, France, the Netherlands, Singapore, Saudi Arabia, Japan, Hong Kong, Malaysia, South Korea, Thailand, Spain, United Arab Emirates, China, Turkey, Mexico, Australia and Poland. They are illustrated with full colour illustrations, many of which have not been published before and form part of an incredible graphic package. This book is essential reading for architects, engineers, planners and environmentalists alike.

ROBERT STEWART has 30 years' experience in the master planning and architectural design of major airports. He has led the aviation and transport sectors in a major architectural practice with a personal emphasis on airport master planning, facilities design, terminal and ancillary projects architectural design.

His work encompasses more than 50 airport projects, including master planning the ancillary facilities for the Heathrow Expansion Project, the design of the New London Hub Airport (Boris Island), Gatwick North Terminal, Bristol, Heathrow T5 and Runway 3 (for British Airways), London City and Medina Airports.

His experience of designing for both airports and airlines has allowed him to write this book with the unique perspective of an 'insider' who understands the drivers of growth and change that have led to the evolution of airport design.

THE EVOLUTION OF AIRPORT DESIGN

ROBERT STEWART

Routledge
Taylor & Francis Group
LONDON AND NEW YORK

Designed cover: Front: Istanbul 2019, one of the largest airports in the world, photograph courtesy of Istanbul Airport IST. Back: Croydon, the oldest airport terminal in the world. Aeroplane/Key Publishing (courtesy of Colin Ambrose).

First published 2024
by Routledge
4 Park Square, Milton Park, Abingdon, Oxon OX14 4RN

and by Routledge
605 Third Avenue, New York, NY 10158

Routledge is an imprint of the Taylor & Francis Group, an informa business

British Library Cataloguing-in-Publication Data
A catalogue record for this book is available from the British Library

ISBN: 978-1-032-34740-0 (hbk)
ISBN: 978-1-003-32362-4 (ebk)

DOI: 10.4324/9781003323624

Typeset in Optima LT Std
by KnowledgeWorks Global Ltd.

Printed and bound in Great Britain by
TJ Books Limited, Padstow, Cornwall

CONTENTS

CHAPTER 1

INTRODUCTION 1

CHAPTER 2

EXPERIMENTAL PERIOD 1904–1919 9

The Wright Field, Huffman Prairie, Dayton, Ohio 1904 9
Prototype interchanges 12

CHAPTER 3

PIONEERING ERA 1920–1939 18

1. *Berlin Flugplatz Johannisthal, 1919 21*
2. *Flughafen Königsberg-Devau (KGD) 1922 22*
3. *Dearborn Ford Airport 1928 25*
4. *Croydon Airport 1920 to 1928 and 1928 to 1959 27*
5. *New Orleans Sushan (Lakefront) (NEW) 1934 39*
6. *London Gatwick (LGW) 1936 43*
7. *Paris Le Bourget (LBG) 1923 and 1937 49*
8. *Berlin Tempelhof (THF) 1928 and 1939 54*
9. *New York LaGuardia (LGA) 1939 68*

CHAPTER 4

EVOLUTIONARY PHASE 1940–1959 77

10. *Chicago O'Hare (ORD) 1947 80*
11. *Washington Friendship (Baltimore) (BWI) 1950 88*
12. *San Francisco (SFO) 1954 90*
13. *London Heathrow (LHR) 1955 92*
14. *St Louis Lambert (STL) 1956 100*
15. *London Gatwick (LGW) 1958 106*

CHAPTER 5

JET AGE 1960–1979 118

16. *New York Idlewild (JFK) 1962 120*
17. *Washington Dulles (IAD) 1962 138*
18. *Los Angeles (LAX) 1962 147*
19. *Amsterdam Schiphol (AMS) 1967 156*
20. *Dallas Fort Worth (DFW) 1973 173*
21. *Paris Charles de Gaulle (CDG) 1974 178*

CHAPTER 6

TRANSFER INTERCHANGE 1980–1999 189

22. *Atlanta, Hartsfield (ATL) 1980 192*
23. *Singapore, Changi (SIN) 1981 204*
24. *Jeddah, King Abdulaziz (JED) 1981 209*
25. *London, Stansted (STN) 1991 215*
26. *Osaka Kansai (KIX) 1994 220*
27. *Denver International (DEN) 1995 228*
28. *Hong Kong, Chek Lap Kok (HKG) 1998 232*
29. *Kuala Lumpur (KUL) 1999 240*

CHAPTER 7

MEGA HUB 2000–2019 244

30. *Seoul Incheon (ICN) 2001 247*
31. *Bangkok Suvarnabhumi (BKK) 2006 253*
32. *Madrid Barajas (MAD) 2006 259*
33. *Heathrow (LHR) T5 2008, T2 2014 and the Heathrow Expansion Programme 264*
34. *Beijing Capital (PEK) 2008 281*
35. *Dubai International (DXB) 2010 285*
36. *iGA Istanbul (IST) 2019 290*
37. *Beijing Daxing (PKX) 2019 295*

CHAPTER 8

FUTURE DEVELOPMENTS 301

38. *Mexico City (MEX) 303*
39. *Dubai World Central (DWC) 307*
40. *Red Sea Umluj Tabuk 312*
41. *Western Sydney International (Nancy-Bird Walton) (WSI) 314*
42. *Centralny Port Komunikacyjny (CPK) 318*

CHAPTER 9

CONCLUSION 323

APPENDIX

AIRPORT MASTER PLANS THROUGH THE AGES 327

Airport master plans, USA 1919 and 1922 327
Airport master plans – USA 1930 and Europe 1932 328
Airport master plans – USA 1939 329
Airport master plans – Europe 1939 330
Post-war airport master plans 1940s 331
Airport master plans 1954–1956 332
Airport master plans 1970s 333
Airport master plans 1980 334
Airport master plans 2016 & 2030+ 335

BIBLIOGRAPHY 336
INDEX 345

CHAPTER 1

INTRODUCTION

Civil aviation is now over a hundred years old, and the airports we both love and hate have evolved greatly over that period. Most of the ideas had their origins many decades ago and had to wait until suitable technology was available to bring them to fruition, but few architects or engineers or indeed writers give credit to the originators of those ideas. This book seeks to correct this and show where the ideas we now take for granted have first been explored and how airport design has evolved in response to the increasing size and performance of aircraft and the greater volume of passengers handled.

A number of airport case studies have been selected to investigate the innovations they instigated. The first airport developed by the Wright brothers at Huffman Prairie, with its grass airfield, in 1904 sets the scene. Subsequent projects have been chosen to illustrate the origin of key step changes in airport design and show how they have influenced subsequent design. Many have been looked at over their lifetime to see how the challenge of growth and change has been addressed and whether the favoured concept has become obsolete.

Airport design is different from other forms of architecture, engineering, planning and environmental design by virtue of the sheer scale of development which is more akin to city planning. Large airports in the United States, such as Dallas–Fort Worth and Denver, are considerably larger than Manhattan Island. Manhattan is admired as a vibrant city precisely because it regenerates itself with every generation.

However, airports also differ from cities because no one lives there, giving rise to the view that airports are non-places, despite major hub airports having city-scale working and transient populations. Airports are also different in the speed of response required when faced with the challenges of growth and changed circumstances. This is in part because, unlike the airlines they handle, when they reach a certain scale, they are hugely profitable organisations that can afford to reinvent themselves on a regular basis.

The third aspect that sets airports apart from other developments is the complexity of the operation that allows an aircraft to land and take off again, often within an hour. This includes uniquely sophisticated passenger and baggage processes and aircraft turn-around logistics not found in other public buildings or complexes.

The airport, like air travel, is an invention of the 20th century. Airports are arguably a unique synthesis of architecture, planning, engineering and environmental design, and no one discipline can create an airport on its own. This book aims to explore the interdependence between these various disciplines and how this influences the evolution of airport design.

Airport design has always been driven by both, a rigorous philosophy and attention to functional and operational requirements. As such airport design can be regarded as the epitome of the modern movement, in that form follows function, despite the variety in the resultant layout and appearance. So airport design brings together an unusual combination of architectural, engineering and planning disciplines, each of which brings its own approach.

DOI: 10.4324/9781003323624-1

Architects have typically focused on satisfying the travelling public with an effortless process through the terminal, despite sometimes contrary commercial pressures. They have also sought to create a look and feel that is in character with the spirit of flight and their concepts of modernity. The Futurists' *Foundation Manifesto* was published in Le Figaro on 20 February 1909, written by Filippo Tomaso Marinetti, the founder of the movement. It was the first design manifesto to consider an attitude of mind that embraced aircraft and aviation while celebrating the twentieth century as a new machine age. The last proposition stated:

> *We will sing of the stirring of great crowds – workers, pleasure seekers, rioters – and the confused sea of colour and sound as revolution sweeps through a modern metropolis......... We will sing ...[of]... the easy flight of aeroplanes, their propellers beating the wind like banners, with a sound like the applause of a mighty crowd.*
>
> (Banham, 1960)

The Futurists therefore envisaged an aesthetic based on technology and machinery, a philosophy that has ever since underpinned much of the best airport architecture and design. Part of the Futurist *Messagio* of 1914, to which Antonio Sant'Elia contributed, includes the following statements:

> *I affirm that the new architecture is the architecture of cold calculation, temerious boldness and simplicity; the architecture of reinforced concrete, iron, glass, textile fibres and all those replacements for wood, stone and brick that make for the attainment in maximum elasticity and lightness.*
>
> *That oblique and elliptical lines are dynamic and by their very nature possess an emotive power a thousand times stronger than perpendiculars and horizontals and that no dynamic architecture can exist that does not include these.*
>
> *That from an architecture so conceived no stock answers, plastic or linear, could arise, because the fundamental characteristics of Futurist architecture will be expendability and transience.*
>
> (Banham, 1960)

These are phrases that would still resonate with airport designers today, although Sant'Elia could not have realised how apposite and prescient the last statement was. The early design concepts developed from that philosophy include Sant'Elia's 1912 Rail/ Air Interchange and Eric Mendelson's 1917 dynamic sketch for an integrated airship and aeroplane interchange (Figures 1.1 and 1.2).

Engineers have largely concentrated on improving the safety and efficiency of the operation and developing internationally recognised standards. In 1919, Wilbur Wright somewhat optimistically argued in a special editorial for Aviation magazine that:

> *The airplane has already been abundantly safe for flight. The problem before the engineer today is that of providing for safe landing.*
>
> (Wright, 1919)

Thus, he attempted to transfer the focus of attention on safety from the aircraft to the airport. The engineering contribution to the evolution of airports has from the start concentrated on safety as the first priority, initially focused on organising fields for the safe take-off and landing of aircraft based on the military experience of the First World War.

Figure 1.1 Futurist Interchange 1912.
SANT'ELIA – SOURCE: COMO MUSEO CIVICA

Figure 1.2 Sketch for an Aerodrome, Erich Mendelson 1917.
SOURCE: MENDELSOHN, STRUCTURES AND SKETCHES (1924)

Henry Ford entered the aviation business in 1924 when he took over the Stout Metal Airplane Company and created one of the earliest American airports to handle passenger services at Dearborn. He recognised that safety and the public's perception of safety were keys to encouraging nervous passengers to fly. He adopted the slogan:

> *Safety – First, Last and Always.*
>
> (Gordon, 2008)

From 1925 to 1931, the Ford Motor Company sponsored annual air tours to promote reliability and safety in commercial aircraft. Planes were rated on the ability to take off and land quickly and maintain consistent speeds and schedules.

Planners have been primarily concerned with an airport's location, with how it might be integrated into the fabric of society and with how passengers might transfer from one mode of transport to another. More recently, they have also sought to minimise any negative environmental impacts.

As early as 1912, Donat-Alfred Agache had proposed the integration of an airport and rail terminus in his masterplan for the Canberra, Federal Capital City design competition. The Aerostatic Station (the name given to the airport), as seen in his master plan, contained all the elements of early airport design: a terminal building, hangars for aircraft and airships and a grandstand surrounding the four sides of an airfield. It was located in the south-east quadrant of the city, immediately adjacent to the main railway station. From an operational viewpoint, surrounding the airfield with buildings and having no regard for the variable direction of the wind would have severely limited its practicality, but Agache was arguably the first to recognise that aviation had a practical use as public transportation and not merely as an aerial spectacle (Figure 1.3).

While civil aviation may have started in 1919, it took some 40 years before the airport community began to recognise the environmental consequences of flying. Although socio/environmental issues such as noise, air quality, safety and property values had been identified as early as 1930 in a study by Harvard University, the environmental impacts of flying were largely ignored or belittled. The new concern was largely provoked by the introduction of noisy jet aircraft in the 1960s and the associated rapid growth of air travel. It is only in the 21st century that global warming has been added to the agenda.

There are now three strands to the story of the evolution of airport environmental sustainability; firstly the influence of environmental protest groups in frustrating airport

Figure 1.3 City of Canberra, Master Plan of 3rd placed competition entry by Donat-Alfred Agache showing the location of the proposed airport adjacent to the rail terminus 1912.

SOURCE: IMAGE COURTESY OF THE NATIONAL ARCHIVES OF AUSTRALIA. NAA: A710.1

development and secondly the development of environmental assessment as a recognised science. Both had their origins in the New York region. The third strand, which is still in its infancy, is the identification of measures required to make airports sustainable, minimise climate change resulting from aviation and address extreme weather events at airports that can result.

However, little has been written about the multidisciplinary nature of airport design and why, when, where and how new airport concepts have evolved, so this is the particular focus of this book. Because airport development is on a city scale, it requires the input of many disciplines with different philosophies.

This book analyses, through a series of case studies, the role of architectural philosophies such as Futurism in capturing the spirit of flight and facilitating growth and change, the engineering approach to safety and operational robustness, the planning focus on connectivity and integration into the city region and the environmentalist's concern with sustainability. So, airport design brings together architectural, engineering, planning and environmental philosophies into a new combined multidisciplinary approach.

Le Corbusier, writing in *Towards a New Architecture* (*Paris 1923*) showed how important aviation and its engineering approach was to him in formulating his architectural ideas by dedicating a whole chapter to airplanes to justify his approach to functional design, stating that:

> *The airplane is the product of close selection. The lesson of the airplane lies in the logic which governed the statement of the problem and its realisation.*
>
> (Corbusier, 1923, English Edition 1927, first paperback edition 1970, p. 100)

He went on to cite the functional engineering design of the airship hangar at Orly by engineers Freyssinet and Limousin to illustrate the new architecture. This emphasis on the interdependence of architects and engineers has characterised the development of airport planning and terminal design.

Competitions such as the RIBA design competition of 1928 and the LeHigh Airport competition in 1929 were sources of innovative ideas. In particular, the LeHigh competition in the United States saw a number of new ideas introduced that can be found in later airport designs. The winning design included an airfield with multi-directional runways and a terminal with an underground walkway linking the terminal with a remote satellite that incorporated retractable telescopic boarding walkways, giving access to the five stands that surrounded it (Figure 1.4 and 1.5).

Rudyard Kipling was among the first to recognise the impact aviation would have on the experience of travel. In 1914, Rudyard Kipling told the Royal Geographical Society:

> *The time is near when men will receive their normal impressions of a new country suddenly and in plan, not slowly and in perspective; when the most extreme distances will be brought within the compass of one week's – one hundred and sixty eight hours' travel; when the word 'inaccessible,' as applied to any given spot on the surface of the globe, will cease to have any meaning.*

Walter Gropius was one of the first architects to recognise that aviation had changed the way that architecture could be viewed. In his book *The New Architecture and the Bauhaus*, first published in 1935, he wrote:

> *With the development of air transport the architect will have to pay as much attention to the bird's-eye perspective of his houses as to their elevations.*
>
> (Gropius, 1935)

This concern with the aerial view has become a particular focus for airport architects and planners. Gropius was also a key proponent of the multidisciplinary approach and organised the Bauhaus design curriculum to foster cross-discipline learning.

In the same year, Le Corbusier published his book *Aircraft*, in which he stated:

> *The bird's eye view: The eye now sees in substance what the mind formerly could only subjectively conceive. It is a new function added to our senses. It is a new standard of measurement. It is a new basis of sensation. Man will make use of it to conceive new aims. Cities will arise out of their ashes.*
>
> (Corbusier, 1935)

In the late 1930s, the design for New York La Guardia brought together architectural, engineering and town planning disciplines but added to them from the early planning stage

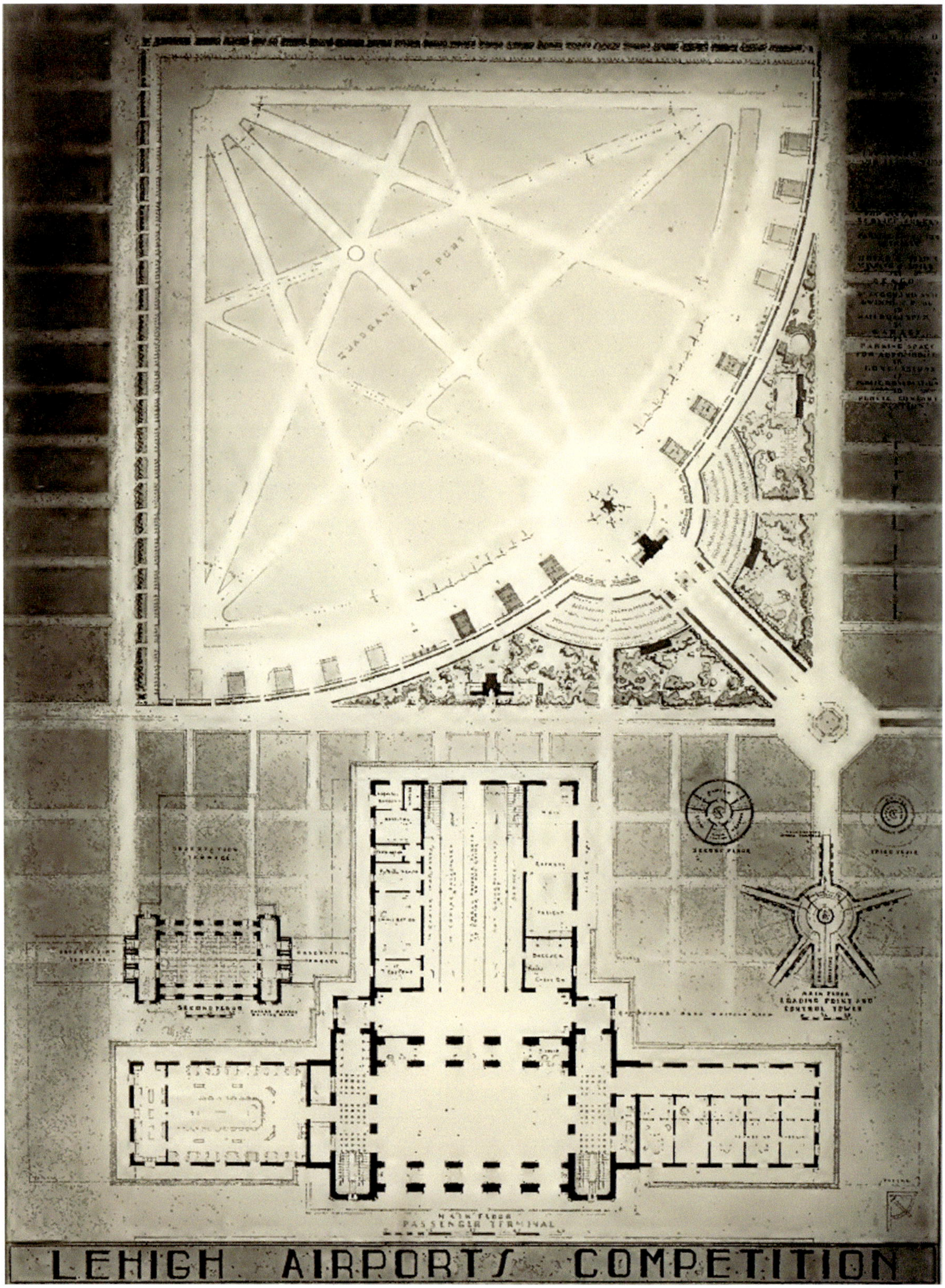

Figure 1.4 First prize-winning design, master plan and terminal layout, by A C Zimmerman and William H Harrison, from the LeHigh Airports Competition 1929.

SOURCE: AMERICAN AIRPORT DESIGNS 1930

Figure 1.5 First prize-winning design, bird's-eye view and cross section, from the LeHigh Airports Competition 1929.
SOURCE: AMERICAN AIRPORT DESIGNS 1930

designers of air navigation systems and communications technology. The air traffic control tower became the focus of the airport design, and a new player was introduced to the airport design team, the Civil Aeronautical Authority, with its overriding concern for safety.

Eero Saarinen later said when designing Washington Dulles Airport:

> *No one asked us to grapple with the problem of a jet-age terminal beyond the question of pure architecture. But I believe the architect has to assume that kind of responsibility. Therefore, together with the team of Ammann & Whitney, engineers; Charles Landrum, airport consultant; and Burns and McDonnell, mechanical engineers, we decided to make a fundamental analysis of the whole problem and we wanted to solve it in a hard boiled way. 'We sent out teams with counters and stopwatches to see what people really do at airports, how far they walk (and) their interchange problems. We analysed special problems of jets, examined schedules, peak loads, effects of weather. We studied baggage handling, economics, methods of operations and so on. We reduced this vast data to a series of about forty charts.*
>
> *We found there were three very critical areas…the tremendous distances passengers…. would have to walk in jet terminals…the heavy cost of taxiing jet planes…(and) the increasing need for the greatest possible flexibility in operations and servicing of aircraft.*
>
> (Federal Aviation Administration, 1962)

Airport designers would recognise the same issues today.

Norman Foster wrote in his 1996 essay on airports about the blurring of distinctions between the various design disciplines:

> *A further shift can be discerned in the nature of such infrastructure projects. The edges between infrastructure and architecture are becoming more blurred. We can see this in structures concerned with information transmission — communication towers and platforms, for example. But we can also see it in structures for physical communication, such as the airport. Is the airport infrastructure or is it architecture?*
>
> *Or is it perhaps inhabited infrastructure? As these edges become less finite, the distinctions between the role of the architect, the engineer and the other professions become similarly blurred. New infrastructure projects are typically becoming more publicly accessible, more multifunctional (and) less unidirectional. Together these trends have the potential to create a new kind of airport building.*
>
> (Foster, 1996)

These studies of commercial airport design conveniently fall into a series of stages of some two decades each, with a total of 42 examples.

Experimental Period	1904–1919
Pioneering Era	1920–1939
Evolutionary Phase	1940–1959
Jet Age	1960–1979
Transfer Interchange	1980–1999
Mega Hub	2000–2019
Future Development	2020–onwards

CHAPTER 2

EXPERIMENTAL PERIOD 1904–1919

Following their successful first flight of 59 seconds and covering 852 ft against a headwind of 20mph at Kittyhawk on December 17, 1903, the Wright brothers returned to their hometown of Dayton, where they created the first permanent airfield. The Wrights began using Huffman Prairie, which they described as a swampy meadow, in 1904 with the permission of the field's owner, Dayton banker Torrence Huffman. The Wrights made about 150 flights in 1904–1905, leading to the development of the Wright Flyer, which they considered to be the first practical airplane.

The Wright Field, Huffman Prairie, Dayton, Ohio 1904

The Wright Field included two features that weren't to make it to later airports: the guide rail launch track and the catapult, both designed to overcome friction during launching. However, both concepts have been adopted by aircraft carriers. The drawing on the following page appeared in a 1940 publication, *Airports,* by John Walter Wood. It was derived from a sketch made by Orville Wright in 1935 (Figure 2.1).

It clearly shows the locations of the guide rails and associated catapults, which were relocated as required to suit the prevailing wind. The plane sat at one end of a 160–195 ft long wooden rail on top of a detachable wheeled trolley. Ropes and pulleys connected the trolley to a 1,200–1,400 pound weight, which fell 16.5 ft from a 20 ft tall wooden derrick. With the plane's engines revving, the weight was dropped, propelling the aircraft down the ramp and into the air. The drawing also shows a feature indicated with an H that was to become a standard feature of all subsequent airports: the aircraft hangar. The hangar measured 44 ft × 16 ft and housed two Wright planes, but only after the front elevators had been removed.

The Wright brothers kept immaculate records of their achievements, which they submitted to the Aero Club of America on 12 March 1906. Key events included their first complete circle on 30 September 1904, in their Flyer II, a flight which lasted 1minute, 36 seconds and covered 4,080 ft. Later on 5 October 1905, Wilbur made a spectacular flight in Flyer III in which he circled the field 30 times in 38 minutes 3 seconds for a total distance of 24.2 miles (38,956 metres) before landing after running out of fuel (Figures 2.2–2.4).

Between 1904 and 1914, a series of aviation firsts took place at Wright Field.

- first controlled turn
- first circle
- first controlled bank
- first figure of eight
- first permanent flying school

DOI: 10.4324/9781003323624-2

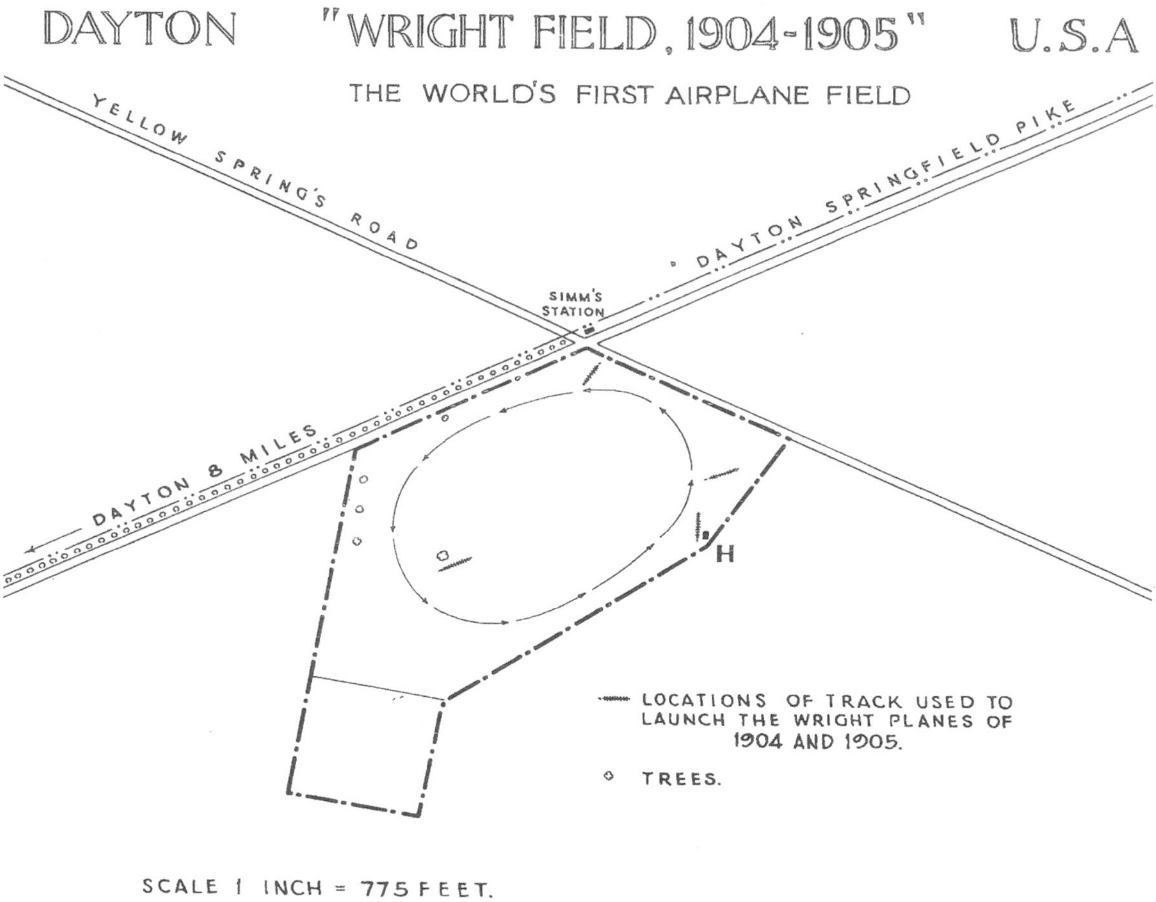

Figure 2.1 Dayton, Wright Field.
SOURCE: AIRPORTS BY JOHN WALTER WOOD FROM A 1936 SKETCH BY ORVILLE WRIGHT

Figure 2.2 Dayton, Wright Field, Flyer 11 & Hangar, May 1904.
SOURCE: US LIBRARY OF CONGRESS

Figure 2.3 Flyer II at the Wright Field, Dayton 1904.
SOURCE: US LIBRARY OF CONGRESS

It is also claimed that the first cargo flight took place at Huffman Prairie as a Wright plane flew several bolts of cloth to Columbus, about 55 miles away, as publicity for a retail shop. Later, Huffman Prairie became the first flying school. The United States Army Signal Corps purchased the field in 1917 and renamed it, along with 2,000 adjacent acres, Wilbur Wright Field. Later, it became part of the Wright-Patterson Air Force Base. The 1905 Wright Flyer III is preserved and on display at Carillion Historical Park, Dayton.

An eyewitness, Amos I. Root, editor of *Gleanings in Bee Culture*, wrote in the 01.01.1905 issue:

> *It was my privilege to see the first successful trip of an airship without a balloon to sustain it, that the world has ever made, that is, to turn corners and come back to the starting point...... When the engine is shut off, the apparatus glides to the ground very quietly and alights on something much like a pair of sled runners, sliding over the grassy surface perhaps a rod or more.*
>
> (Root, 1905)

Wilbur Wright provided a vivid description of Huffman Prairie in a letter to Octave Chanute, the scientist and pioneer aviator who had flown a series of man-carrying gliders before the Wright brothers:

> *We are in a large meadow of about 100 acres. It is skirted on the west and north by trees. This not only shuts off the wind somewhat, but gives a slight downward trend. However, this is a matter we do not consider anything serious. The greater troubles are the facts that in addition to the cattle there have been a dozen or more horses in the pasture and as it is surrounded by barbwire fencing we have been at much trouble to get them safely away before making any trials. Also, the*

Figure 2.4 Start of the first flight of Flyer III, June 23, 1905, Orville at the controls. The catapult tower, which they began using in September 1904, is on the right.
SOURCE: US LIBRARY OF CONGRESS

> *ground is an old swamp and is filled with grassy hummocks some six inches high, so that it resembles dog town.*
>
> (Wright, 1908)

An electric interurban rail line between Dayton and Osborn ran by the field. Simms Road Station, the penultimate stop, was conveniently located fifty yards away. It took less than 30 minutes for the brothers to make the trip from their home in West Dayton. This might be regarded as the first rail-air interchange.

The Wright field at Dayton showed the importance of having a level field with the minimum of obstructions around the edge to make it suitable as an airfield. It also had good public transport access via the train service from Dayton, which is how the Wright brothers reached the field. The only design element there, which can still be found in modern airports, is the provision of a hangar to house the aircraft. Other features, such as the catapult and guide rail for assisted take-off, are used in aircraft carriers but not at airports.

Prototype interchanges

The airport terminal is an unusual building type, as it is rarely a destination that the user chooses to experience. Rather, its raison d'être is to facilitate the efficient transfer of passengers from one mode of transport to another, or, in the case of the larger hubs, between journeys on the same mode. No longer is air travel the preserve of an exclusive, wealthy, and adventurous elite who were prepared to brave the dangers and discomforts of early flying, but it is an everyday experience for many people.

The first interchanges preceded the advent of scheduled services and were designed to enable large numbers of spectators to attend the air shows that were held from 1908 onward. The concept of an airport as an interchange can be found in the organisation of these earliest air shows. The entry ticket to the Frankfurt air display of 1909 included a map below, which showed the layout of the site with many of the key elements of the modern airport interchange: a railway station (rail line in red), a tram terminus (pink), a runway (green) and a transit system (yellow) to take visitors around the site to the hangars and Zeppelin mooring point (Figure 2.5).

The same year, 1909, many of the 500,000 visitors to the famous air show at Reims, Grand Semaine d'Aviation de Champagne, during the week of 22–29 August, reached the venue by rail via a purpose-built extension to the railway with a new station, Gare de Fresnois (Figure 2.6).

The difference between developing an urban site with good transport links, as at the Frankfurt Air show, or creating a new site away from the city, as at the Rheims Air show and providing new transport links has been the subject of controversy ever since then and has fueled many debates on the appropriateness and sustainability of these alternative approaches to airport siting.

The airport from which the first domestic scheduled services began later in 1919 was Berlin Johannisthal, which also had its origins in air displays and as a manufacturing base for aircraft. The first aircraft to land there on 27 September 1909 was an Antionette flown by Hubert Latham, a Briton.

The airfield attracted a number of aircraft manufactures to the site. It also hosted several flying shows and air races, for which two spectator stands were built, one for 2,300 people and a second one for 1,750 people. One of the best-known events is the Rund um Berlin (around Berlin) race on the 30 September 1913 (Figure 2.7).

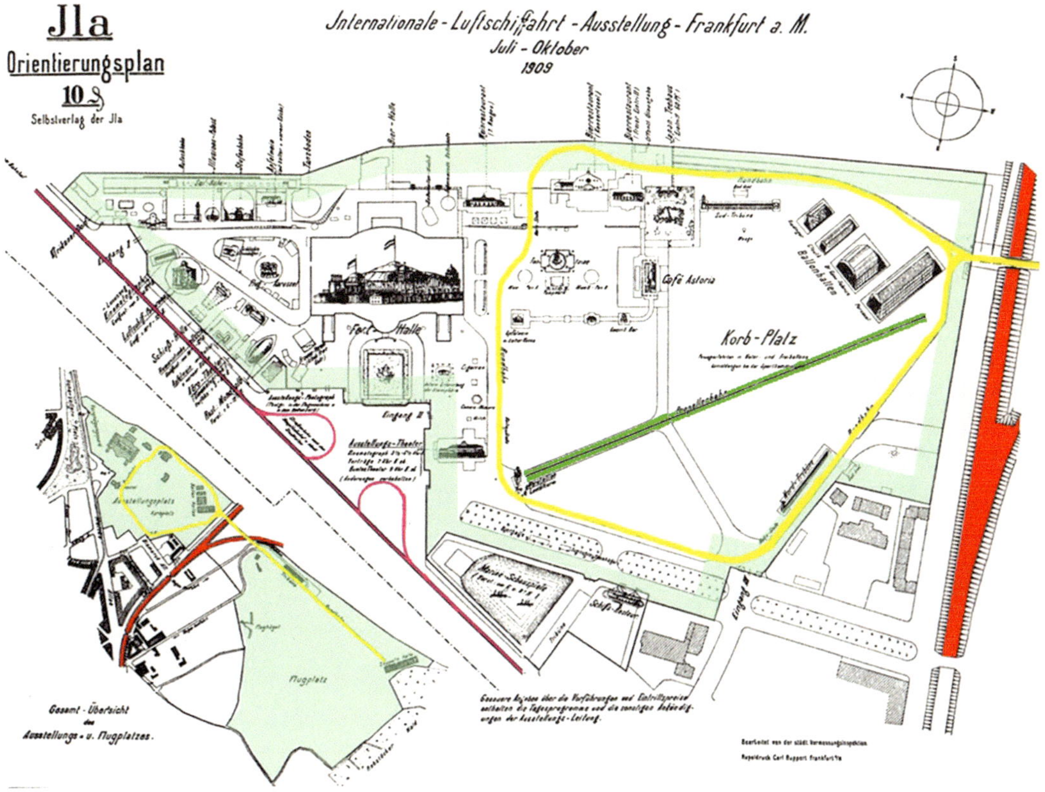

Figure 2.5 Frankfurt Internationale Luftschiffahrt Augglellung (ILA) site plan, 1909.
SOURCE: AUTHOR'S COLLECTION

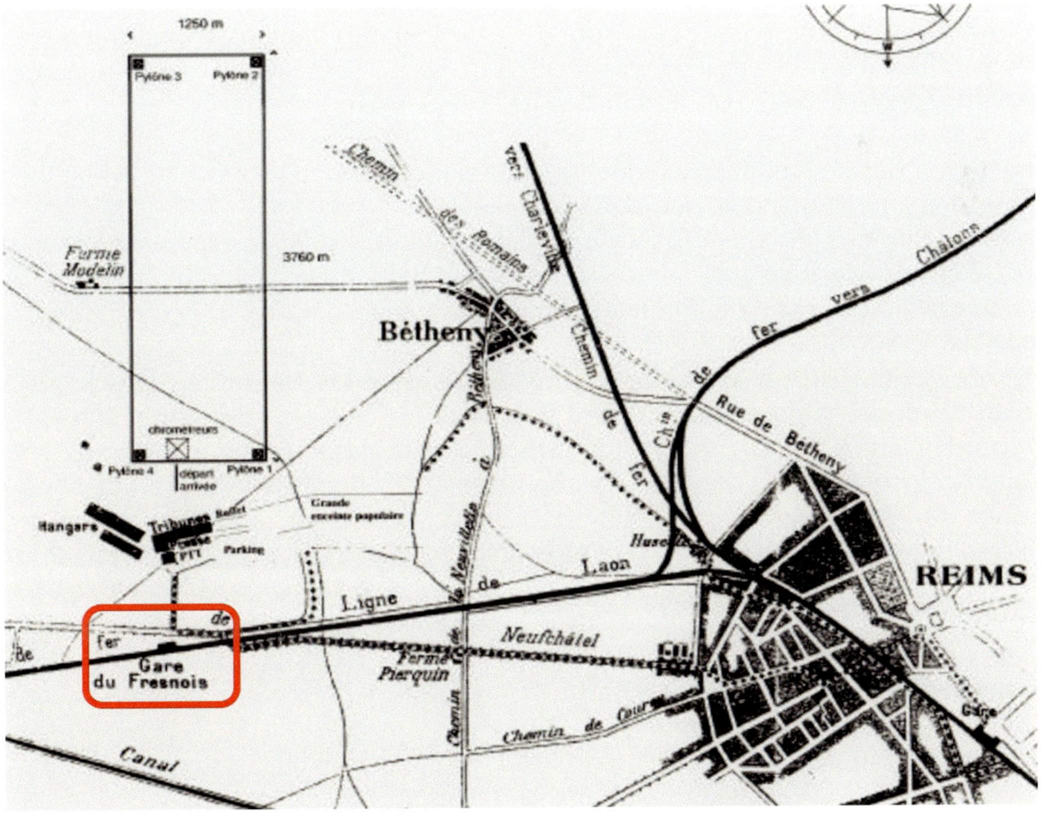

Figure 2.6 Grand Semaine d'Aviation de Champagne 1909 – site plan – New Station highlighted in red.
SOURCE: CLICHÉ ILLUSTRATION VIA AVIATION.MAISONS-CHAMPAGNE.COM

Figure 2.7 Berlin Johannisthal, 'Rund um Berlin' race 1913.
SOURCE: CLEMENS KURZ – WORDPRESS

For a long time, it was not clear whether heavier than air aircraft or lighter than air airships would prevail, so early airports like Berlin Johannisthal hedged their bets and made provision for both types of transport. So, two airship hangars were built there; the first for the Parseval Company was erected in April 1910, while a second airship hangar, the Zepplin-Halle, was completed for the German Imperial Navy's Zigarre airship in 1911.

The airport became militarised with the outbreak of the First World War, when the focus became aircraft manufacture in the hangars.

Images of airports and the role they can play as a public transport interchange between a planned city and its transport network existed long before it was possible to put those ideas into practice, as can be seen in the competition entry for the Canberra Federal Capital City competition (Figure 2.8).

The idea of the airport as an interchange has evolved with concepts such as the multi-modal transport hub, airport city and aerotropolis emerging. These developments recognise the increasing importance of commercial aviation to contemporary society and the need for the airport to be woven into the built environment. Allied to planning issues are environmental sustainability concerns that have grown in importance in recent years, focusing on noise, emissions, traffic impact and global warming. There have been a series of small steps in design development to reach this point and many blind alleys as architects, engineers and planners have sought to define a new type of built environment, part architecture, part engineering infrastructure and part urban design, with each discipline bringing very different perspectives to the task.

The drawings in the 1914 Futurist Manifesto included Sant'Elia's dramatic vision, 'Statione per Treni e Aerei' (station for trains and aeroplanes), an imaginary multi-level interchange between road, rail and air linked by dramatic external escalator enclosures. It

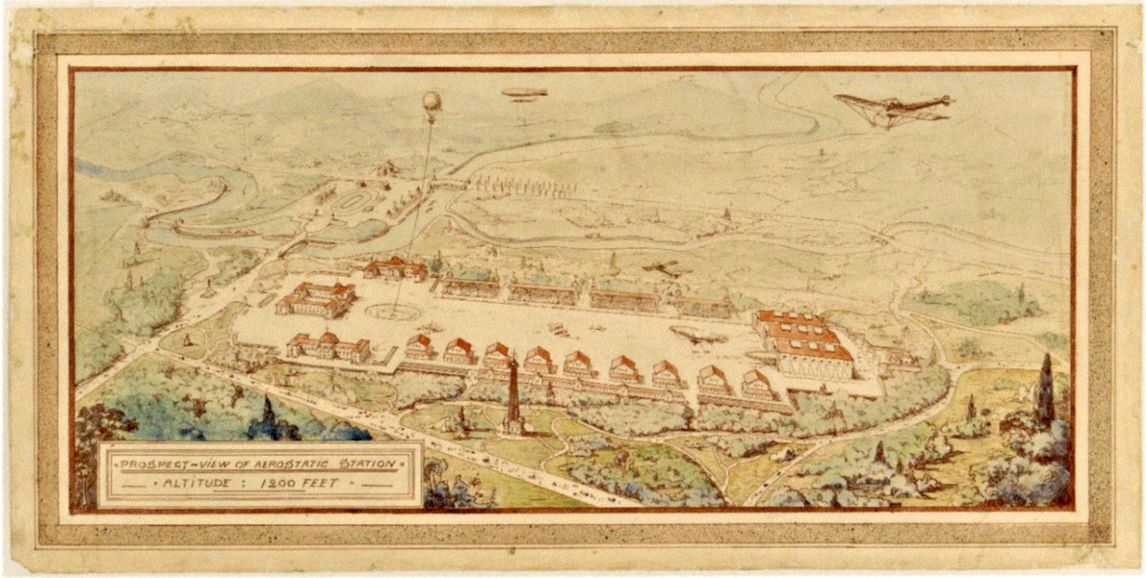

Figure 2.8 Canberra Aerostatic Station 1912. Prospect view of the Aerostatic Station. Altitude 1,200 ft.

SOURCE: IMAGE COURTESY OF THE NATIONAL ARCHIVES OF AUSTRALIA, NAA: A710, 11
ENTRY NO. 4, FEDERAL CAPITAL CITY DESIGN COMPETITION, DONAT-ALFRED AGACHE

was his architectural concept for the redevelopment of Milan station and part of his vision for La Citta Nuova (The New City) (Figure 2.9).

It was arguably a more grandiose, monumental and static vision of the future than his manifesto suggested, but the Futurist Manifesto and Sant'Elia's words and drawings did provide an intellectual basis and inspiration for generations of architects, planners and engineers who designed subsequent airports.

In contrast to these grand ideas of the airport for the future, the very first experimental scheduled service, by the St. Petersburg Tampa Airboat Lines, took place in the USA for a short period between January and April 1914, between St Petersburg and Tampa in Florida, with a Benoist flying boat crossing Tampa Bay, with the pilot and one fare-paying passenger. This service, however, made use of the existing slipway to launch the seaplane and did not require any permanent infrastructure that would be recognised as an airport. Two flying boats carried 1,205 passengers and flew over 11,000 miles (17,702 kilometres) (Figure 2.10).

The first flight's pilot was Tony Jannus, an experienced test pilot. The first paying passenger was Abram C. Pheil, former mayor of St. Petersburg. Their 21-mile (34-kilometer) flight across the bay to Tampa took 23 minutes. They flew in a flying boat designed by Thomas Benoist, an aviation entrepreneur from St. Louis.

Figure 2.9 'Statione per Treni e Aerei'- Antonio Sant'Elia – 1914.
SOURCE: WORDPRESS

Figure 2.10 SPT Airlines' Benoist Type XIV flying boat takes off on the first scheduled commercial passenger flight, St. Petersburg, Florida, 1 January 1914.

SOURCE: STATE ARCHIVES OF FLORIDA, FLORIDA MEMORY

CHAPTER 3

PIONEERING ERA 1920–1939

While commercial flying started in 1919, the few services that operated at this date were all from former military airfields that had become redundant at the end of the First World War. Scheduled domestic services began on 5 February 1919 between Berlin Johannisthal and Weimar. Directly after the war commercial aviation was banned in the UK so one of the first international flights from France to England on 8 February 1919, which publicised the merits of the Farman Goliath aircraft, carried former pilots who all travelled in uniform carrying mission orders. This aircraft had been converted to handle passengers from its wartime role as a heavy bomber. The design was admired by Le Corbusier and figured in his book *Aircraft*. Later, the aircraft's shape informed the plan of the terminal at Le Bourget. Scheduled international services followed a few months later, in August 1919, between London Hounslow and Paris Le Bourget.

Airports of this first period of commercial aviation typically had grass airfields, and the earliest permanent buildings were hangars. As aircraft were so flimsy, they had to be stored indoors when not in use to protect them against bad weather.

The first purpose-designed terminal was built at Konigsberg Airport (1922), but there is no evidence that it influenced any subsequent designs. The new terminal at London Croydon (1928) was however highly influential and widely published and analysed in specialist aviation publications of the late twenties, thirties and forties. It contained the first control tower, introduced the first airport hotel and included most of the elements of a modern terminal except security control. Security measures did not emerge until the seventies in response to hijackings and were then made more rigorous to counter terrorist attacks. The major limitation of Croydon was that it was only designed to handle one departing and arriving flight at a time.

At the beginning of this period, it was not clear whether aircraft or airships would dominate, and for a time, some believed that aircraft would prove to be more suitable for short-range flights while airships would hold sway in the long-haul market. So, many airports, such as London Croydon, Berlin Tempelhof and Dearborn Ford, were designed to handle both aircraft and airships. However, after a series of setbacks to airships culminating in the loss of the Hindenburg at Lakehurst, New Jersey, USA, in 1937, it became clear that the aeroplane would dominate future air travel.

Art movements had a limited effect on the design of terminals during this period, and none of the earlier Futurist or Expressionist themes featured in terminal design. While Croydon looked backwards to a Beaux Arts model for its aesthetic and Le Bourget followed a contemporary Art Deco aesthetic, the International Style can be found first at the Leipzig airport restaurant and later at Rio de Janeiro Santos Dumond, designed in the late thirties but not completed until after the war. However, the international style did not dominate terminal design until after the Second World War (Figures 3.1, 3.2, 3.3 and 3.4).

DOI: 10.4324/9781003323624-3

Figure 3.1 London Croydon 1928.
SOURCE: PHOTO WIKIMEDIA

Figure 3.2 Paris Le Bourget – 1937.
SOURCE: GENEANET CARTE POSTALE ANCIENNE

Figure 3.3 Leipzig – Halle Restaurant 1929.
SOURCE: POSTALES INVENTADAS

Figure 3.4 Rio de Janeiro, Santos Dumont 1947.
SOURCE: WIKIPEDIA PHOTO – MARIO ROBERTO DURAN ORTIZ

The first terminal at Berlin Tempelhof demonstrated that terminals could be built in phases while still maintaining business continuity at all times. It also introduced the idea of linking the terminal to the underground Metro system to provide an integrated transport network and pioneered runway lighting to permit night flights.

An early design theoretically capable of handling multiple flights simultaneously can be found at Gatwick Airport 'Beehive' (1936). The circular plan owed much to the winning design for the 1929 Lehigh Airport Competition which had been widely published. The latter design boasted a hexagonal satellite with telescopic walkways providing covered access to the aircraft parked around its perimeter and an underground walkway linking the building to the main terminal. These features were incorporated into the circular Gatwick design.

Gatwick was also the first airport to be linked to the mainline railway via its own station and pioneered the integration of rail and air travel as a single offer. Despite these innovations, the boggy landing conditions on Gatwick's grass field meant that this capability was rarely needed.

There is evidence of cross-fertilisation of design ideas between Europe and the USA in books, periodicals and lectures. This can be seen in books like Steadman Hanks' survey of European airports for an American audience or the Norman and Dawbarn lecture to the Royal Aeronautical Society following their visit to new airports in the USA. This also led to numerous articles in architectural and engineering journals.

While hard-surfaced landing strips can be found as early as 1922 in Boston, USA, it was the concrete runways introduced at Ford Airport Dearborn that brought hard-surfaced runways into the mainstream of airport design in the USA. Most of the major airports in the USA were constructed with hard-surfaced runways in the late twenties and were progressively introduced in Europe in the late thirties. New Orleans Sushan illustrates the different direction taken in the USA where hard-surfaced runways were generally adopted a decade before Europe. Instead of the perimeter development associated with the omni-directional grass fields in Europe, the Art Deco Terminal at New Orleans Sushan projects into the gap between the runways.

Throughout this pioneering period, commercial aviation needed financial subsidies, either directly from governments in Europe, or indirectly via the postal service in the USA. Airports did make many efforts to supplement their income by holding flying shows as major spectator events and by increasingly sophisticated catering and leisure offers. Sushan airport in particular was a pioneer in conceiving of the airport as a leisure destination.

The first signs of evolution from these simple beginnings can be found at the second terminal at Berlin Tempelhof and the domestic terminal at LaGuardia, where designs were developed so that multiple aircraft could be turned around simultaneously. With so many new ideas and technological innovations in this period, airport design fits conveniently within Banham's characterisation of this period as 'The First Machine Age', but it certainly does not fit with his description of aviation being in a 'Pastoral Phase'. Many of the ideas we see in modern airports have their origins in this most creative of design periods.

The airports selected to illustrate the key characteristics of the era are:

- Berlin Flugplatz Johannisthal, 1919
- Flughafen Königsberg-Devau (KGD) 1922
- Dearborn Ford 1926
- London Croydon 1928
- New Orleans Sushan (NEW) 1934
- London Gatwick (LGW) 1936
- Paris Le Bourget (LBG) 1937
- Berlin Tempelhof (THF) 1928 and 1939
- New York LaGuardia (LGA) 1939

1. Berlin Flugplatz Johannisthal, 1919

Berlin Johannisthal was the airport from which the first scheduled services flew on 5 February 1919, initially with flights to Weimar, home of the new German National Assembly, by Deutsche Luft Reederei (the forerunner of Lufthansa). The first flight took off from Flugplatz Johannisthal with two pilots, and carrying 40 letters and some 65 kilos of newspapers bound for Weimar (Figures 3.5 and 3.6). In the following months, the average journey time on this route was 2 hrs 11 mins, with a record trip of 1 hr 15 mins. 538 flights were recorded by the end of April. A second route between Berlin and Hamburg was opened on 1st March, with 262 flights by the end of April. Further destinations were quickly added by the summer, including Breslau, Frankfurt, Leipzig, Hamburg, Hanover, Munich, Warnmünde and Westphalia. Longer flights were added later in the year, including a six-hour flight to Konigsberg to link East Prussia with Germany.

Passengers were handled through an existing hangar, and a flying kit and motor transport to and from the aerodrome were provided as an inclusive part of the fare. Luggage was also carried free of charge, but the total weight of the passenger and baggage could not exceed a certain weight. Transferable serial tickets could also be purchased.

Berlin Johannisthal had been in existence for some ten years before scheduled flying started. It opened in 1909 as the first civilian airfield and was known at the time as *Motorflugplatz Johannisthal-Andershof* as it was on the border between the two communities. The airfield had earlier been used for the round Berlin air races in 1909, for aircraft manufacture and then for military use during the First World War. Over time, a series of hangars and aircraft manufacturing facilities had grown up around the perimeter of the grass airfield, which became the model for European airfield layouts in the 1920s and 1930s. The use of the airfield sharply diminished after 1923 when Berlin Tempelhof opened, and after civilian flying was eventually banned in 1931, it became an important test installation for the secret development of the Luftwaffe.

The enduring legacy of Berlin Johannisthal is the idea of handling passengers from the hangar. This approach later reappeared in the lean-to concepts that were prevalent in the USA during the 1930s and more recently in executive terminals that support business aviation and at the Virgin Galactic Spaceport.

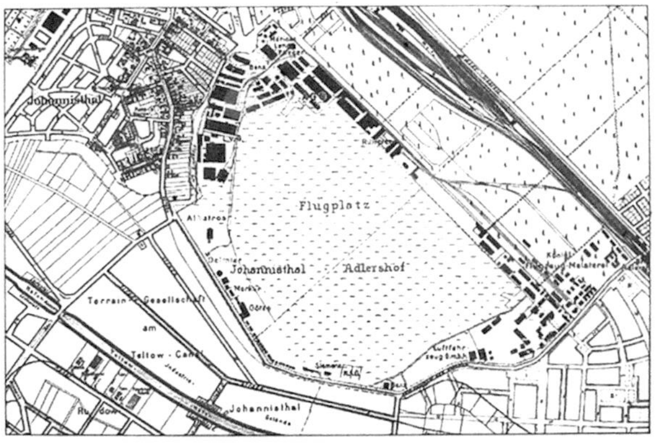

Figure 3.5 Flugplatz Johannisthal Plan 1916.
SOURCE: WIKIMEDIA 05.02.1919

Figure 3.6 Berlin Johannisthal – First scheduled flight.
SOURCE: WIKIPEDIA

2. Flughafen Königsberg-Devau (KGD) 1922

Königsberg Airport is recognised as the first airport that was laid out specifically for civilian flying and the first example of a new building type, the passenger terminal which opened in 1922, although it was at this time simply known as the administration building. Königsberg received special attention in the 1920s and 1930s as an otherwise isolated German enclave because of its strategic location as the capital city of East Prussia (now Kalingrad, a Russian enclave). It was separated from Germany by the Danzig corridor that linked Poland with the Baltic Sea. Flying was seen as a way of overcoming this territorial split. The airport master plan included a terminal building designed by the young Hans Hopp (1890–1971) in a neo-classical style and located in the corner of a grass airfield between two arched hangars set at right angles to one another (Figures 3.7–3.10). The city's tram service was extended from the railway station to the airport in 1924 soon after the opening of the airport.

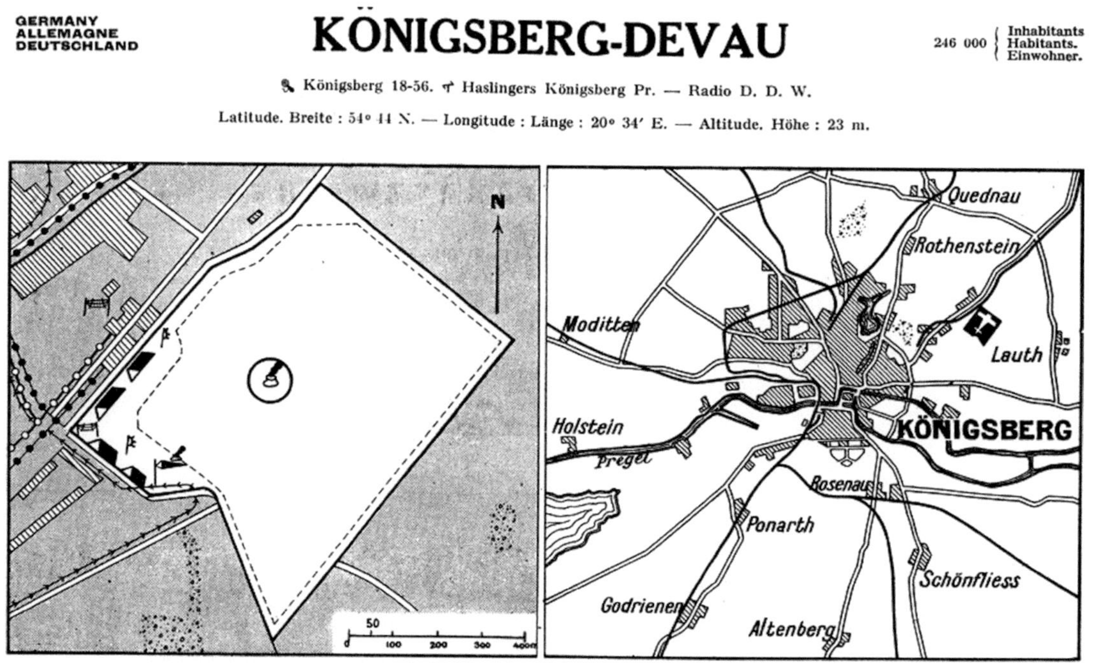

Figure 3.7 Königsberg site plan and location plan.
SOURCE: HTTPS://KALININGRAD-CITY24.RU/

Figure 3.8 Königsberg seen from the airfield.
SOURCE: HTTPS://KALININGRAD-CITY24.RU/

Figure 3.9 Konigsberg 1922, main entrance – Architect, Hans Hopp.
SOURCE: HTTPS://KALININGRAD-CITY24.RU/

While this might have been the first purpose-designed terminal, there is little evidence that it influenced subsequent designs and was largely unknown until it was rediscovered by Wolfgang Voight and published in *Buildings for Air Travel* in 1996. The terminal was short-lived and was replaced by a new, larger and more modern terminal in the early 1930s, which was in turn demolished by the occupying Russian Force in 1945 (Figures 3.11 and 3.12).

Figure 3.10 Königsberg showing spectator facilities and rooftop airport identification.
SOURCE: HTTPS://KALININGRAD-CITY24.RU/

Figure 3.11 Königsberg terminal, rebuilt in the 1930s.
SOURCE: HTTPS://KALININGRAD-CITY24.RU/

Figure 3.12 Königsberg, interior view of second terminal.
SOURCE: HTTPS://KALININGRAD-CITY24.RU/ & MUSEUM DER STADT KÖNIGSBERG (KALINGRAD)

3. Dearborn Ford Airport 1928

Ford Airport pioneered, in 1928, one of the most important airport engineering innovations which contributed to both the safety and reliability of the operation: the hard-surfaced concrete runway (Figures 3.13 and 3.14). This innovation would later enable the development of a whole new approach to airport masterplanning that puts the terminal at the centre of the airport between the runways, rather than at the periphery of the all-directional grass field.

Figure 3.13 Ford Dearborn airfield layout showing concrete runway in the background.
SOURCE: DETROIT HISTORICAL SOCIETY

Figure 3.14 Ford Dearborn, showing the final runway layout with vehicle test track around the perimeter.
SOURCE: ABANDONED AND LITTLE KNOWN AIRFIELDS: MICHIGAN NORTHWEST DETROIT AREA – MACSMOTORCITYGARAGE.COM

Civil aviation started later in the USA than in Europe, but Ford Airport pioneered a number of other advances in the States and was noted in *Flight Magazine* in 1926 as *'one of the biggest and most active civilian airports in America'*. Key dates are identified below:

- 1924: November, Ford Airport is dedicated as the first modern airport, equipped with two grass runways and flood lights for night landings.
- 1925: The Ford Motor Company Airplane Division starts manufacture of aircraft.
- 1925: The first in the USA to have an airline terminal for passenger use.
- 1925: The only privately owned permanent airship mooring mast in the USA becomes operational.
- 1925–31: the Annual National Air Tour (to demonstrate the safety and reliability of commercial aviation) started and ended at Ford Airport.
- 1926: Pioneered the world's 1st flight of a commercial airliner guided by radio.
- 1926: First US airport to operate a scheduled passenger airline service between Detroit (Dearborn) and Chicago and first commercial postal service to Cleveland.
- 1928: Construction of the world's first hard-surfaced concrete runway.
- 1931: First hotel in the USA (the Dearborn Inn) designed and built for the air traveller.

The terminal building at Ford airport was designed by Albert Kahn, architect to Henry Ford, who later went onto design a number of important airport buildings in the USA (Figure 3.15).

Figure 3.15 Ford Dearborn terminal building.
SOURCE: MACSMOTORCITYGARAGE.COM/

4. Croydon Airport 1920 to 1928 and 1928 to 1959

The first Croydon Airport was chosen as the Air Port of London in 1920 (customs entry point for international flights) in succession to Hounslow Heath, partly because it was no longer required by the Royal Flying Corps after the First World War. It was also considered better to have an aerodrome to the south of London so that pilots would not have to fly over the city. It was well connected to London by both road and rail and was less affected by fog than Hownslow. It comprised a hotchpotch of buildings, serving a split airfield originally individually known as Beddington and Waddon Aerodromes, connected by a level crossing allowing aircraft to taxi from one field to another but became better known as Croydon Aerodrome. (Figures 3.16 and 3.17).

Croydon pioneered the manned aerodrome control tower to direct the movement of aircraft on the ground and in the air, as seen below right. Initially it was a simple shack on an elevated platform overlooking both parts of the airfield (Figure 3.18). The plan prepared by the London Borough of Sutton, Library Service shows the layout of the first Croydon airport (Figure 3.19).

Figure 3.16 Aerial view of the first Croydon Airport in 1925.
SOURCE: NATS IN CONJUNCTION WITH HISTORIC CROYDON AIRPORT TRUST COLOURISED BY MARINA AMARAL

Figure 3.17 London, Croydon 1920, main entrance-1920.
SOURCE: UK AIRFIELDS AND AIRPORTS

Figure 3.18 Original control tower and customs facility.
SOURCE: NATS IN CONJUNCTION WITH HISTORIC CROYDON AIRPORT TRUST COLOURISED BY MARINA AMARAL

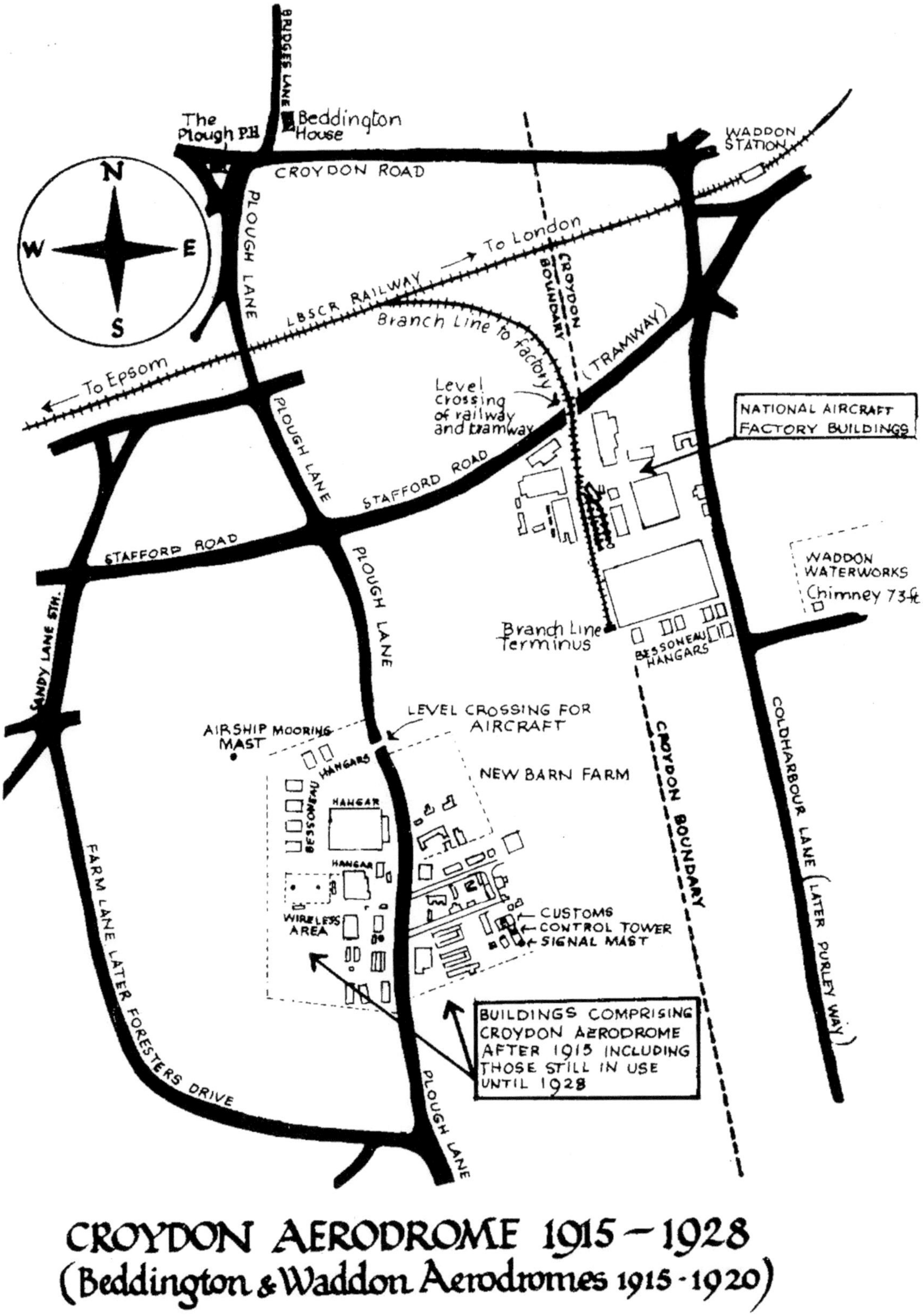

Figure 3.19 Croydon Aerodrome plan 1920–1928.

SOURCE: LONDON BOROUGH OF SUTTON, LIBRARY SERVICE

In 1927, Lowell Thomas embarked, with his wife, on a tour of Europe by air, starting from Croydon Airport. He described his experiences in a book titled European Skyways. He was picked up from his hotel on the Thames Embankment by Imperial Airways and travelled to the airport at Plough Lane which he described as *'The Aerial Crossroads of the British Empire.'* His experience started with breakfast at the Trust House Hotel, followed by a tour of the dispersed airport buildings which he referred to as a *'Harbour in the Fields'*. He described the airport as follows:

> *The flying-field itself occupies a vast area two or three times the size of a university athletic field. Along one side are the cavernous hangars, or sheds, where the aircraft are protected from the rough weather and groomed by expert mechanics for the next flight. Hemming in the field on the other sides are the hotel and restaurant, machine shops, wireless plant, with its three lofty masts, police headquarters and a score of other buildings. Not the least important is the weather bureau, because without complete meteorological information along the air routes, commercial aviation would be a farce.*
>
> *Dominating the converted cow-pasture, figuratively and literally, rises a structure forty feet above the surrounding buildings. This is the Control Tower. And that man sitting up there is the aerial harbour-master. To visit the throne room of this czar of the heavens we climb, ship fashion, a narrow iron ladder. Aloft in this crow's-nest, with an uninterrupted view of the flying-field, sits this celestial ruler, whose official title is Civil Aviation Transport Officer of the London Terminal Aerodrome. But the pilots, with their instinct for short cuts, call him the C.A.T.O.*
>
> *In front of him spreads a huge map, showing all the air routes that radiate out from Croydon. At its side are signal button, telephones, and speaking tubes, by means of which he can keep in touch with the near-by offices, shops and hangars. His windows overlook the aerodrome. A great map painted on the top of his flat desk shows him the position of every plane that flies over the commercial air routes between London and the Continent. His head is adorned with wireless ear-phones, and he can talk even to the pilots of planes a hundred miles away. The CATO directs the movements of all aeroplanes entering or leaving Plough Lane and supervises the management of the whole Croydon Aerodrome.......*
>
> *I saw nothing in Europe that so intrigued and amazed me. The CATO in that Control Tower at Croydon is in direct communication with the pilot of every commercial liner within a radius of two hundred miles, and sometimes he can carry on a conversation with a pilot three hundred miles away. All passenger planes are equipped with the latest wireless sets, and the pilots can send messages as well as receive them. At any hour the pilots can talk directly to the men in Croydon Tower and hear their voices as clearly as if they were sitting in the crow's-nest.*
>
> (Thomas, 1927)

It was quickly apparent that the layout inherited from the RFC (RAF) was unsatisfactory. Plans were made to replace the dispersed facilities with a single terminal building, handling all the necessary processing and administration functions for handling

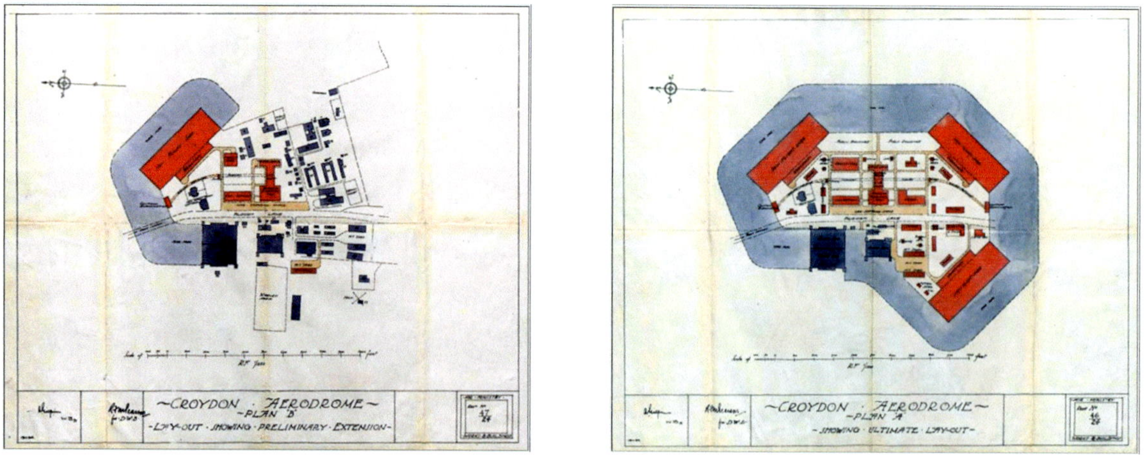

Figure 3.20 Croydon Aerodrome, early re-development plans of 1924.
SOURCE: AIR MINISTRY NATIONAL ARCHIVE VIA HISTORIC CROYDON AIRPORT TRUST

passengers on the existing Plough Lane site. This was a considerable advance over Le Bourget which had continued with a dispersed layout when it was rebuilt in the previous year. Croydon was also the first airport to prepare a master plan for phased development that ensured business continuity during reconstruction (Figure 3.20).

However, the plans were changed because of a fatal air crash on Christmas Eve 1924, which resulted in an inquest and the first Public Inquiry in aviation history. The Inquiry looked into the shortcomings of the First World War airfield and suggested safety could be improved by extending the grass runways. To extend the runways, the airport would need enlargement and improvement, requiring an Act of Parliament to enable the closure of Plough Lane and the amalgamation of the two airfields.

An early revised version of the master plan, to the right, appeared in an article by Archibald Black in the *American City Magazine* in March 1926, titled *London Makes a Mistake.* He commented (Figure 3.21):

> *'Unsatisfactory arrangement of buildings and equipment necessitates expensive removals and reconstruction.'*
>
> (Black, 1926)

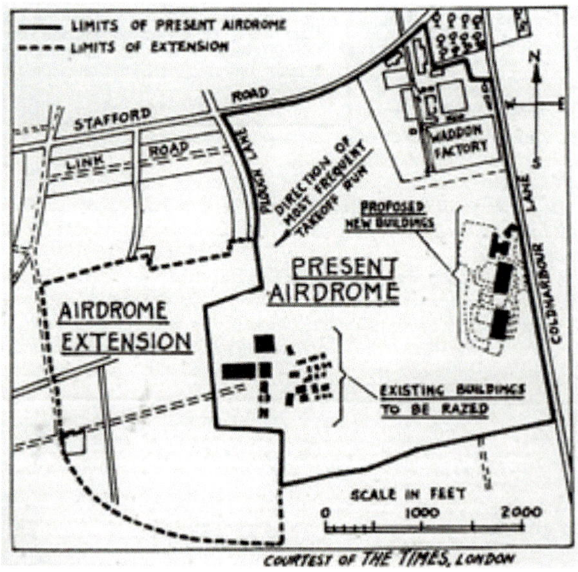

Figure 3.21 Croydon revised master plan.
SOURCE: AMERICAN CITY MAGAZINE, MARCH 1926

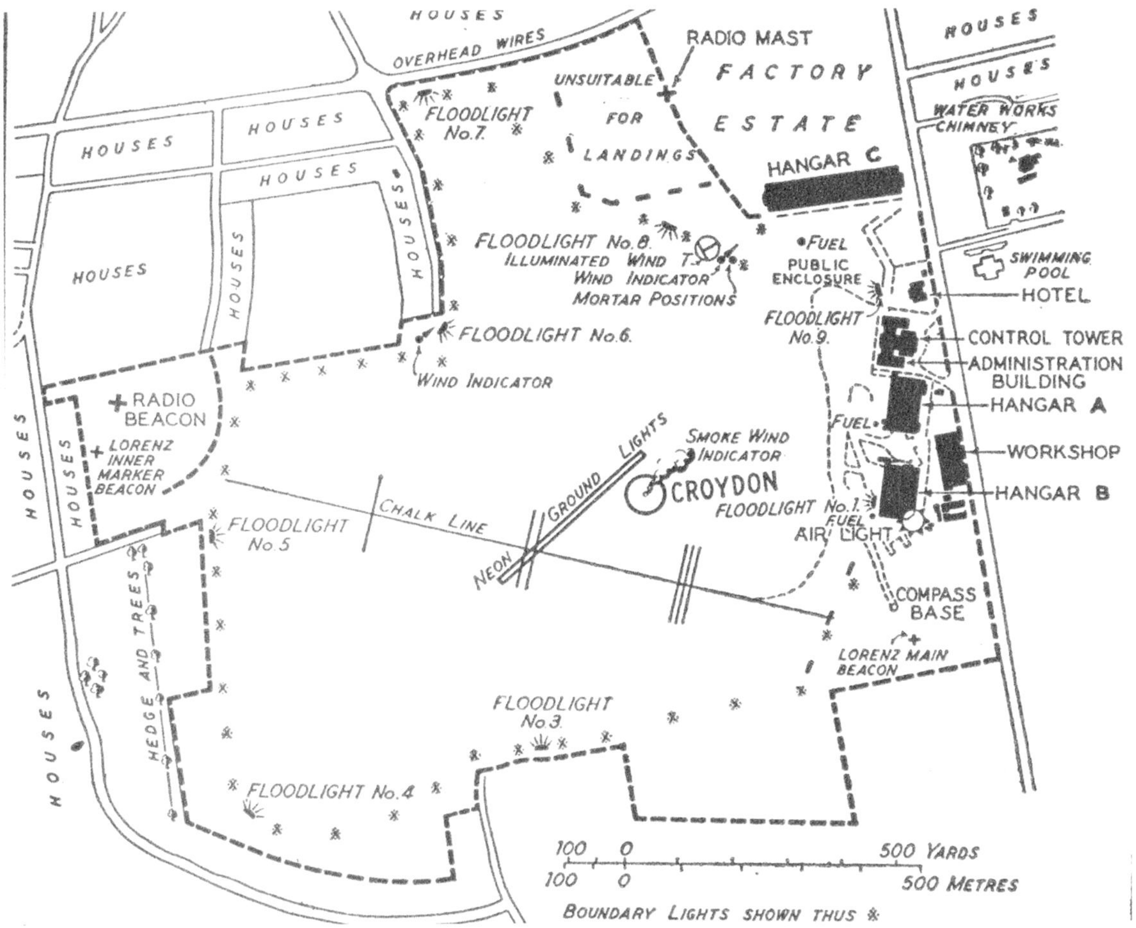

Figure 3.22 The second Croydon airport saw the amalgamation of the eastern and western sections of the airfield by the closure of Plough Lane and its level crossing, and the construction of the new terminal building, next to the Croydon By-pass, Purley Way, and opening in 1928. This is rare example of an airport being expanded and improved primarily for safety reasons.

SOURCE: UK AIRFIELDS GUIDE 1938

The master plan shows an evolution of the terminal plans of 1924, but with the terminal/hangar complex relocated to the eastern boundary, adjacent to the newly opened Croydon By-pass, Purley Way (Figure 3.22).

The new terminal had most of the facilities of the modern terminal: a set-down and pick-up forecourt, check-in desks, immigration and customs areas, simple baggage facilities, a shop and catering outlet, support offices and an early flight information system. However, it did not have two facilities that are found in the modern terminal – passenger and baggage security search and an airside lounge.

Ground-breaking in design, Croydon was the first terminal to integrate the major airport service functions into one structure. The new terminal featured a rational, symmetrical layout to facilitate the logical sequence of departing and arriving passenger movement. Considerable care was taken to introduce daylight to the main double-height concourse through an octagonal skylight (Figures 3.23–3.26).

Despite its conservative neo-classical appearance, which offered a sense of solidity and security from familiar heavyweight construction, the terminal employed a number of novel construction techniques, including a steel frame clad with concrete blocks finished to imitate stone and precast concrete floors and roofs. However, Mr John Dower, Chairman and Secretary of the RIBA Aerodromes Advisory Committee in the 1930s, took a dim view of Croydon Airport's architectural merit, stating that *'Croydon is a dull and grim sobriety and does not express aviation at all!'*

Figure 3.23 Model of the terminal at the Croydon Airport Micro-Museum – note the glazed roof lights that lit up the main concourse and the route through immigration control.
SOURCE: PHOTO BY 'THE CROYDONIST'

During its short life as a civilian airport, Croydon introduced a number of innovations that make it important in the evolution of airport design, including:

- the integration of the major airport service functions into a single structure, the terminal designed from first principles for sequential passenger processing
- the invention of the control tower to direct aircraft movement
- the use of radio and telephone for communication and navigation
- the first airport hotel and restaurant
- the provision of aircraft direction-finding equipment and lighting for night flying
- the foundation of Imperial Airways as their base for flights to the Empire
- the original Airport City development along the Purley Way
- the invention of the Mayday distress call (m'aidez in French)
- the first complaint and legal action about aircraft noise nuisance 1924.

Croydon Airport was by far the most publicised pre-war airport, in books, newspapers, magazines, periodicals and films. The terminal, with its logical sequential layout based on the requirements for passenger processing, was a major innovation and became the benchmark against which all subsequent pre-war airport designs have been compared. The control tower was physically the world's largest, most technologically advanced and comprehensive air traffic control (ATC) operation at the time. Its format remains essentially unchanged in the control towers of today (Figures 3.27–3.30).

Ultimately, Croydon airport was too small and located too close to London. The grass airfield was unsuited to the new larger and heavier aircraft and was surrounded by residential areas, making expansion impossible. It was not considered for post-war use and scheduled services were reduced with the opening of Heathrow in 1946. However, the airport remained in operation for some scheduled services and general

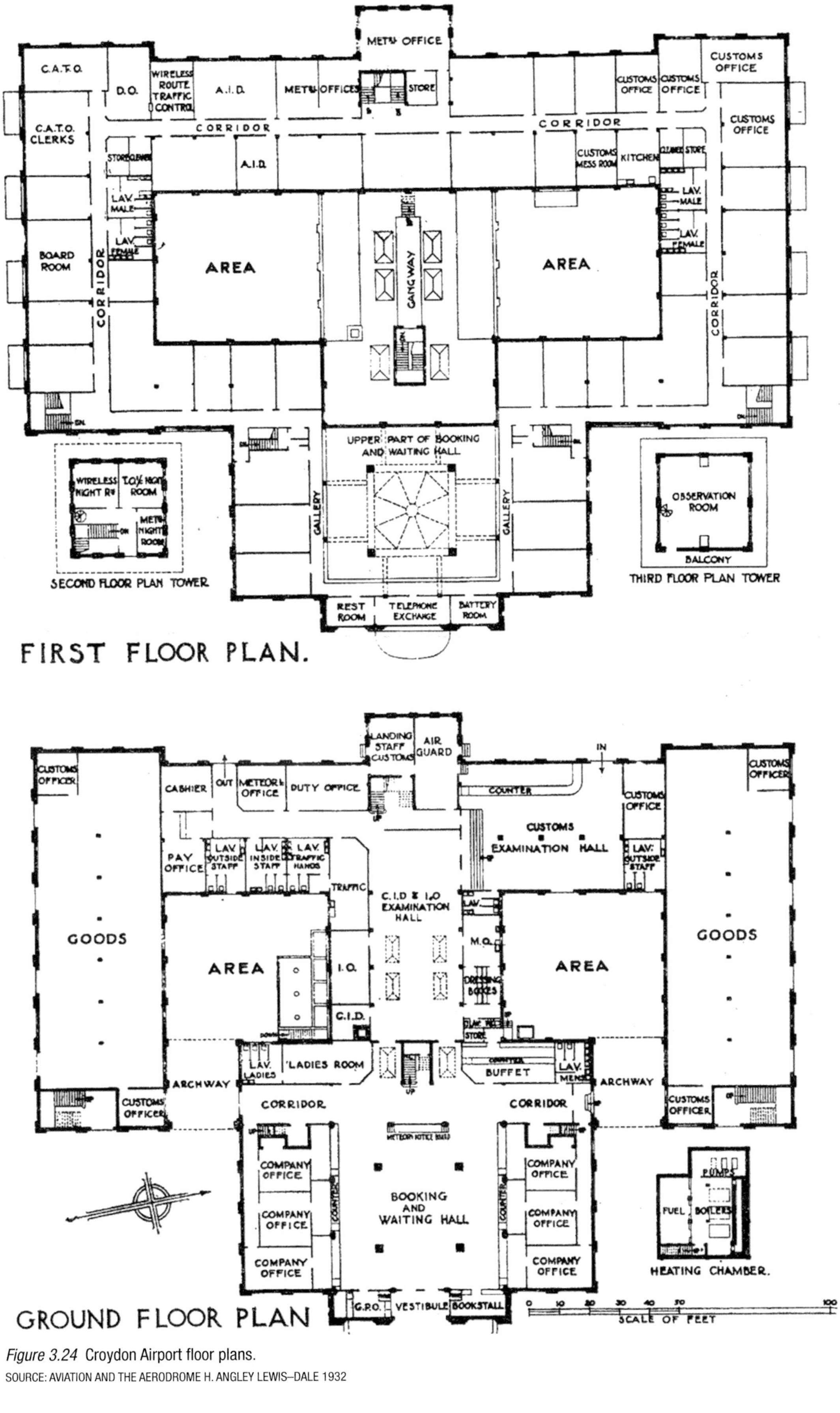

Figure 3.24 Croydon Airport floor plans.

SOURCE: AVIATION AND THE AERODROME H. ANGLEY LEWIS–DALE 1932

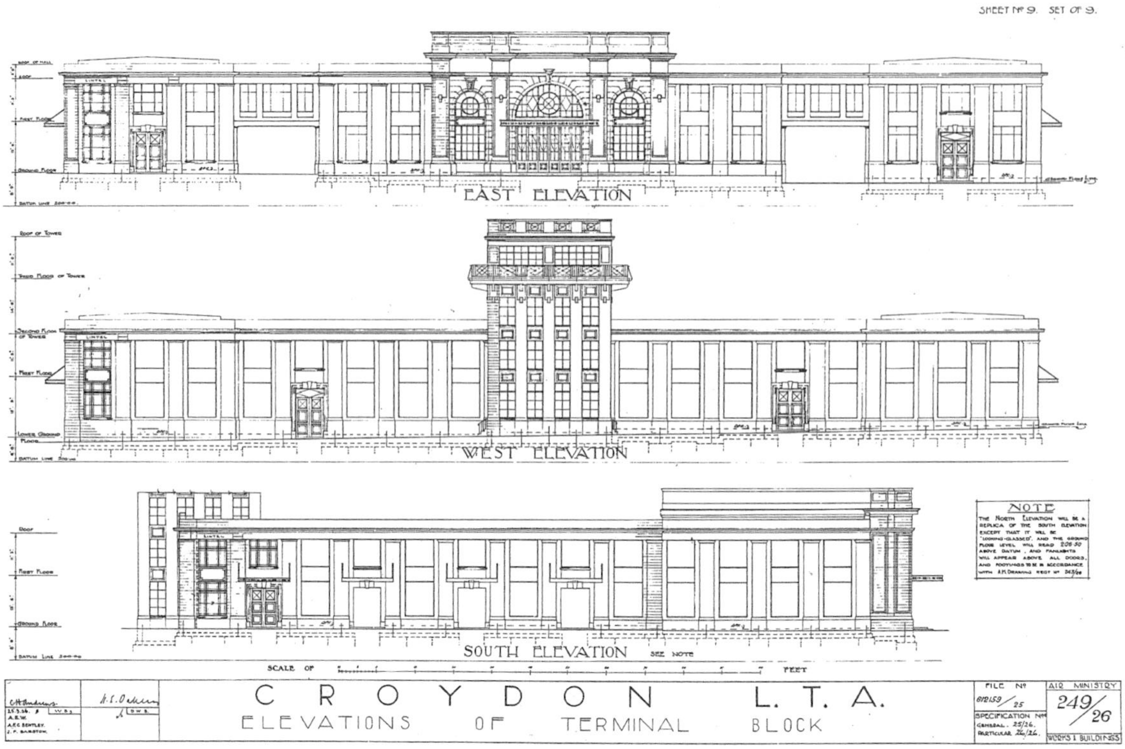

Figure 3.25 Croydon Terminal 1928 – elevations.

SOURCE: AIR MINISTRY ARCHIVE VIA HISTORIC CROYDON AIRPORT TRUST

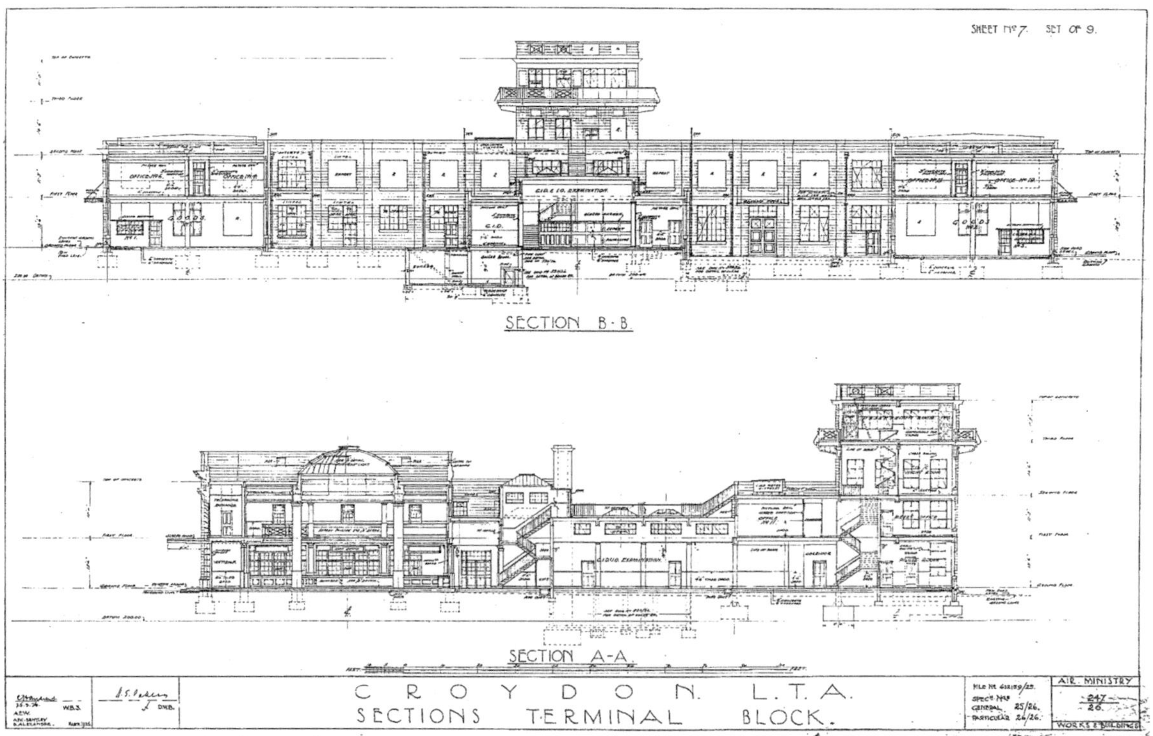

Figure 3.26 Croydon Terminal 1928 – sections.

SOURCE: AIR MINISTRY ARCHIVE VIA HISTORIC CROYDON AIRPORT TRUST

Figure 3.27 The de Havilland DH 91 Albatross Fortuna in front of the control tower at Croydon Airport in 1939. The integrated control tower formed the central element of the airside of the new terminal, influencing a generation of subsequent terminal designs. The top level contained the visual control room and radio communications, while the floor below was dedicated to support accommodation, with a meteorological office on the floor below that again. The control tower at Croydon was the nerve centre of the airport's operation and housed an unrivalled communications and navigation system far in advance of any other airport of the period.

SOURCE: AEROPLANE/KEY PUBLISHING (COURTESY OF COLIN AMBROSE)

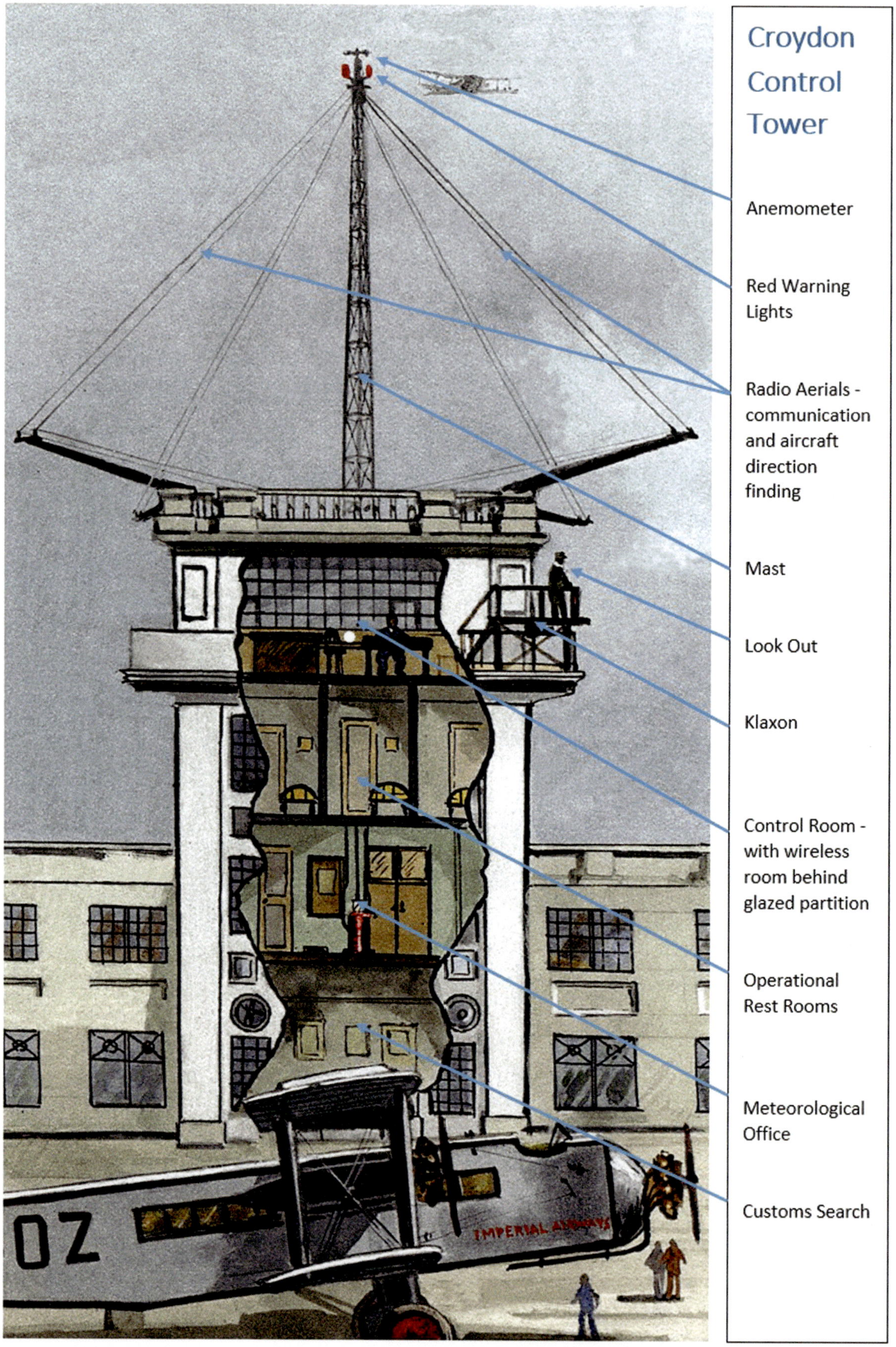

Figure 3.28 Croydon control tower.
SOURCE: SCIENCEPHOTO.CO.UK – POSTER BY SHEILA TERRY

Figure 3.29 Croydon, 'Radio Officers' in the control room with the map room beyond. The role of the first controllers was split between 'Radio Officers' who relayed messages to pilots and the 'Civil Aviation Traffic Officers' (CATOs) who plotted their progress using paper maps and pins.

SOURCE: NATS IN CONJUNCTION WITH HISTORIC CROYDON AIRPORT TRUST, COLOURISED BY MARINA AMARAL

Figure 3.30 Croydon – 'Civil Aviation Traffic Officer' (CATO) recording flight the progress of a in the map room.

SOURCE: NATS IN CONJUNCTION WITH HISTORIC CROYDON AIRPORT TRUST COLOURISED BY MARINA AMARAL

Figure 3.31 Croydon 1928 with the flight information system in the form of clock faces to give arrival and departure times to the left with a news kiosk in the background.

SOURCE: HISTORIC CROYDON AIRPORT TRUST

aviation until its final closure in 1959. By contrast, many of the terminal planning, safety, operational and commercial innovations first seen at Croydon have influenced subsequent airport design.

Croydon also pioneered the introduction of flight information systems and the ability to check in remotely at a town check-in (Figures 3.31 and 3.32).

Figure 3.32 Imperial Airways and Air France coaches arriving at Croydon Airport. The former came from London Victoria Hotel the original town terminal for Imperial Airways from 1924.

SOURCE: HISTORIC CROYDON AIRPORT TRUST

5. New Orleans Sushan (Lakefront) (NEW) 1934

The New Orleans Sushan was built on land reclaimed from Lake Pontchartrain and was the first combined land and seaplane air terminal in the world (Figure 3.33). It was the largest airport in the USA at the time and an early example of multi-directional asphalt runways (Figure 3.34). The airport was one of the first new airports in the USA to be designed to meet the highest A1A standard of the Aeronautics Branch of the Department of Commerce. The standards, which had been published as a circular in 1929, required an airport to have:

- administration, cargo and passenger facilities together with fully equipped hangars, maintenance buildings and refuelling facilities,
- an airfield with a landing area of 2,500 ft in all directions, with increased area if the airport was at a high altitude,
- a comprehensive lighting system of beacons, airfield floodlighting, boundary and obstruction lighting.

The shape of the airport economised on construction costs by virtue of its triangular-shaped airfield, with east and west boundaries converging towards the lake to reduce the amount of fill required with the increasing depth of the lake and to minimise the area of retaining structure.

The airfield can be seen as an exemplar of the American Airport design philosophy of the 1930s. It safeguarded the provision of additional parallel runways to segregate arriving and departing flights when traffic built up. The four runways, orientated to suit the prevailing wind directions, vary between 3,100 ft (945 m) and 4000 ft (1220 m) and are constructed of asphalt. The airfield layout, with its multi-directional asphalt runways, is important in the evolution of airport planning (Figure 3.35).

The terminal is now recognised as a prime example of Art Deco design (Figures 3.36–3.38). Sushan is also significant in developing the concept of the airport as a leisure destination rather than for any innovation in the terminal design. Following the Second World War, the airport lost commercial flights to Moisant Field (now Louis Armstrong International Airport), but it continues to handle personal and company flights, flight schools and some military flights. It is now categorised as a general aviation-reliever airport.

Figure 3.33 New Orleans Sushan, location plan.
SOURCE: FLIGHT 1932

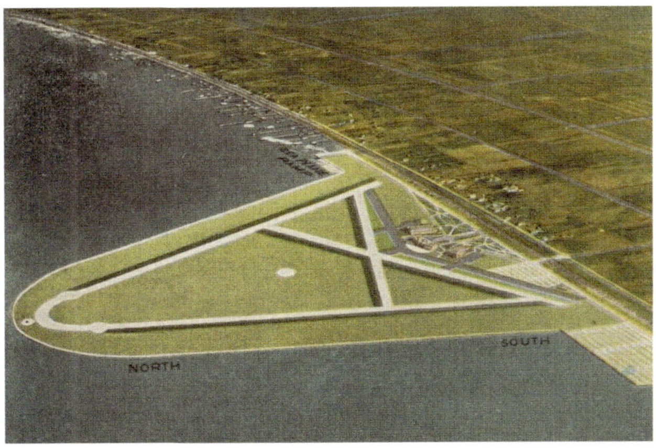

Figure 3.34 New Orleans Sushan, postcard of aerial view.
SOURCE: WIKIPEDIA

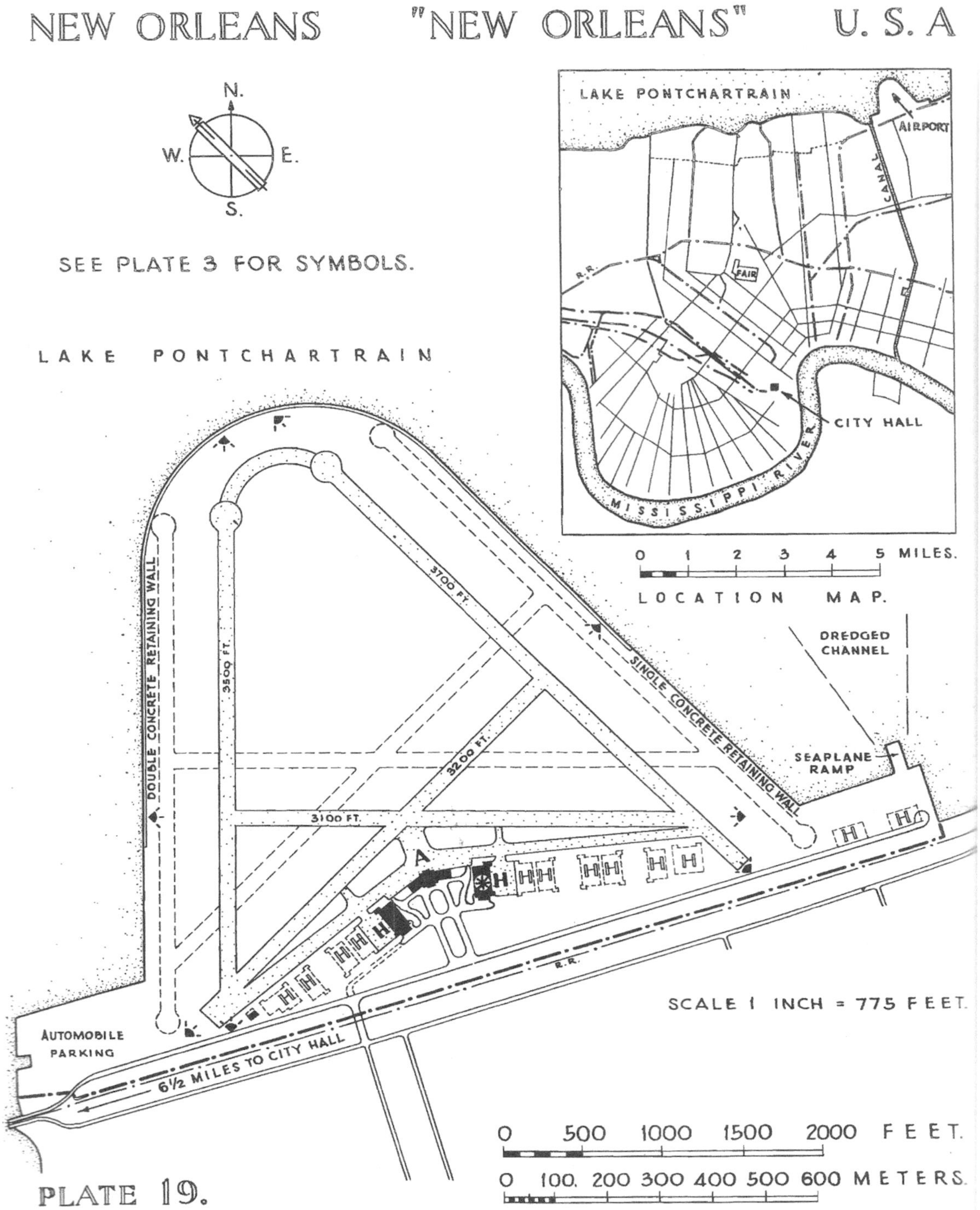

Figure 3.35 New Orleans Sushan (Lakefront) masterplan. The terminal 'A' can be seen projecting into the airfield in the angle between the two shorter cross-wind runways, in contrast to the European airports of the period where terminals followed the perimeter of the airfield, setting a new trend that would eventually culminate in the later mid-field terminal layouts we see today.

SOURCE: AIRPORTS – SOME ELEMENTS OF DESIGN AND FUTURE DEVELOPMENT, JOHN WALTER WOOD & CHARLES L. LAWRANCE

Figure 3.36 New Orleans Sushan, airside view of the terminal with the glazed control tower as the focus and with one of the symmetrical pair of hangars in the background. The sculptural relief on the terminal's façade traces the progress of aviation and its influence on society. The terminal design made no provision for extension, but land was reserved beyond the hangars to allow up to ten additional hangars to be built.

SOURCE: NOLA.COM

Figure 3.37 New Orleans Sushan – main entrance – above the main entrance is a figure of man as a flying machine, flanked by window panels with similar representations; on the side walls are sculptures representing the four winds. Among the sculptors who created these works were Enrique Alferez, William Proctor and John M. Lachin.

SOURCE: SUZASSIPPI.WORDPRESS.COM

Figure 3.38 The Art Deco Terminal and Hangar buildings have now been lovingly restored and can be regarded as one of the prime examples of the style.

SOURCE: ATLASOBSCURA.COM – PHOTOGRAPHY: SANDRA COHEN-ROSE & COLIN ROSE/CC BY-AA 2.0 HTTPS://WWW.FLICKR.COM/PHOTOS/73416633@N00/16294445211/

6. London Gatwick (LGW) 1936

Gatwick Airport, 1936, is recognised as a pioneer in its terminal design, introducing a number of innovations, including the development of the circular form and the use of multiple retractable passenger boarding canopies. The airport was also the first to connect to the mainline railway network and build an underground link from the station to the centre of the terminal. These features have all influenced the subsequent evolution of airport design.

Gatwick was the first realisation of a circular terminal building which allowed aircraft to park all around it, putting into practice ideas that had earlier been put forward in the Lehigh and Stockholm Bromma competitions. This early interpretation of the circular design was developed by architects Hoar, Marlow and Lovett, from a concept provided and patented by the owner Morris Jackaman, which they described as a *Martello Air Station* and later became more colloquially known as the *Gatwick Beehive* (Figure 3.39).

Gatwick was also one of the first airports to introduce dry boarding using simple retractable canopies at each of the six gates. A single canopy had first been experimentally deployed at Oakland Airport USA in 1928, and while multiple canopies were a feature of the winning competition, entry to the Lehigh Airport competition in 1929, Gatwick was the first to deploy multiple retractable boarding canopies in practice.

Although its inability to expand was recognised from the outset, many subsequent airports have followed Gatwick's example and have constructed circular satellites. The excellent connectivity to the main-line railway is also a feature of Gatwick that is now commonplace in modern airport designs.

Gatwick is located some 30 miles south of London adjacent to the London–Brighton main line railway and just south of the then popular Gatwick racecourse. The site was chosen, in part, because the London–Brighton railway was one of the main navigation landmarks that pilots followed on their way from the continent to Croydon airport, making Gatwick a very good location for a diversionary airport (Figures 3.40 and 3.41).

Figure 3.39 'The Beehive' Gatwick 1936 – Model displayed at the Airports and Airways Exhibition at the RIBA 1937.
SOURCE: JANE MARLOW

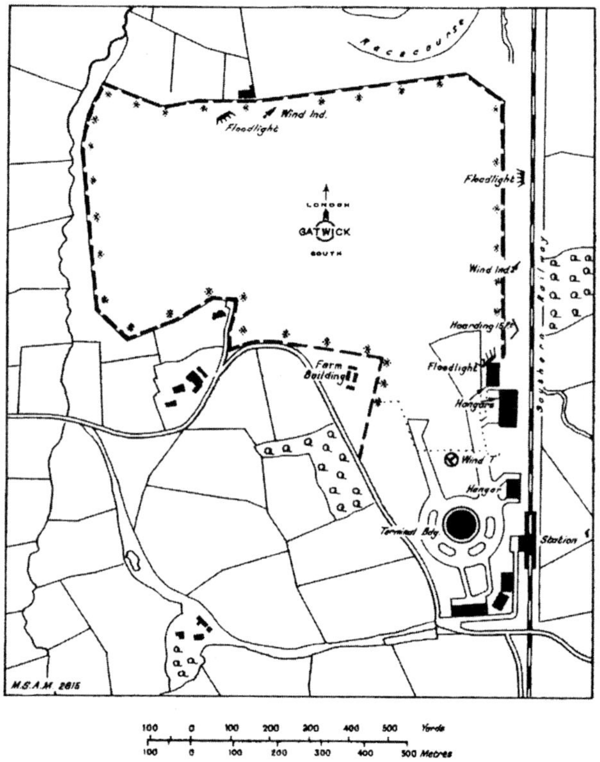

Figure 3.40 Gatwick 1936, site plan.
SOURCE: UK AIRFIELDS AND AIRPORTS

Creating the site involved some challenges: the diversion of the river Mole and drainage of land that lay in the river's floodplain and the construction of a new railway station. It became a public aerodrome in 1934 and the terminal was completed and opened in 1936.

The airport site was small even by the standards of the day, occupying a total area of 196 acres (79.3 Ha) and the triangular southern section of the airport where the terminal, apron and hangars were located occupied about a quarter of the site. The resultant landing area restricted take-off lengths to about 2,200 ft (670 m) east–west and 1,500 ft (457 m) north-south, much less than the UK Air Ministry recommendation, at the time, of 3600–5100 ft.

The terminal building rises from a single storey in the exterior ring to three in the centre. This central

Figure 3.41 Gatwick 1936, aerial view of the southern section of the airfield, with the terminal apron and hangars in the foreground and the railway station to the right.
SOURCE: FLYERTALK.COM (BA)

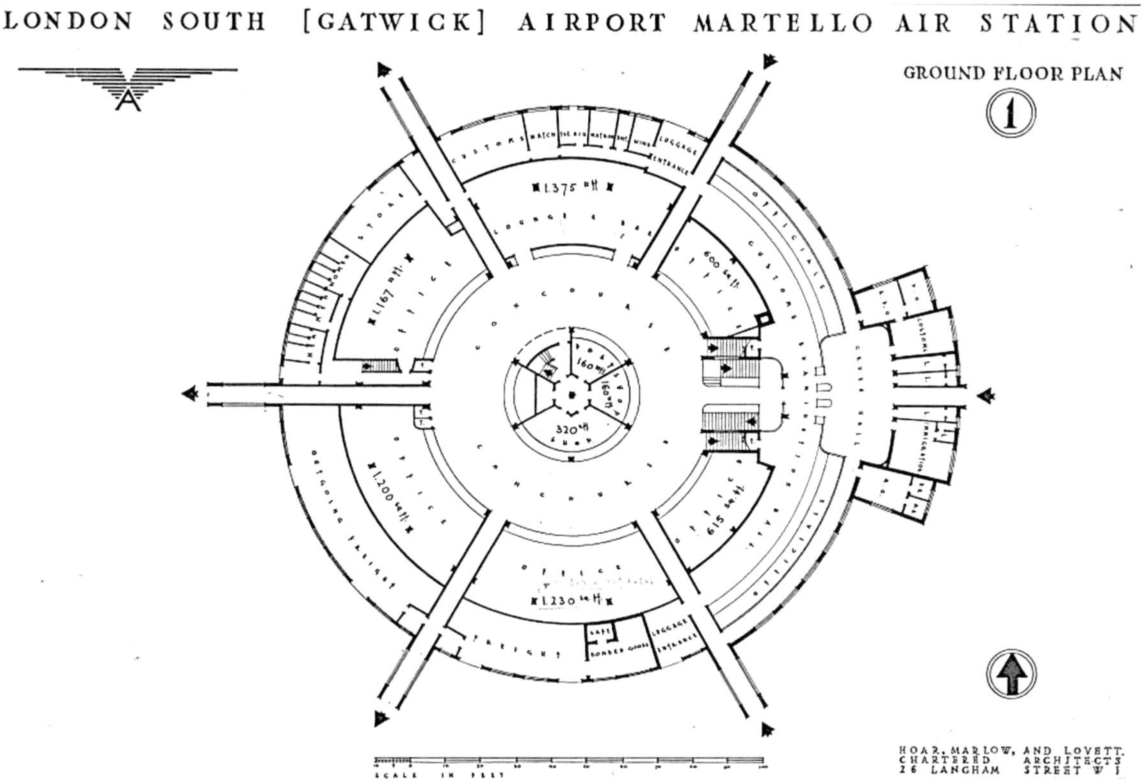

Figure 3.42 Gatwick, Martello Air Station 1936, terminal plan.
SOURCE: JANE MARLOW

section contained a control tower and weather station, with glazing that sloped back at 60 degrees. There was a restaurant with an outdoor terrace and airport offices in a middle ring on the first floor accessed by a grand staircase from the ground floor which contained passenger and baggage processing facilities. The terminal was surrounded by six gates, three for arrivals and three for departures. Unlike today, after the passengers had disembarked, the aircraft had to be moved from one gate to another before the next flight could board the aircraft (Figures 3.42 and 3.43). The sections through the terminal show the underground link to the station 130 yds (119 m) away highlighted in red, ensuring that passengers arriving by train from London stayed undercover from the time they arrived at Victoria station until the time their aircraft reached its destination (Figures 3.44 and 3.45).

Telescopic retractable boarding canopies led from the main concourse, allowing six aircraft to be accessed under cover at one time, while protecting passengers from both the elements and from propeller wash (Figure 3.46).

Figure 3.43 Gatwick 'Beehive' terminal 1936.
SOURCE: UK AIRFIELDS AND AIRPORTS

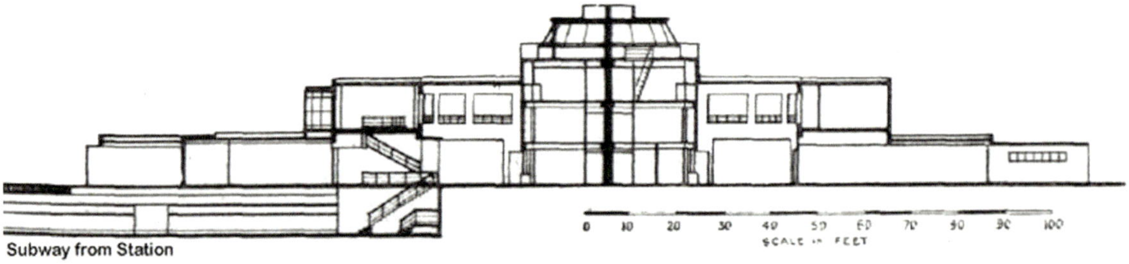

Figure 3.44 Gatwick 1936 section through terminal.
SOURCE: BRITISH CALEDONIAN

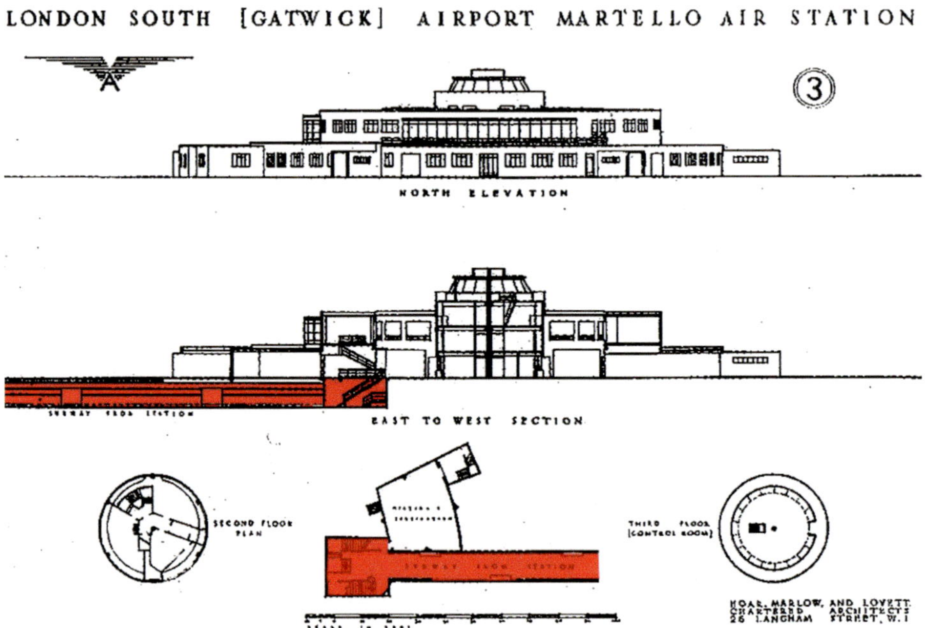

Figure 3.45 Gatwick 1936 section highlighting the underground link.
SOURCE: JANE MARLOW

Figure 3.46 Gatwick 1936 – telescopic boarding canopies.
SOURCE: GATWICK AIRPORT, THE FIRST 50 YEARS, CHARLES WOODLEY

Figure 3.47 Gatwick, interior view of check–in desks and weigh scales.
SOURCE: RIBAPIX

The interior views show how the main concourse has been adapted from its original purpose as a check-in hall to modern day use as a foyer to serve office accommodation (Figures 3.47 and 3.48).

Figure 3.48 Gatwick, interior of the refurbished 'Beehive'.
SOURCE: ALCHETRON.COM

Figure 3.49 British Airways Advertisement.
SOURCE: FLIGHT 12.11.1936

The Bristish Airways advertisement promoted the advantage of the integrated railway station and complementing it by offering combined rail/air tickets that are rare even now. The journey time from London to Paris was better in 1936 than can be achieved today (Figure 3.49).

The aerial view of 1964 shows the relationship of the 1936 terminal (top right) and new 1958 terminal (top left) – the former 'Beehive' terminal has been listed as a historic building and has been refurbished as offices and occupied by a number of airlines (Figure 3.50).

Figure 3.50 Gatwick aerial view.
SOURCE: AUTHOR'S OWN COLLECTION (COURTESY OF KEITH NEWTON) PHOTOGRAPHED IN 1964

7. Paris Le Bourget (LBG) 1923 and 1937

Paris Le Bourget, along with London Hounslow Heath, was the first international airport with passenger services, starting in August 1919 and using the facilities of the former military airfield (Figure 3.51).

The first purpose-designed passenger processing buildings at Le Bourget, which were completed in 1923, were unusual in perpetuating the dispersed layouts of the former military airfield (Figures 3.52 and 3.53). Charles Lindbergh landed at Le Bourget in 1927, making the airport world-famous, after his first solo transatlantic flight in the Spirit of St. Louis.

Figure 3.51 London Hounslow to Paris Le Bourget 1919, the first international scheduled flight.
SOURCE: BA

Figure 3.52 Views of Le Bourget, 1923 showing the dispersed facilities. At this stage Le Bourget missed the opportunity to consolidate all the operational and passenger processing facilities into one integrated terminal which was becoming the norm elsewhere. This separation of services continued until 1937 when the new terminal opened.
SOURCE: TAYLOR EMPIRE AIRWAYS

Figure 3.53 Le Bourget in 1927, showing the facility rebuilt in the neo-classical style, retaining the layout of the original dispersed administration buildings, set amongst a series of aircraft hangars and serving a grass airfield. Passengers had to present themselves in turn to several separate buildings before boarding their flight. The hangars are in the foreground and the dispersed passenger facilities beyond.

SOURCE: AÉROPORTS DE PARIS SA

The location of the first dispersed passenger complex (1923) is highlighted in green, the hangars in blue, while the replacement new linear terminal (1937) is shown in red in the airport plan of 1939 (Figure 3.54). The second terminal, however, was the first purpose-designed linear terminal and a leading example of Classic Moderne or Art Deco Style (Figures 3.55 and 3.56).

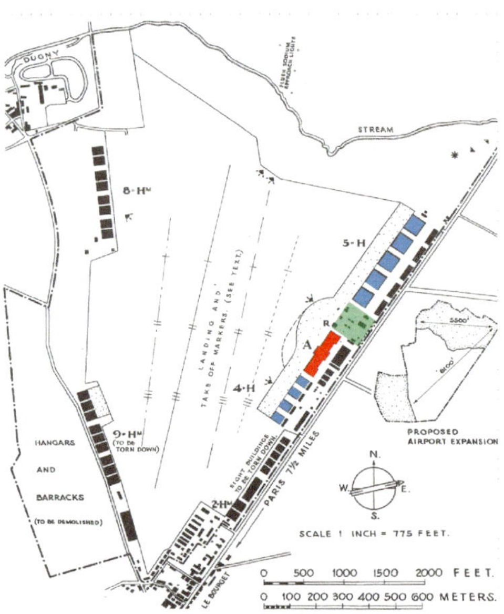

Figure 3.54 Paris Le Bourget, site plan 1939.

SOURCE: AIRPORTS – JOHN WALTER WOOD 1940

Figure 3.55 Le Bourget landside view.

SOURCE: HTTPSPARIS-BISE-ART.BLOGSPOT.COM

Figure 3.56 Airside view of the control tower.

SOURCE: DEL CA

The new terminal, designed by architect Georges Labro, was due to open in 1936 in time for the June International Exhibition, but was not completed in time and only opened in November 1937. In an article published in L'Architecture in January 1938, Labro described his design as follows:

> *My project respected the desire for a simple composition, developing in a straight line slightly in front of the alignment of the hangars, and keeping the depth of the building to a minimum. It also made the most of the heights authorised for the building, the profile of which is intentionally low. In this way I could obtain the largest possible surfaces for the various parts of the terminal, maximising its potential. Furthermore – and this is very important – my design, articulated around a central feature, or if you prefer, around a kind of prow with long wings on each side, allows for the inevitable extensions of the future, with no obstacles placed at the ends of the wings. This makes future changes possible without any fear of compromising the overall order of the composition. As regards the structure, reinforced concrete made rapid construction possible, its simple skeletal framework offering various possibilities for its interior organisation with different and practical combinations of partitions.*
>
> (Labro, 1938)

The centralised linear terminal at Le Bourget can be regarded as a linear extension, on a larger scale, of the same basic form as the simple terminal at Croydon. It provided a far greater number of check-in and airline ticketing desks and a larger reclaim area than earlier designs (Figure 3.57). Horizontal separation of arriving and departing passengers, which can be seen very clearly in Le Bourget, is a common feature of subsequent terminal designs such as Gatwick (1958) and Stansted (1991).

His description of his design having a 'prow with a long wing' is compared with the shape of the Farman Goliath Aircraft overleaf and shows a remarkable similarity of form, but his inspiration was an aircraft that was by then long obsolete (Figures 3.58 and 3.59).

The terminal interior continues the Art Deco theme (Figure 3.60).

Le Bourget has now been restored as a museum, Musée de l'Air et de l'Espace. The original signs show the horizontal separation of arriving and departing passengers (Figure 3.61).

The Art Deco interior *Salle aux huit Colonnes* provided a backdrop to *Death in the Clouds*, the episode of the ITV series, where Hercule Poirot visits both Shoreham Airport in the UK and Le Bourget in France.

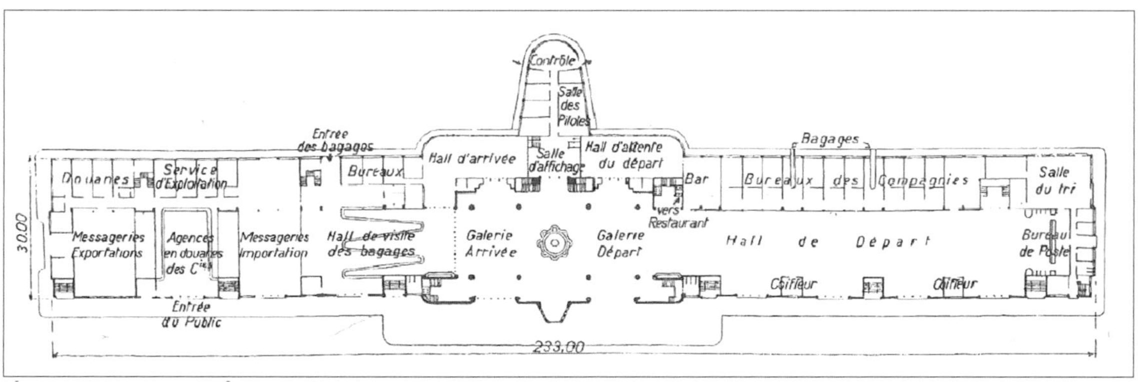

Figure 3.57 Paris Le Bourget, 1937.
SOURCE: PENCIL POINTS

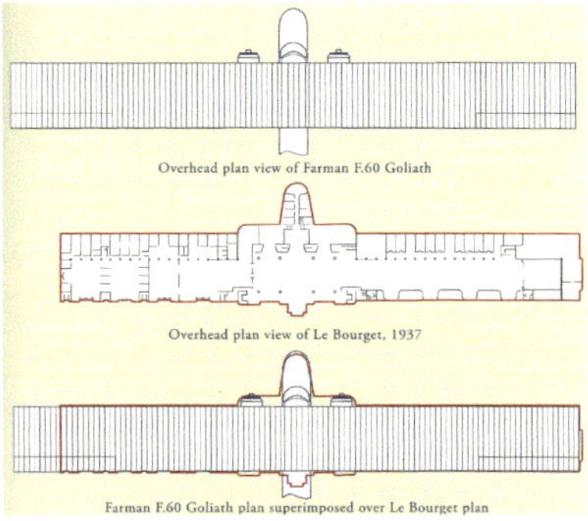

Figure 3.58 Comparison of Le Bourget.
SOURCE: ART DECO AIRPORTS – TERRY MOYLE

Figure 3.59 Farman Goliath Aircraft.
SOURCE: VEHICLEPAD.COM

While Georges Labro, the architect, in theory embraced the idea of future extension to facilitate growth and change, the terminal was only designed for a single departing and a single arriving flight at one time. However, after the war Labro proposed modifications to the terminal to segregate domestic and international passenger flows and extend the airside face of the building. Further airside extensions were introduced as part of a odernization programme in 1964 to increase the airside waiting space (Figure 3.62).

Since 1975, the extensions have been demolished and the terminal has been restored to its original design. It now houses the *Musée de l'Air et de l'Espace*, France's main aviation museum, and hosts an international air show bi-annually alternating with Farnborough in the UK. This restoration shows that airport architecture can be re-purposed when it ceases to function as originally designed.

Figure 3.60 Paris Le Bourget, interior view of the Art Deco main concourse entrance.
SOURCE: HTTPSPARIS-BISE-ART.BLOGSPOT.COM

Figure 3.61 Paris, Le Bourget restored as a museum, showing the horizontal separation of arriving and departing passengers.
SOURCE: ANDREI VIA WORDPRESS

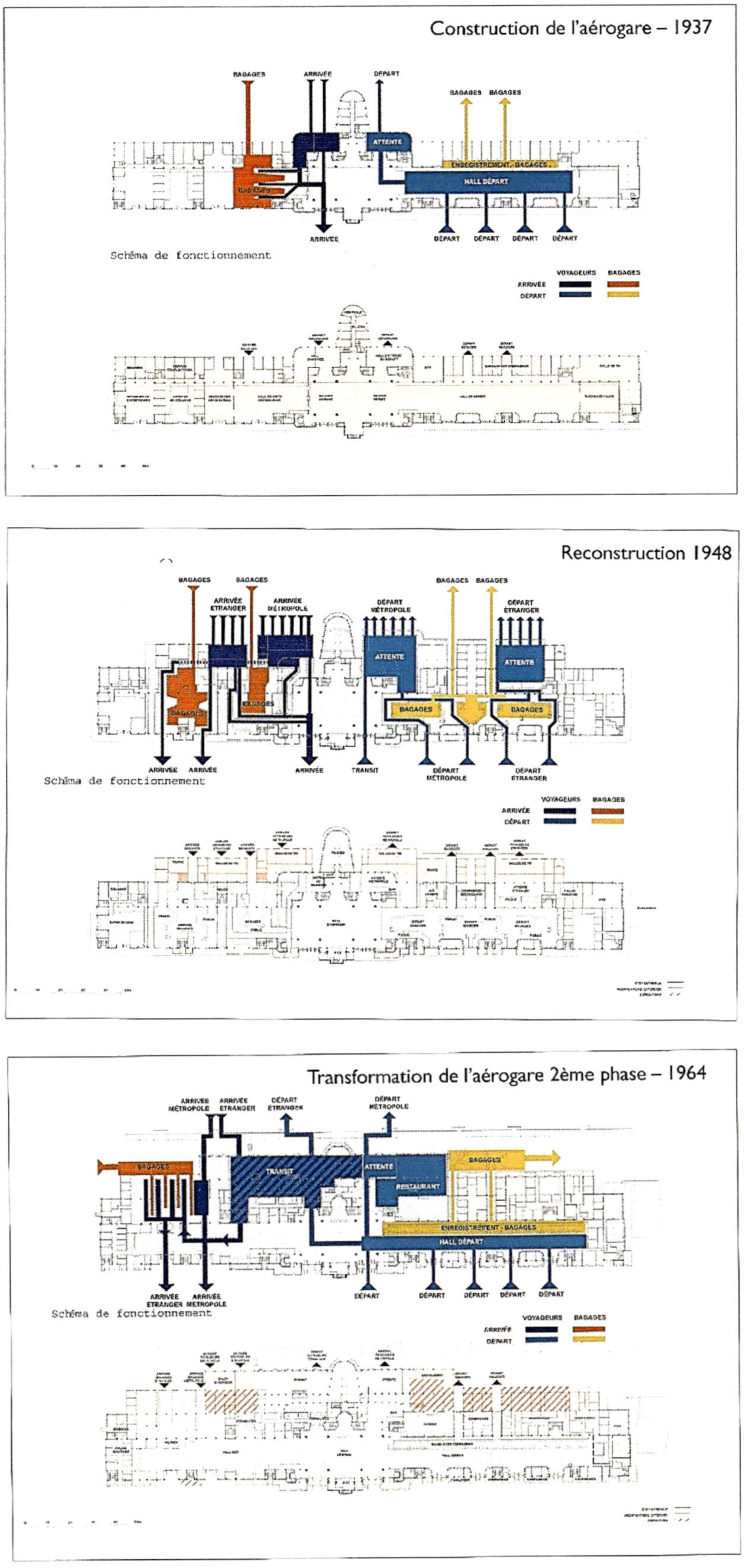

Figure 3.62 Growth and change at Le Bourget: 1937, 1948, 1964.
SOURCE: HISTORIC AIRPORTS

8. Berlin Tempelhof (THF) 1928 and 1939

Designed as the successor to Berlin Johannisthal, pre-war Berlin Tempelhof was the busiest international airport in the world in the 1930s. This was, in part, because of its strategic location as the focus of the Lufthansa network and because Germany used the development of civil aviation as a cover for re-armament, heavily subsidising the development of the national airline and the creation of a range of airports, of which Berlin Tempelhof was the most important.

The original masterplan shows the angled hangars on the periphery of the airfield, framing the central terminal and embracing the airfield (Figure 3.63). The hangars were designed by Heinrich Kosina and Paul Mahlberg and were completed first. The terminal, designed by architects Paul and Klaus Engler, followed soon after and was designed to be built in phases, though only phases 1 and 2 of a three-phase project were completed (Figures 3.64 and 3.65). The horizontal brick-built terminal was designed with a slight curvature that reflected the circular shape of the airfield (Figure 3.66).

Photographs of the landside and airside of the terminal show its appearance when it was newly completed in 1928. The airport was also used for spectator events such as air shows or Nazi rallies and the slight curve of the terminal emphasised the airport's role as an amphitheatre (Figure 3.67). Uniquely the airport was designed with its own underground station – Flughafen. The airport and terminal remained in use for civilian flights during the war, with the last scheduled flight leaving as late as April 1945.

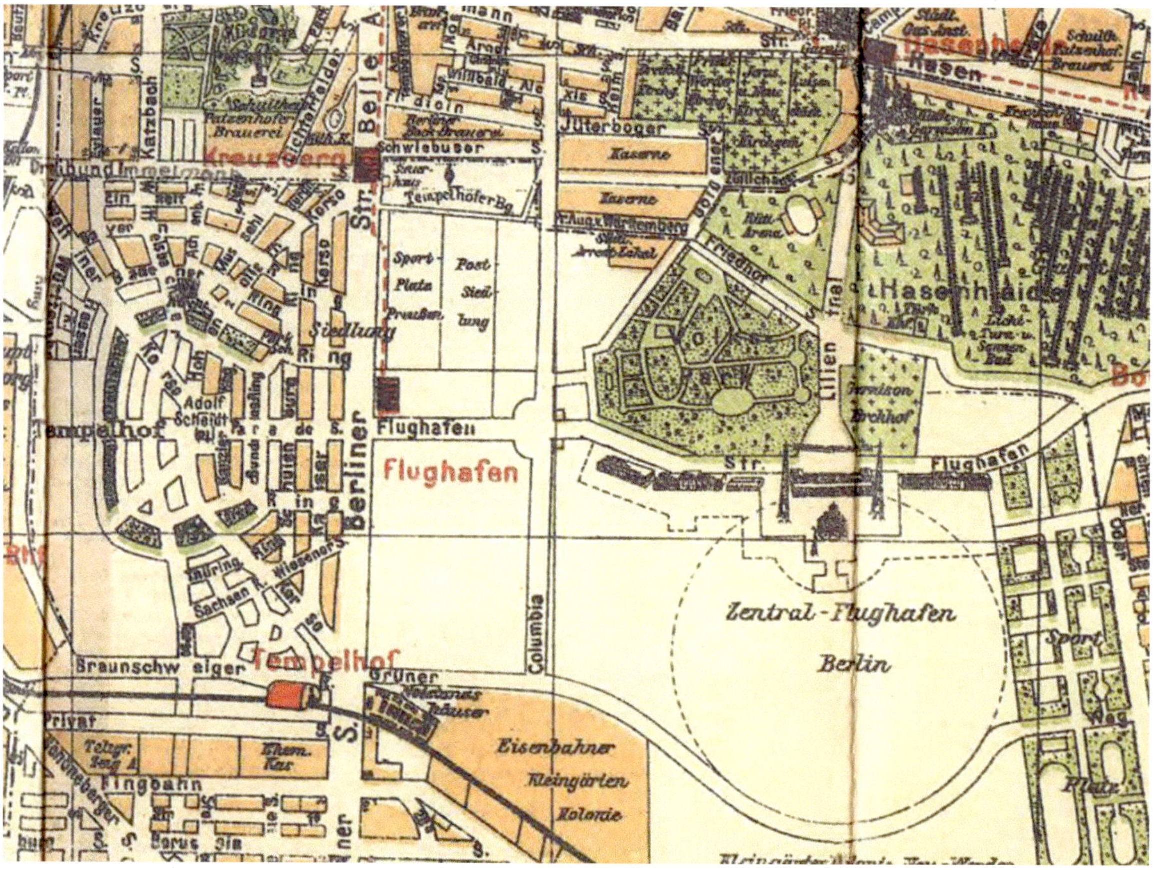

Figure 3.63 Berlin Tempelhof masterplan with Flughafen the underground station serving the airport 1928.
SOURCE: A MULTIFACETED MONUMENT, THE COMPLEX HERITAGE OF TEMPELHOF CENTRAL AIRPORT

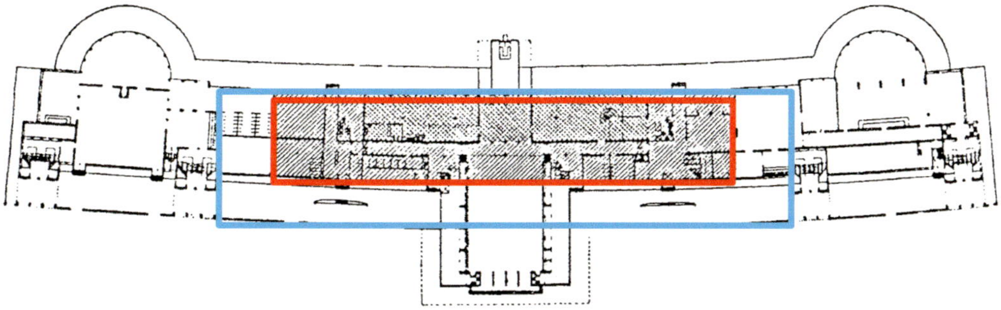

Figure 3.64 Berlin Tempelhof, floor plan, phase 1 in red, phase 2 blue and unbuilt phase 3 with author's added colour.
SOURCE: DIE DEUTSCHE LUFTFARHT, GESCHICHTE DER DEUTSCHEN VERKEHRSFLUGHAFEN

Figure 3.65 Berlin Tempelhof, phase 1 of terminal development.
SOURCE: DIESELPUNKS.ORG

Figure 3.66 Berlin Tempelhof, phase 2 of the terminal development, 1928. Note the anachronistic donkey and cart in front of the terminal.
SOURCE: WORDPRESS.COM

Figure 3.67 Berlin Tempelhof 1928 terrace facing the apron used for large-scale dining and to complement the spectator facilities at the air shows.
SOURCE: POSTCARD -SAMPOR.DE

Lowell Thomas wrote about his experience of landing there in 1927 when only phase 1 of the terminal had been completed:

> *We make a perfect landing at the finest commercial aerodrome in the world, the Tempelhof Field within the city limits of Berlin...... We taxi up to a concrete platform in front of the spacious main offices of the Deutsche Luft Hansa, where a roomy open car with motor running is waiting to speed us to our hotels.*
>
> (Thomas, 1927)

On leaving for Moscow they took one of the earliest night flights describing the experience as follows:

> *After a midnight supper we taxied to Tempelhof, arriving a half-hour before the time to take off. At ten minutes to two our baggage was weighed and we were weighed. Then we were led across the field to the concrete platform where a mighty three-engined Junkers plane stood ready to whirl us through the night across the hills and forests of Prussia to Danzig, our first stop.*
>
> (Thomas, 1927)

When Stedman Hanks described Berlin Tempelhof in 1929, after completion of phase 2 of the terminal, he particularly recognised the excellent quality of its connectivity to the city centre:

> *The airport buildings are placed along the northern boundary of the airport and on the main artery of travel between the airport and the city. An underground station and tram line are near enough to the administration building so that transportation facilities are rapid and at frequent intervals.*
>
> (Hanks, 1929)

He recognised that the terminal had been designed with extension in mind, as can be seen from the Phase 1 and phase 2 photographs and plan, and noted that, unlike Croydon, passenger waiting facilities were airside:

> *The main airport administration building, which has been materially extended during the past year, forms the centre structure of the group. Air passengers enter this building for examination by customs officials, and then pass on to the waiting room to wait for the departure of their airplanes.*
>
> (Hanks, 1929)

By the late 1930s, airline networks were sufficiently developed to allow passengers to plan a multiple sequence of flights, though often with an overnight stop. The drawing below shows the Lufthansa flight network from Berlin and the frequency and times of services. As can be seen, Berlin was the centre of a major network by 1939 (Figure 3.68).

The airport and first terminal remained in use for civilian flights during the war, with the last scheduled flight leaving as late as April 1945. During the 1930s, Berlin Tempelhof was by some distance the busiest international airport in the world. In 1939, its timetable showed that it scheduled some 20 aircraft movements an hour between 9 and 10 am. All traffic was handled through the original 1928 terminal, as the new 1939 terminal, though largely complete, was not yet operational.

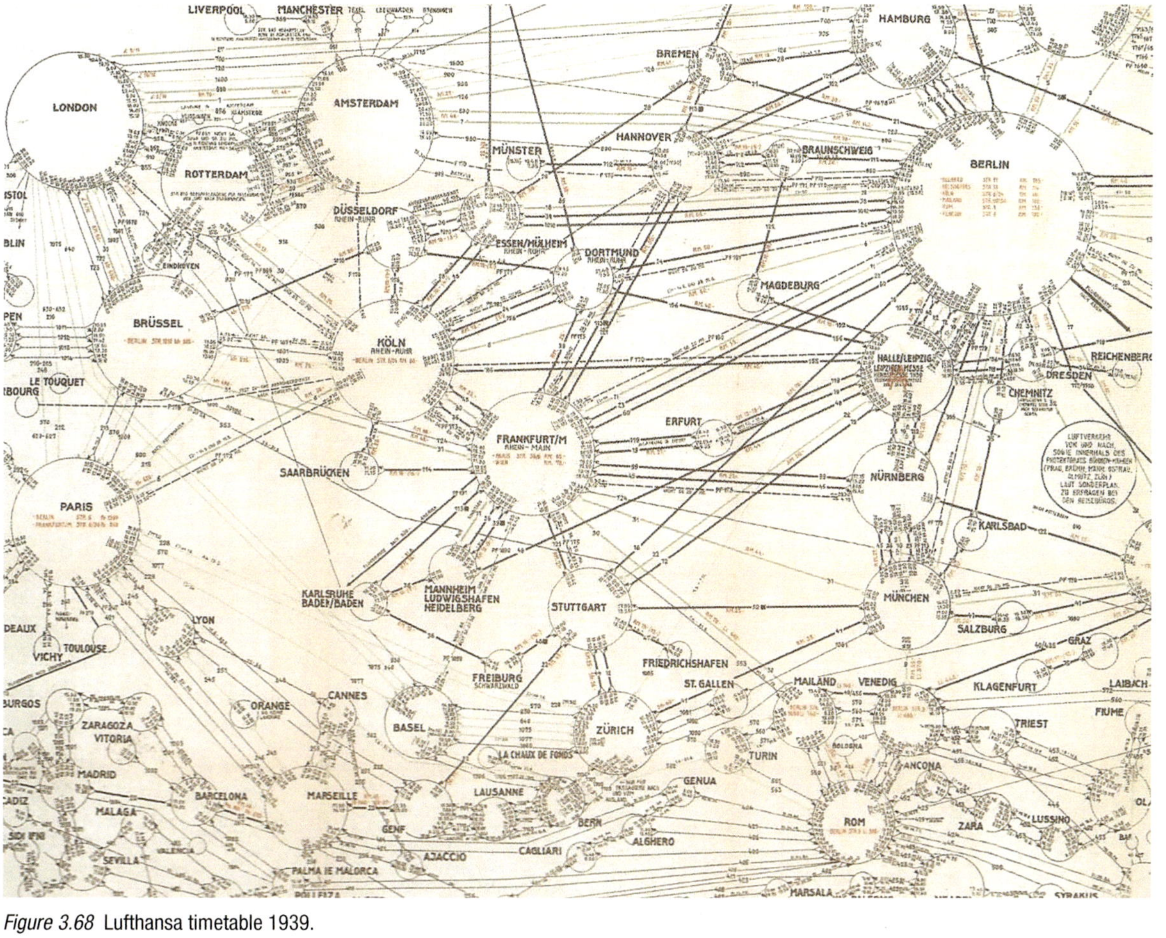

Figure 3.68 Lufthansa timetable 1939.

SOURCE: HISTORIC AIRPORTS (COURTESY OF JOHN KING)

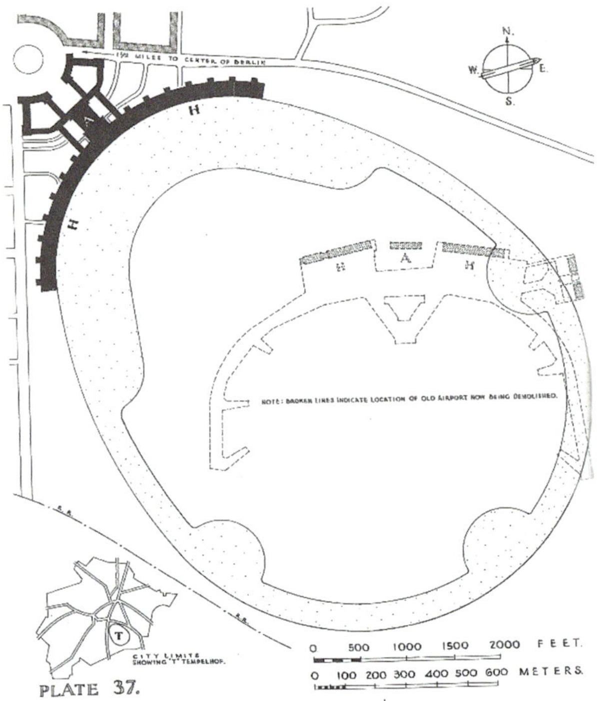

Figure 3.69 Berlin Tempelhof site plan showing the 1st and 2nd terminals.
SOURCE: AIRPORTS – JOHN WALTER WOOD

Figure 3.70 Berlin Tempelhof 2 1955, view under cantilever canopy.
SOURCE: HTTP://PETERPAPKE

Figure 3.71 Berlin Tempelhof 1 1928.
SOURCE: DEUTSCHE LUFTHANSA AG

The site plan illustrates the location of the first terminal (1928), marked A and shown with a broken outline, and the new terminal and hangar complex in the north–west corner of the site (1939) shown in solid. As can be seen, the airfield was extended but was still a grass field, while the new terminal was radically larger and designed to handle traffic that was 30 times greater than the existing demand (Figures 3.69–3.71).

The second terminal at Tempelhof, designed in the 1930s, incorporated a much more far-sighted and ambitious interchange, while the airport still retained the obsolescent concept of a grass landing field. If the first terminal was modest and extendible, the second terminal was the reverse, a grandiose statement of Nazi ambition over a kilometre long and designed to last sixty years without needing extension with a dramatic cantilever canopy to provide shelter to boarding passengers (Figures 3.72 and 3.73).

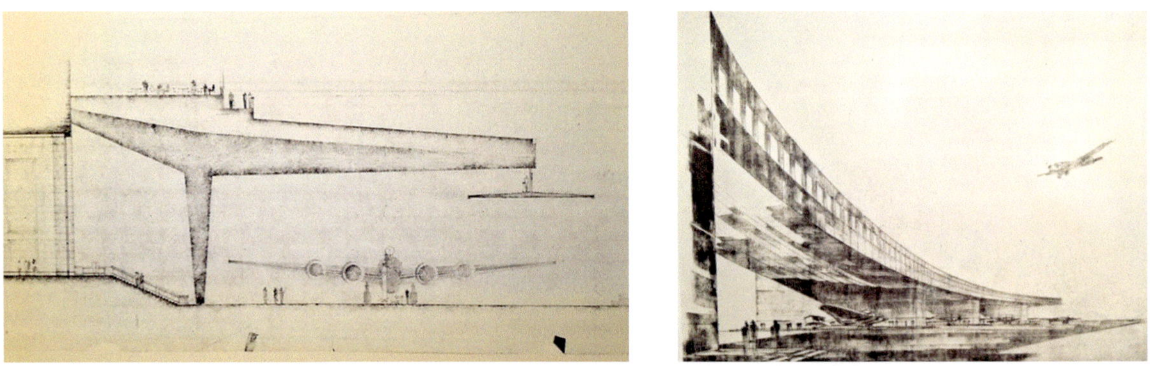

Figure 3.72 Berlin Tempelhof, Ernst Sagebiel early drawings of the cantilever roof concept in 1934.
SOURCE: TECHNISCHE UNIVERSITAT BRAUNSCHWEIG

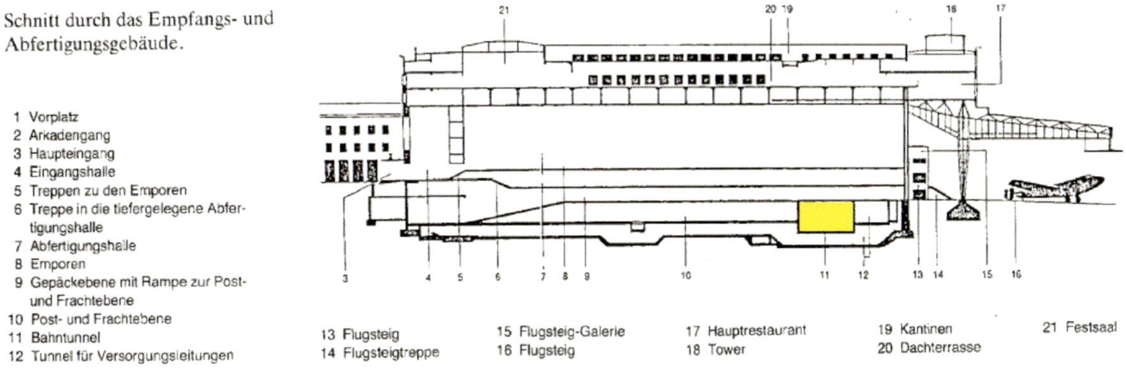

Figure 3.73 Berlin Tempelhof 1939-section through terminal, featuring the interchange functions, with the rail track highlighted in yellow.
SOURCE: DIE DEUTSCHE LUFTFAHRT, GESCHICHTE DER DEUTSCHEN VERKEHRSFLUGHAFEN

It was planned to be integrated into the master plan for Germania, designed by Albert Speer, the intended capital of a new empire (Figures 3.74–3.76). It was also remarkably innovative. For the first time, multiple aircraft could be disembarked, turned around and boarded simultaneously from the gate rooms and under the shelter of an enormous canopy. It was the only comprehensively integrated terminal and hangar complex handling scheduled services. The terminal, with its rooftop seating and the airside apron, also acted as an amphitheatre for major Nazi rallies (Figure 3.77).

The Reich Air Ministry, under the leadership of Hermann Göering, undertook the financing and enlargement of the airport and the construction of the second terminal. In 1935, Ernst Sagebiel, the architect, received the order to design the new airport. Sagebiel, whose work in the Reich Air Ministry began at the end of 1933, had made a rapid advance in his career there with his designs for the Air Force. He especially made his mark because of his planning of the Reich Air Ministry. Many large scale models were made during the design process to show how the airport could be integrated with the Germania master plan (Figure 3.78).

The construction of the new airport terminal began in the spring of 1936. After the initial speedy progress – the first building section was already constructed in the year of the topping-out ceremony in 1937 – construction work came to a halt due to the war. The opening intended for 1939 could no longer take place, though parts of the administration areas were brought into use.

The terminal employed a number of innovative features that helped its role as an interchange (Figure 3.79):

- Vertically segregated passenger and visitor movement was built on the concept first seen in Hamburg. Here passengers remain at a single level, one floor above the apron level where the aircraft parked, while visitors were led up to a gallery level where they could view departing passengers checking in below without interfering with their movement.
- The multiple aircraft boarding gates were strung out as curved arms from the central terminal, allowing up to 20 aircraft to be handled simultaneously, radically increasing capacity and allowing passengers to transfer easily from one flight to another.
- A baggage sorting hall was located immediately below the check-in hall. The mechanical movement of baggage was by lifts, a first for any airport.
- A new station was built to integrate the existing underground railway, which ran along Berliner Straße, with the terminal.
- A new rail track was integrated into the terminal at a lower basement level to transport freight and post to the terminal. The postal facility was linked to the landside road network on Berliner Straße, while the freight facility was linked to Columbiadamm.

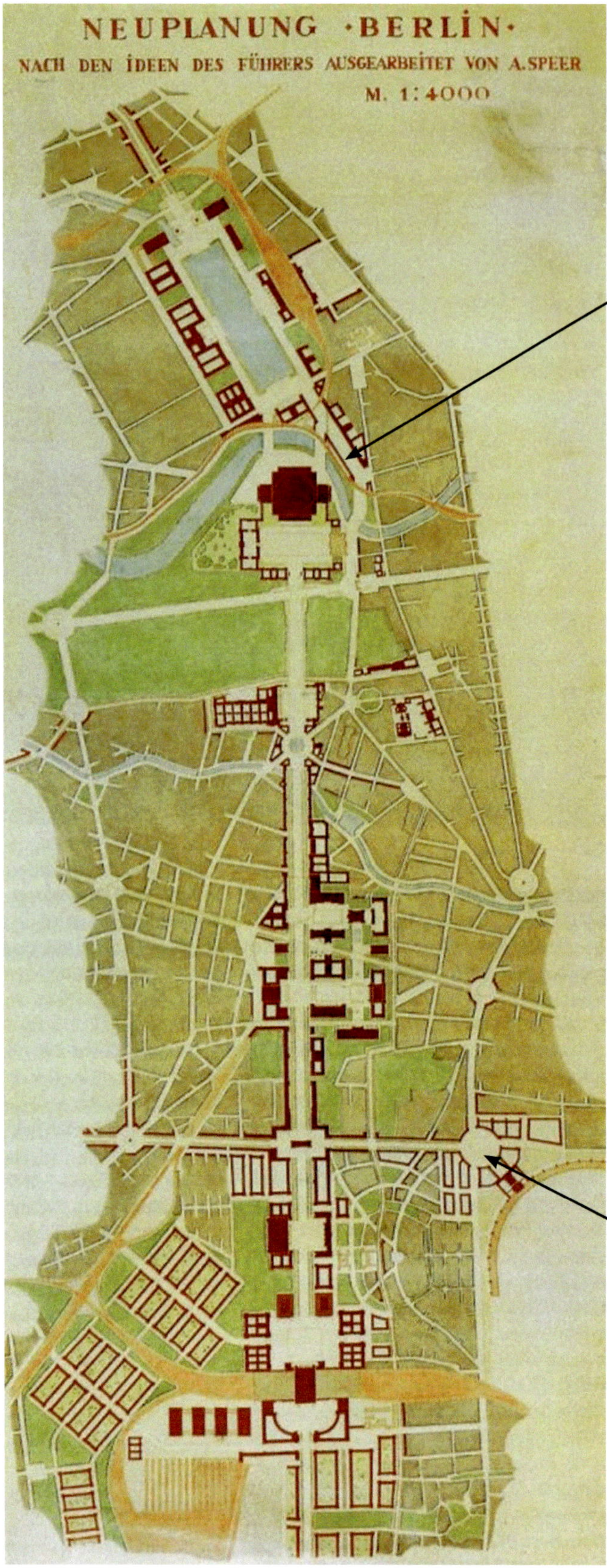

Figure 3.74 Germania master plan 1937.
ARCHITECT: ALBERT SPEER
SOURCE: LANDESARCHIV, BERLIN

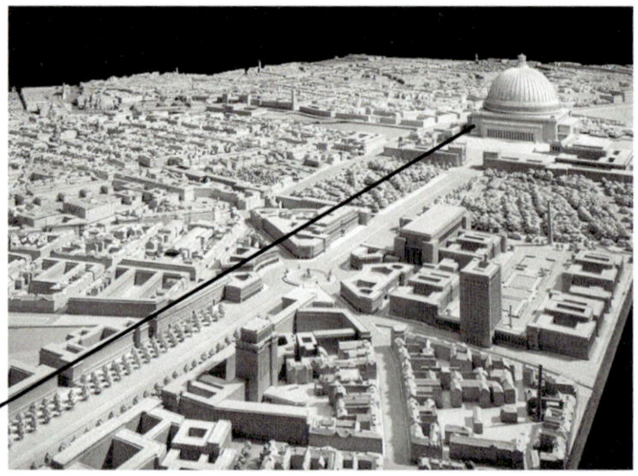

Figure 3.75 Volkshalle/Große Halle, Berlin model 1937.
ARCHITECT: ALBERT SPEER
SOURCE: TWITTER

Figure 3.76 Tempelhof Airport New Terminal model showing original concept for a circus at the entrance 1937 – architect: Ernst Sagebiel.
SOURCE: TUMBLR

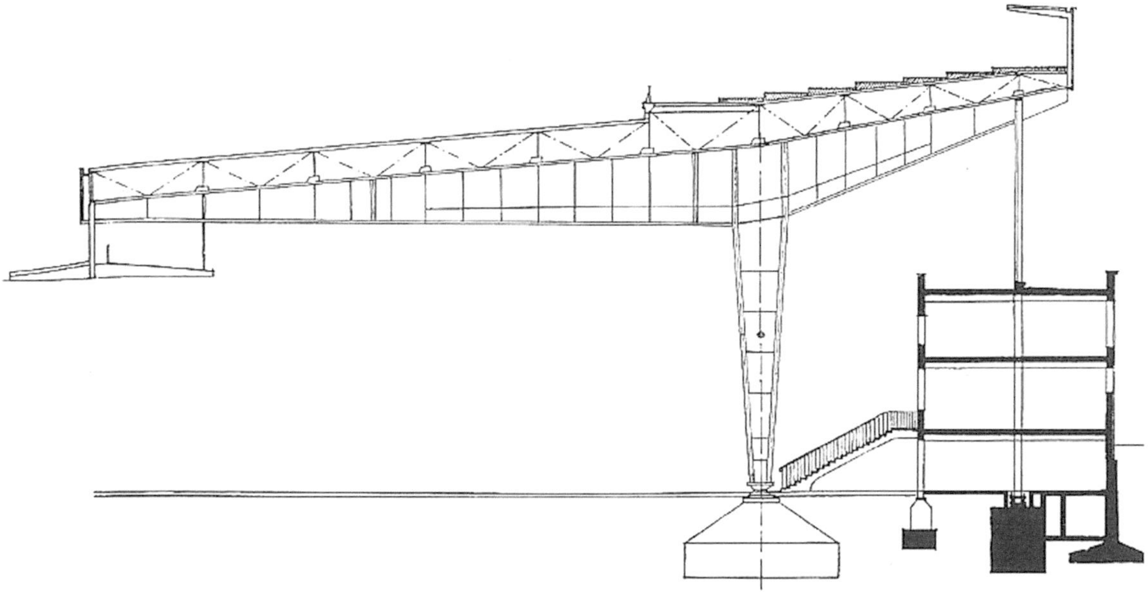

Figure 3.77 Berlin Tempelhof section though 40-metre overhang of the canopy by A. Schleusner showing the cantilever roof canopy over the passenger boarding area and spectator terrace over it. Access to the spectator gallery was provided by a series of staircases around the perimeter of the building in the manner of stadia.

SOURCE: DER FLUGHAFEN TEMPELHOF, ELKE DITTRICH

Figure 3.78 Berlin Tempelhof, model of second terminal building 1937.

SOURCE: THF-BERLIN.DE ©TEMPELHOF PROJECT

Figure 3.79 Berlin Tempelhof layout of the gaterooms that allowed multiple aircraft to be enplaned and deplaned simultaneously under cover of an enormous cantilever canopy.

SOURCE: HISTORIC AIRPORTS

The cantilever canopy of the second terminal allowed aircraft to be boarded under cover, but this concept fell out of favour with the advent of the passenger boarding bridge. The second terminal was the largest building in the world when designed in the 1930s and has been described by Norman Foster as the '*Mother of all Airports*.' It was also designed for long-term growth to last until AD 2000. However, its unique location close to the centre of the city, which was regarded as a virtue when it opened, was ultimately the reason for its closure when the resultant environmental impact was no longer considered acceptable.

Writing in 1940, John Walter Wood identified the two key features that were open to criticism – the extravagance of the cantilever roof and the proximity of the hangars:

> *It seems unnecessary and wasteful of space to build vast roof spans to bring large transport planes under cover to the loading platform when a covered passenger platform would accomplish the same purpose at less expense and without limiting the dimensions of planes and their free circulation at the loading points. Similarly, there are obvious disadvantages in bringing the loading platform and the passengers so closely in contact with the hangars. A more complete separation of incoming and outgoing passenger traffic would have been an advantage to airport operation.*
>
> (Wood, 1940)

His perceptive comments were to become the guidelines for subsequent airport design.

The airfield was less well thought out for retaining a grass field when others had begun to adopt concrete runways. Even the airport director of the time recognised that this feature was obsolescent before it was completed. However, because it had been built with so much surplus capacity, the terminal continued to be used virtually unchanged and was only closed in 2007 due to adverse noise and other environmental impacts.

The proposal for the second terminal at Tempelhof was developed in parallel with Albert Speer's concept for Berlin, but its neo-classical diagonal symmetry was readily assimilated into the master plan for Germania. This vision of integrating an airport with the community has not been reproduced since, but the sophistication of the design concept that integrated passenger, spectator, baggage, freight, mail and VIP circulation suggests that it was indeed a precursor to the modern interchange terminal.

The main structure and appearance of the terminal has remained largely unchanged throughout its lifetime (Figures 3.80–3.82).

The US Air Force occupied the terminal after the war and then built hard-surfaced runways to replace the obsolete grass field. This enabled the airport to be used in the Berlin Airlift when the city was blockaded by the Russians. Aircraft took off and landed three minutes apart, employing new air traffic control strategies which continue to be used today. The airspace system that we enjoy today was developed and perfected as a result of the Berlin airlift in 1948–1949, with modern ground approach control evolving directly from that experience (Figures 3.83 and 3.84).

Figure 3.80 Berlin Tempelhof, second terminal aerial view 1984.
PHOTOGRAPH TSGT JOSE LOPEZ JR., US AIR FORCE

Figure 3.81 Berlin Tempelhof, landside aerial view.
SOURCE: BERLINER FLUGHÄFEN ARCHIV

Figure 3.82 Berlin Tempelhof 1939, airside view with the canopy to the terminal boarding area in the foreground and part of the hangar complex in the background.
SOURCE: PHOTOGRAPHER, DANICA O. KUS

Figure 3.83 British and American officers consult a 3D map of Berlin airspace, with lengths of string tied between wooden pegs representing the vertically separated flightpaths.

SOURCE: IMPERIAL WAR MUSEUM

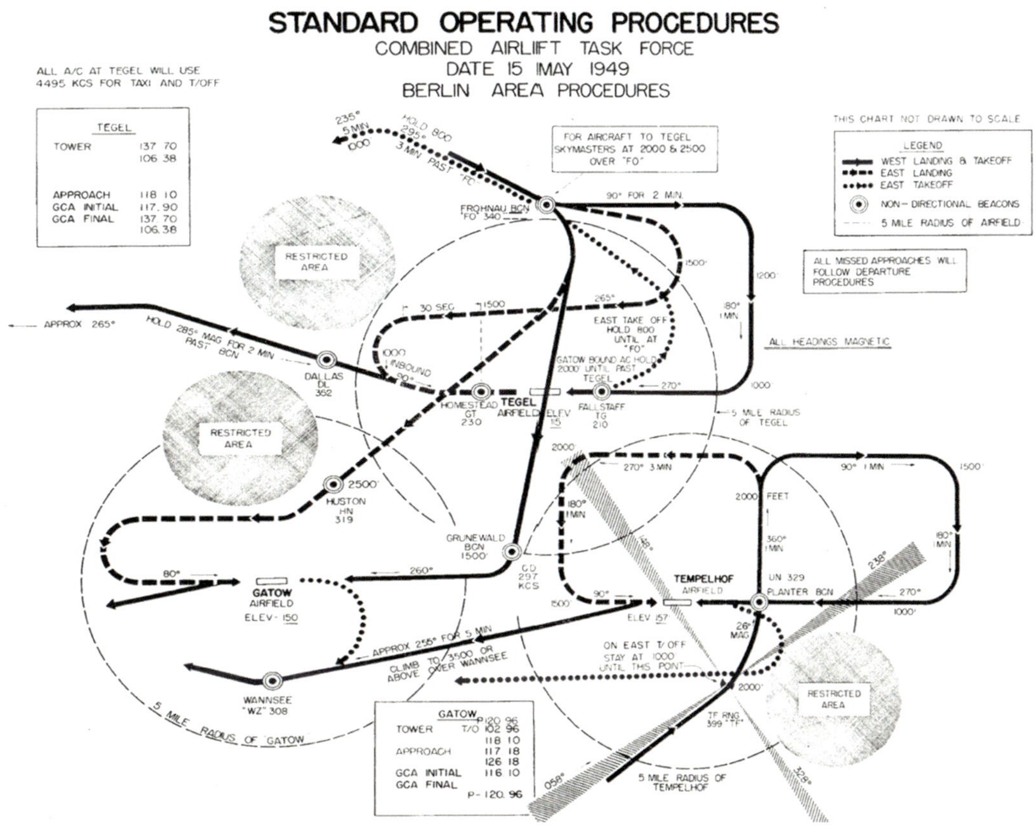

Figure 3.84 Berlin Airlift airspace layout.

SOURCE: US DEPARTMENT OF DEFENCE

Figure 3.85 Berlin Tempelhof, aerial view of final airport layout with twin runways.
SOURCE: GOOGLE MAPS

The photograph of the 3D map of Berlin airspace shows the vertical separation of aircraft, while the standard operating procedures diagram shows the multiple arrival and departure routes in plan. In combination they maximise aircraft separation and optimise capacity.

The standardisation of aircraft type to the C-54 Skymaster during the airlift aided the pit stop performance in aircraft turnaround, a factor that would later be rediscovered by the Low Cost Carriers.

While the terminal remained largely unchanged during its lifetime, the airfield was redeveloped after the war with twin parallel concrete runways to bring it up to date (Figure 3.85). However environmental concerns of noise and safety forced its eventual closure of the airport in 2007, with the airfield becomming re-purposed as a park and the terminal as a museum (Figure 3.86).

Figure 3.86 Berlin Tempelhof, check-in hall re-used as a museum in 2019 – the airport has taken on a new role as a museum and park, following the closure of the airport.

SOURCE: TAGESSPIEGEL.DE, PHOTO: KITTY KLEIST – HEINRICH

9. New York LaGuardia (LGA) 1939

LaGuardia (LGA) included several innovative features and is recognised as a key transition between the earlier simple terminal and later linear terminal designs (Figures 3.87–3.89).

The initial idea to develop an airport for commercial flights and the selection of the airport's location began as a publicity stunt in 1934. New York Mayor, Fiorello La Guardia, in a well-publicised event, refused to disembark from a TWA flight from Pittsburgh to Newark, declaring that his ticket showed his destination as New York. With reporters conveniently already aboard the airplane, the press conference that followed almost immediately, made it clear that the time had come for a new, modern facility closer to Manhattan. The preferred site was identified at North Beach, on reclaimed land next to Bowery Bay on Long Island (Figure 3.90).

On the day the airport officially opened in 1939, it was known as the New York Municipal Airport but was later renamed LaGuardia after the mayor in 1947. The construction of the new airport was supported by the Works Progress Administration (WPA) as part of the 'New Deal' to create new jobs during the Depression and partly to develop America's commercial transportation system.

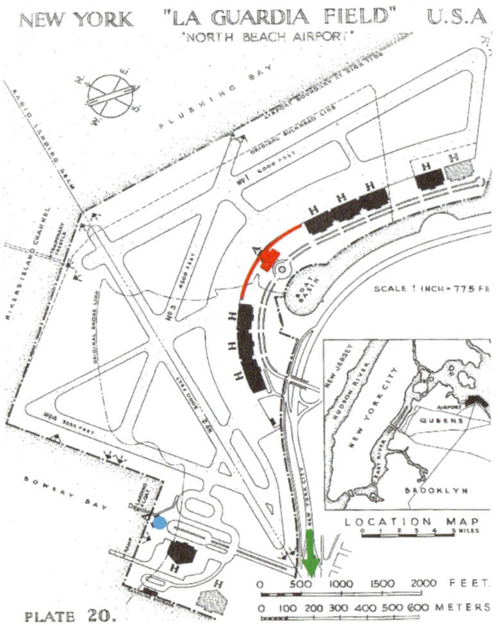

Figure 3.87 LaGuardia master plan 1939.
SOURCE: AIRPORTS – JOHN WALTER WOOD 1940

Figure 3.88 LaGuardia, airside view with 'Skywalk' in the foreground.
SOURCE: PANYNJ VIA PINTEREST

Figure 3.89 LaGuardia, aerial landside view.
SOURCE: PANYNJ VIA STUCKATTHEAIRPORT

The terminal is notable for a number of innovations (Figures 3.91–3.93):

- the Main Terminal (red) for land planes, projected into the airfield,
- a second circular Marine Air Terminal (blue) catered for international flying boats
- direct access to Grand Central Parkway (green arrow) linking with Manhattan
- a two-level forecourt separating departing and arriving vehicles,
- a curved 'Skywalk' to access aircraft, a precursor to the linear pier.

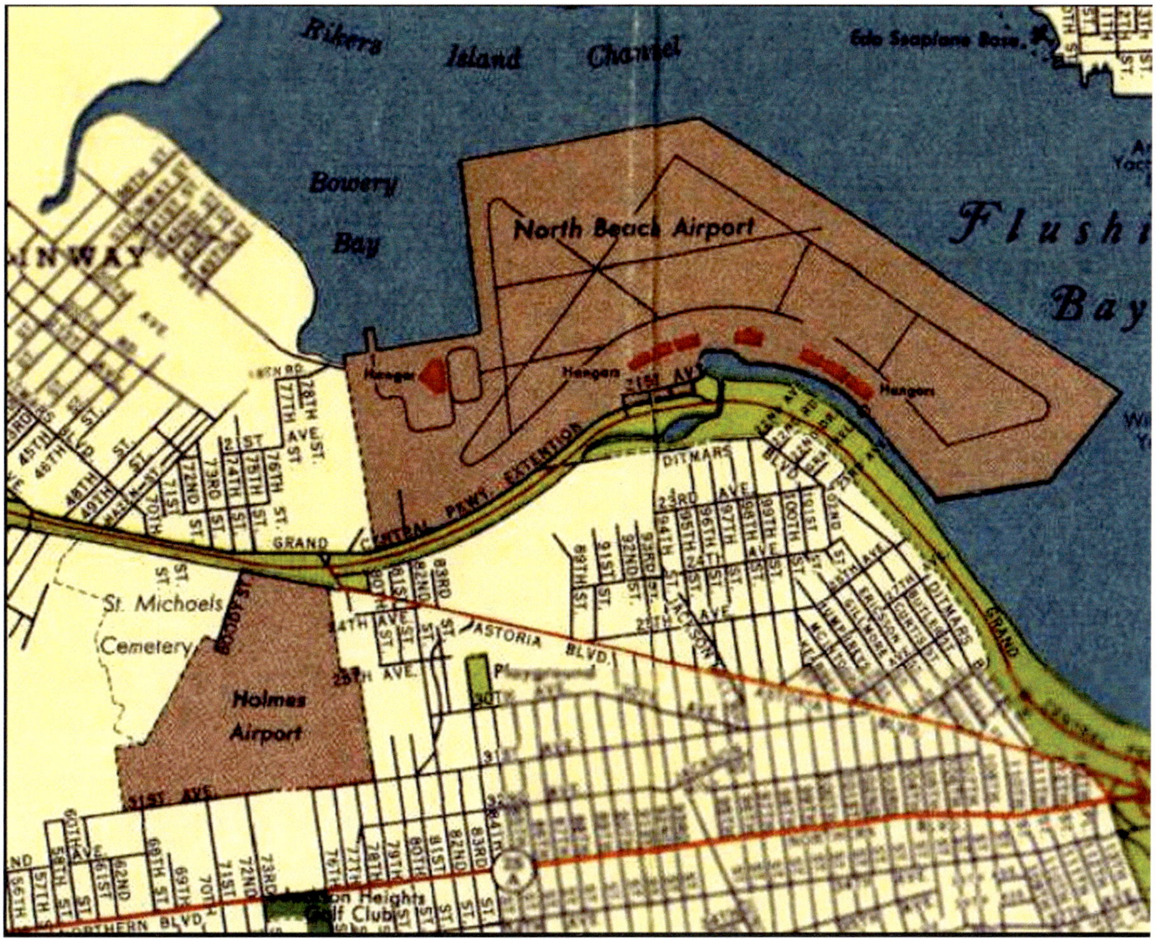

Figure 3.90 Original 1936 masterplan for North Beach airport (later LaGuardia).
SOURCE: NEW YORK POLICE DEPARTMENT

Delano & Aldrich, the architects, designed the circular terminals in Art Deco style, characteristic of the period, incorporating elaborate friezes, murals and sculptures that illustrate flight (Figures 3.94–3.96). They also developed a simple approach to multiple aircraft enplaning and deplaning by incorporating a curved and covered walkway extended either side of the domestic terminal to provide access to several aircraft.

Robert Moses, the infamous New York planner and power broker, built 416 miles of Parkways, which were closed to all commercial traffic. This included the Grand Central Parkway which opened in 1936, linking Manhattan Island to Long Island and passing directly in front of the land which was then being reclaimed to construct LaGuardia Airport. Parkways were conceived as *'ribbon parks'* for pleasure driving. Moses went to great lengths to prevent the use of Parkways by public transport and wrote legislation prohibiting the use of Parkways by buses or other commercial vehicles and, to ensure a

Figure 3.91 LaGuardia, New York 1939 showing the integrated planning concept with interchange between road, airplanes, large seaplanes, light seaplanes and yachts.
SOURCE: PANYNJ VIA MACAULAY HONORS COLLEGE

simple reversal of legislation couldn't undo his strategy, built bridges over the parkway with only 11 ft (3.3 metres) clearance.

Seaplanes became obsolete soon after the war as the range and performance of landplanes increased. So the Marine Air Terminal, now obsolete for seaplanes, became the airport's international departure point for land planes, but larger aircraft and a need for more space prompted other carriers to follow Pan American's lead and move to Idlewild Airport by the end of the decade. However, the postwar growth in domestic air traffic more than compensated for these losses as LaGuardia cemented its reputation as the city's "close-in" airport of choice for short- to medium-haul domestic flights.

Figure 3.92 New York LaGuardia 1939, Art Deco style Landplane Terminal showing the two level forecourt that segregated arriving and departing traffic. Departures vehicles set down at the upper level and arrivals vehicles picked up at the lower level. While the photograph shows a coach in the lower forecourt it would have been prevented from taking the direct Parkway route to the airport.

SOURCE: PANYNJ VIA BETTMANN ARCHIVE

Figure 3.93 New York LaGuardia 1939 with the curved Skywalk in the foreground.

SOURCE: POSTCARD – PHOTO BY W. HOFF OFFICIAL PHOTOGRAPHER LA GUARDIA FIELD

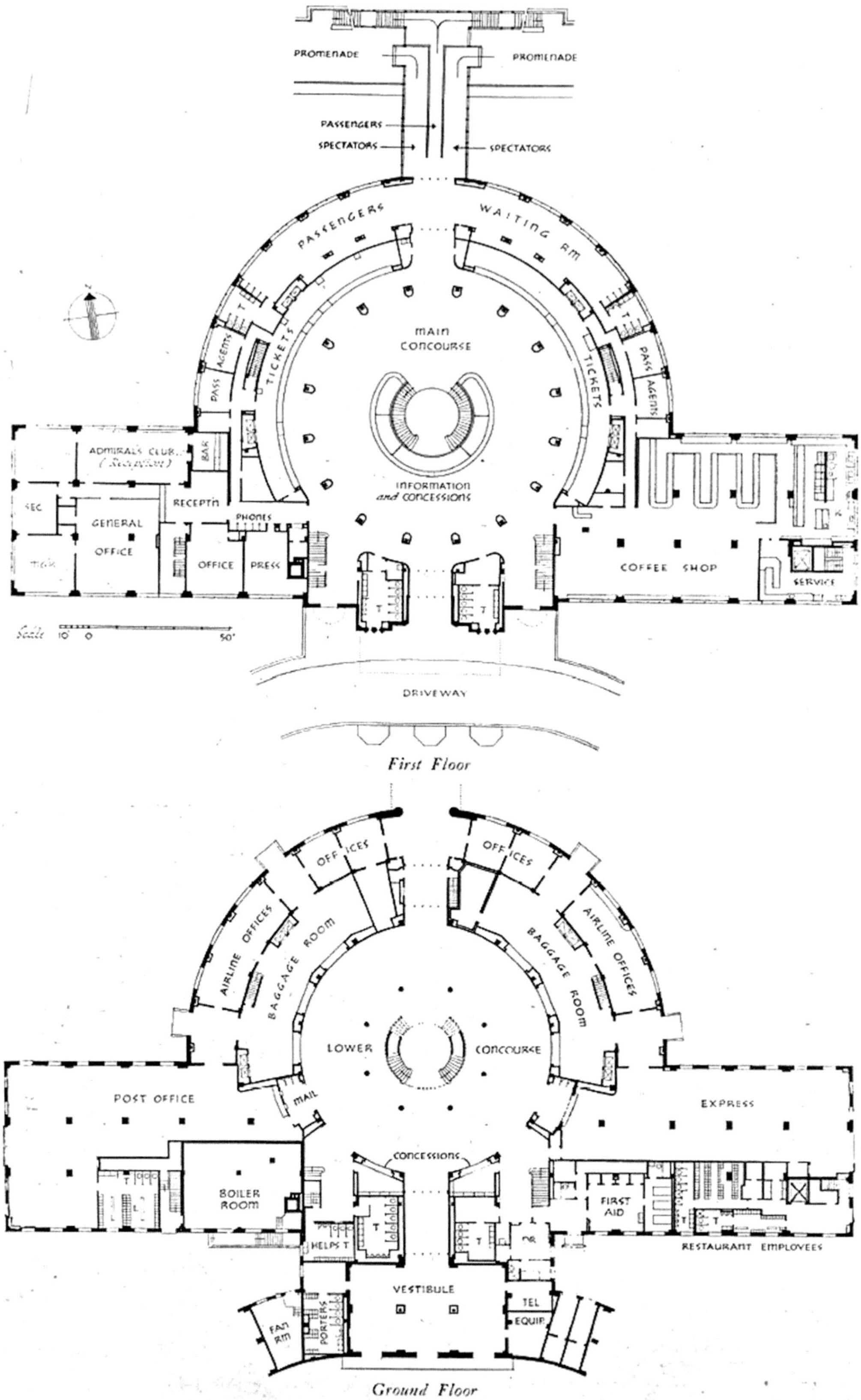

Figure 3.94 New York LaGuardia, 1939, ground & first floor plans of Domestic Terminal.
SOURCE: PENCIL POINTS

Figure 3.95 New York La Guardia 1940, Seaplane terminal.
SOURCE: WIKIPEDIA © HANS GROENHOFF (SMITHSONIAN INSTITUTION)

Figure 3.96 New York LaGuardia, summer 1940 – A Pan Am B314 at anchor in Bowery Bay. At upper right is Rikers Island Channel, take-off and landing area for the Boeing Clippers.
SOURCE: PANYNJ

Figure 3.97 New York LaGuardia, International (Seaplane) Terminal 1939 – circular check-in concourse with 'Flight', Mural by James Brooks.
SOURCE: PANYNJ VIA TWITTER

Inside the Marine Air Terminal at New York's LaGuardia Airport is a piece of transport art called '*Flight*' by American painter James Brooks: it's a massive mural 3.7 m tall and 72 m long, and it depicts the human obsession with taking to the skies, from our very first desire to emulate birds, right up to the latest form of air travel. '*Flight*' was completed in 1940, and the most modern aircraft it depicts is that pinnacle of stylish transport, the flying boat' (top right in the photo) (Figures 3.97 and 3.98).

Figure 3.98 New York LaGuardia 1939, International (Seaplane) Terminal – restored 'Flight' Mural by James Brooks.
SOURCE: PHOTO BY ATDISNEYAGAIN (OWN WORK) [CC BY-SA 4.0], VIA WIKIMEDIA COMMONS

Figure 3.99 Airlines Terminal 42nd Street New York, Postcard 1941. Opening in 1941, it continued in operation until 1978.
SOURCE: POSTCARD – PANYNJ VIA WWW.6SQFT.COM

The design for the five-story Airlines Terminal at New York's 42nd Street by Architect John B. Peterkins (1886 – 1969), developed the idea of a town terminal to support LaGuardia, by consolidating the reservations, ticketing, and baggage handling for the five major American airlines (American, Eastern, TWA, United and Pam Am) (Figure 3.99). The design included a number of technical innovations not previously seen in airport terminals including escalators, car lifts and welded steel frame construction. A limousine and bus service linked the Airlines Terminal with LaGuardia, Newark and later Idlewild (JFK) Airports, from the basement of the Terminal (Figure 3.100).

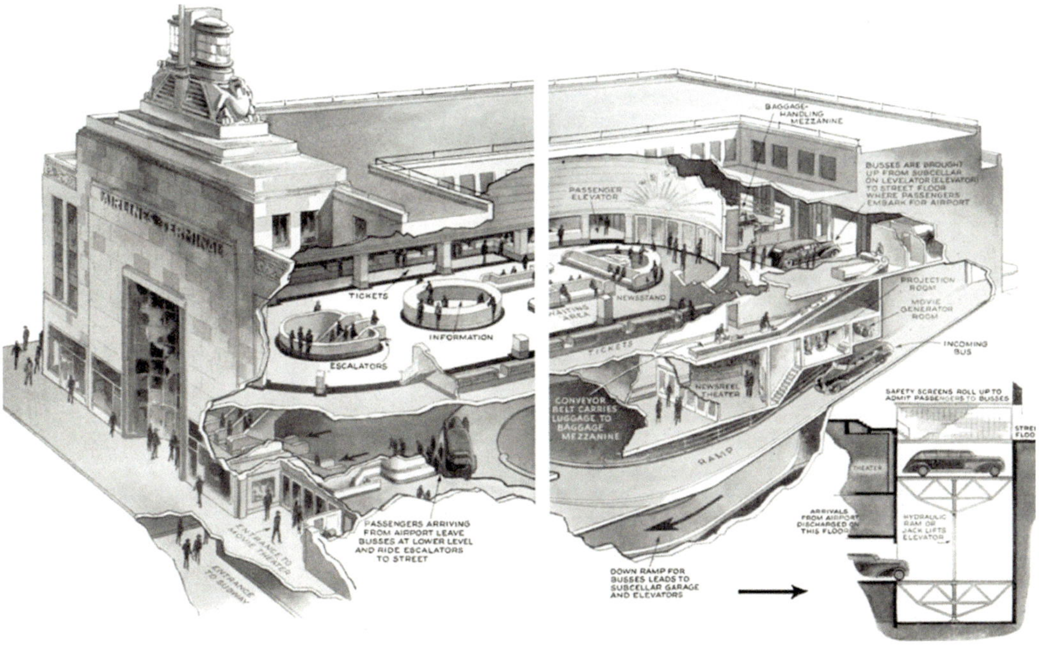

Figure 3.100 Airlines Terminal, New York 42nd Street 1941 sectional perspective.
SOURCE: PANYNJ – LAGUARDIA AIRPORT

Figure 3.101 Aerial view of the redeveloped airport in the 1960s with rebuilt terminal and added piers. Changes to the master plan included lengthening the runways and infilling the original water basin.
SOURCE: WIKIPEDIA

Despite rebuilding in the 1960s the airport has more recently come in for a lot of criticism, mostly due to the condition of the central terminal, with one example being President Joe Biden's comment that his experiences at LGA make him feel as though he were 'in a *third world country*.' The evolution of LaGuardia shows that ideas that were once considered forward-looking and avant-garde can easily become out of date with the growth of traffic, requiring the airport to be reinvented for every generation (Figures 3.101 and 3.102).

Figure 3.102 New York LaGuardia, proposed reconstruction illustrating the plans put forward by the Governor of New York in 2019 to rebuild the entire terminal complex. The entire project is set much closer to the Grand Central Parkway in order to create additional space for aircraft operations and movement, in what will still be a very small and congested airport.
SOURCE: ARCHITECTURE DAILY, COPYRIGHT GOVERNOR ANDREW CUOMO

CHAPTER 4

EVOLUTIONARY PHASE 1940–1959

During the Second World War, a number of airport master plans were developed to be ready for peacetime when construction could start again. From this point onwards, all new major airports incorporated hard-surfaced runways, either concrete or asphalt, to respond to the increasing weight and performance of aircraft.

Some featured an important innovation, placing the terminal zones at the centre of the airport, rather than on the periphery, and were surrounded by multi-directional runways. This idea later became commonplace for the largest international airports. This was the first time central terminal zones appeared on master plans of actual airports, although an earlier similar concept had been proposed in the entry to the Lehigh Airport Competition in 1929 by H.L. Gogerty. It had been relegated to an honourable mention, on safety grounds, as at that time safe operation was deemed incompatible with building in the centre of the airfield. Other master plans focused on integrating terminals for landplanes (short haul) and seaplanes (long haul) with rail and road interchanges, as at New York, La Guardia and the early masterplan for New York, Idlewild (JFK). Seaplanes became obsolete soon after the war as the range that land-based aircraft could travel without refuelling expanded and seaplanes were progressively retired.

Airport design during this period got off to a slow start as design efforts were directed towards the Second World War. Commercial flying was largely curtailed, though scheduled services regularly flew from Berlin Tempelhof right up to April 1945. Several masterplans were produced during the war that looked forward to a peaceful future. Some were more farsighted than others, notably plans for London Heathrow, New York Idlewild and Chicago O'Hare. They all included multidirectional concrete runways, with terminals in the middle of the runway system, rather than on the periphery of the airfield, an idea that had been dismissed as unsafe in previous decades, but was now to become the norm (Figure 4.1).

The 1940s saw the introduction of aircraft with tricycle undercarriages rather than the tail-dragging types of the pioneering era, which made the possibility of airbridges more practicable as aircraft now parked on an even keel. Four-engine aircraft models such as the Douglas DC4, permitted these larger, heavier and faster aircraft designs to fly with increased range. This in turn made hard-surfaced runways a necessity and progressively made flying boats obsolete, particularly as land planes were now capable of flying the Atlantic with a reasonable payload, albeit with intermediate stops in Newfoundland, Iceland and Ireland.

Airlines were responsible for a number of initiatives during the period, particularly in the United States. Eastern Airlines proposed both the finger pier concept and the apron drive airbridge, while United Airlines pioneered the latter's introduction at Chicago O'Hare. Delta Airlines introduced the hub and spoke concept as early as 1955.

DOI: 10.4324/9781003323624-4

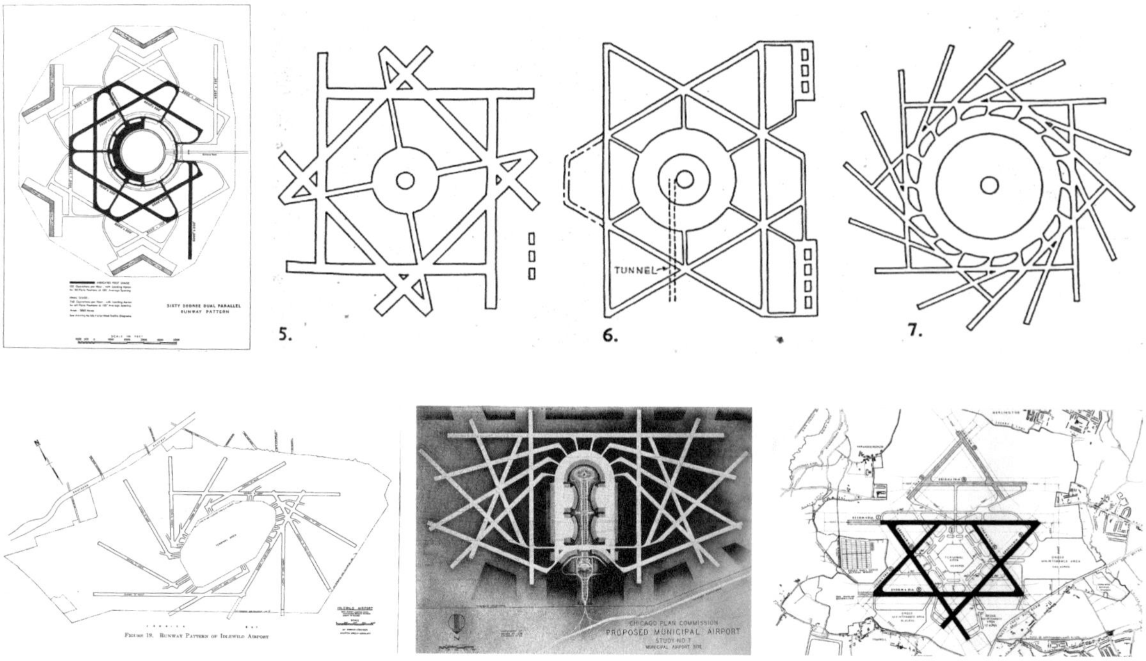

Figure 4.1 Post WW2 Airport master plans – Chicago, New York and London.
SOURCE: AUTHOR'S COLLECTION

Meanwhile, Pan American Airways introduced economy fares in 1953, dramatically increasing the volume of traffic. Many airlines introduced lounges for frequent flyers, following the example of American Airlines which first opened one at La Guardia in 1940.

The airport city idea first emerged at New York Idlewild, initially as an expedient concept where individual airlines designed and funded their own terminals because the original centralised terminal concept could not be afforded by the airport authority. This led to a hotchpotch of individualistic terminal designs which may have originally been seen as innovative, but have not survived the test of time and are now largely either demolished or rebuilt.

The far-sighted Chicago O'Hare master plan of 1947 was arguably the most influential master plan of the period, and its radiating finger piers and centralised terminal concept were later widely imitated at airports such as Schiphol and Frankfurt. The masterplan was the subject of numerous updates, with the airport finally becoming fully operational in 1962, after revising the terminal geometry and increasing runway lengths, to suit the ever-larger aircraft types that had been introduced in the intervening period. The original master plan estimated that 22 million passengers would fly through the airport in 1970, but even this prediction, which was considered wildly optimistic at the time, proved to be a substantial underestimate of growth in demand, with the actual throughput in 1974 being 37.6 million passengers, making it by some margin the busiest airport in the world at that time. The airport, however, has proven adaptable to even higher levels of traffic, handling some 80 million passengers in 2018. It is now undergoing a radical expansion programme, made possible by changing the space-consuming radial runway layout to multiple parallel runways, thereby releasing substantial areas for new aircraft stands and expansion of the terminals.

A number of terminal planning innovations appeared during this period. Heathrow introduced the short-lived reversible unit terminal concept and developed town check-in at the West London Air Terminal. The latter was rendered obsolete by the later introduction of underground metro and mainline rail links. Finger pier concepts first became operational at Washington Friendship in 1950, followed by San Francisco in 1954, St. Louis Lambert 1955 and Gatwick 1958.

Aesthetic developments, seeking to capture the spirit of flight, can be seen at St. Louis, Lambert, with its expressionist design, featuring a lightweight vaulted shell roof to the terminal and freestanding cabins inside. This strategy was later adopted by Eero Saarinen for the TWA terminal at New York JFK and Washington Dulles. At Gatwick, a Miesian version of the International Style provided an aesthetic framework that allowed the airport to remain functionally and visually coherent for many years despite extensive growth and change.

It was an era of mechanisation due to the increasing size of terminals. St. Louis, Lambert, for example, introduced mechanical baggage handling using conveyors and moved people on escalators between levels, introductions described by the architectural press at the time as *'new-fangled'*. All these innovations have become the norm in subsequent terminal designs.

The airports selected to illustrate the key characteristics of the era are:

- Chicago O'Hare (ORD) 1947
- Washington Friendship (BWI) 1950
- San Francisco (SFO) 1954
- London Heathrow (LHR) 1955
- St Louis Lambert (STL) 1956
- London Gatwick (LGW) 1958

10. Chicago O'Hare (ORD) 1947

Chicago O'Hare was designed to replace Chicago Midway, as the latter was reaching capacity, due to the growth in traffic following the Second World War. The City of Chicago appointed Ralph Burke, the City's former chief engineer, to plan a future airport on Orchard Field, later renamed O'Hare, extending the former Douglas Aircraft Company Factory and Airbase. He prepared a very ambitious master plan for the airport to replace Midway airport. His master plan was formally adopted by the City in December 1947. Burke's plan called for a central terminal around which runways would be arranged tangentially. The terminal itself would consist of a main concourse building with 'fingers' spreading out into the apron area. The original centralised terminal plan later gave way to a less ambitious concept with a series of smaller terminals arranged around a central parking lot and serving a series of finger piers.

The finger pier concept, first conceived at O'Hare, has inspired many generations of terminal design, including Washington Friendship, St Louis Lambert, Gatwick South Terminal, Amsterdam Schiphol and Frankfurt airports. The airport also pioneered the development of the aircraft boarding bridge. The airport was subject to continuing updates, both during construction and subsequently in use, with runway extensions and terminal improvements incorporated into evolving masterplans to cater for new aircraft and traffic growth (Figure 4.2).

However, the concept at O'Hare has evolved several times to suit increasing aircraft size, the introduction of jet aircraft and changes from parallel aircraft parking to nose-in parking. Chicago O'Hare took a long time to realise and was redesigned several times in the intervening period, eventually opening fully in 1962 (Figures 4.3 and 4.4).

The first update to the master plan was published in 1961, with the terminal complex update following in 1963, when the original domestic terminal built in 1955 was renovated to become a new International Terminal (Figure 4.5). The radial runway pattern has not had the same longevity, giving way after a series of steps to multiple parallel runways as newer generations of aircraft have become less sensitive to cross winds.

Figure 4.2 Chicago O'Hare, master plan 1948.
SOURCE: CHICAGO TRIBUNE GRAPHICS

Figure 4.3 Chicago O'Hare 1948, aerial view and master plan.
SOURCE: CHICAGO MASTER PLAN (BURKE, 1948)

The story of Chicago O'Hare is, then, one of repeated innovation to accommodate the growth and change that comes with increasing passenger volumes and aircraft movements. Its ability to adapt is demonstrated by the fact that it was the busiest airport in the world in 2018 measured by the number of aircraft movements at 904,000 annual air traffic movements.

Most recently, as seen in the United Terminal and the still evolving future master plan, the advantages of the satellite system in providing unobstructed aircraft movement and an increase in overall capacity within the same site are considered to outweigh the easy passenger connectivity of the finger pier.

The development of the airbridge was made much easier with the widespread introduction of tricycle undercarriages to aircraft such as the Douglas DC4, which

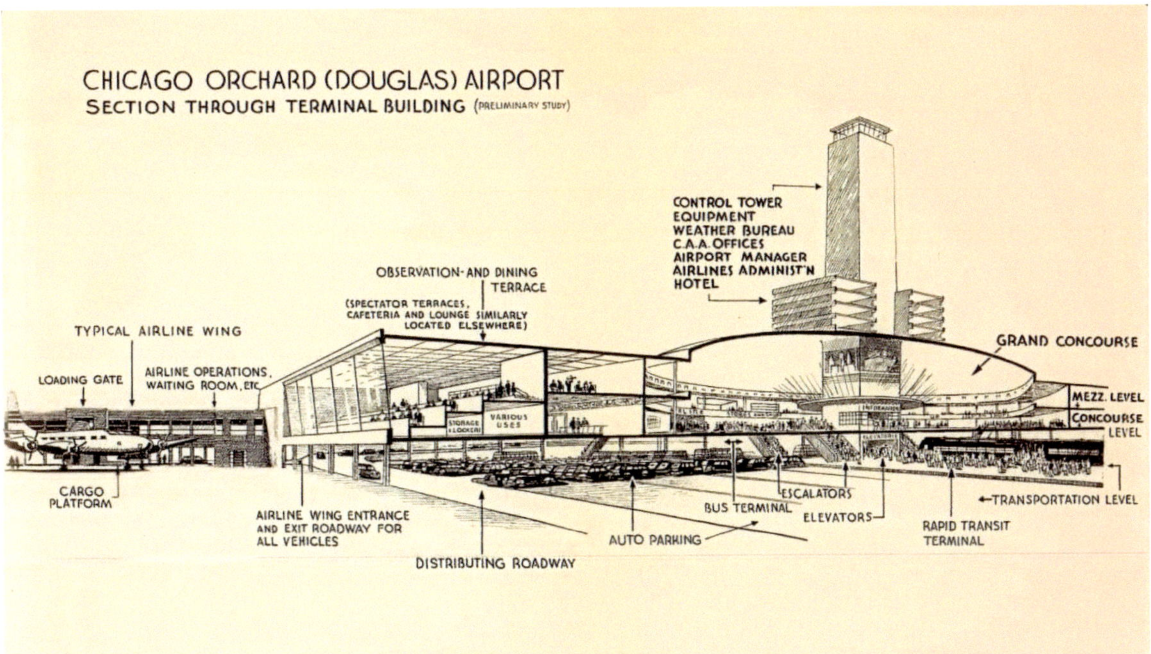

Figure 4.4 Chicago O'Hare, original centralised terminal concept 1948.
SOURCE: CITY OF CHICAGO

Figure 4.5 Chicago 1962 – The early terminals were designed by Naes & Murphy (later C.F. Murphy) with the 7 Continents restaurant in the Rotunda (1962) in the foreground.

SOURCE: POSTCARD – PENROD

meant that aircraft were approximately horizontal when parked rather than the previous tail-dragging type of aircraft such as the Douglas DC3. The first masterplan to illustrate airbridges was Chicago O'Hare (1948) (Figure 4.6). However, the arrangement illustrated was rather impractical, relying on a hinged arrangement and without a visible means of support. The concept did, however, fundamentally alter subsequent terminal design raising passenger movement to the first floor level instead of ground level, a major step in the evolution of airport design. There were numerous patents relating to

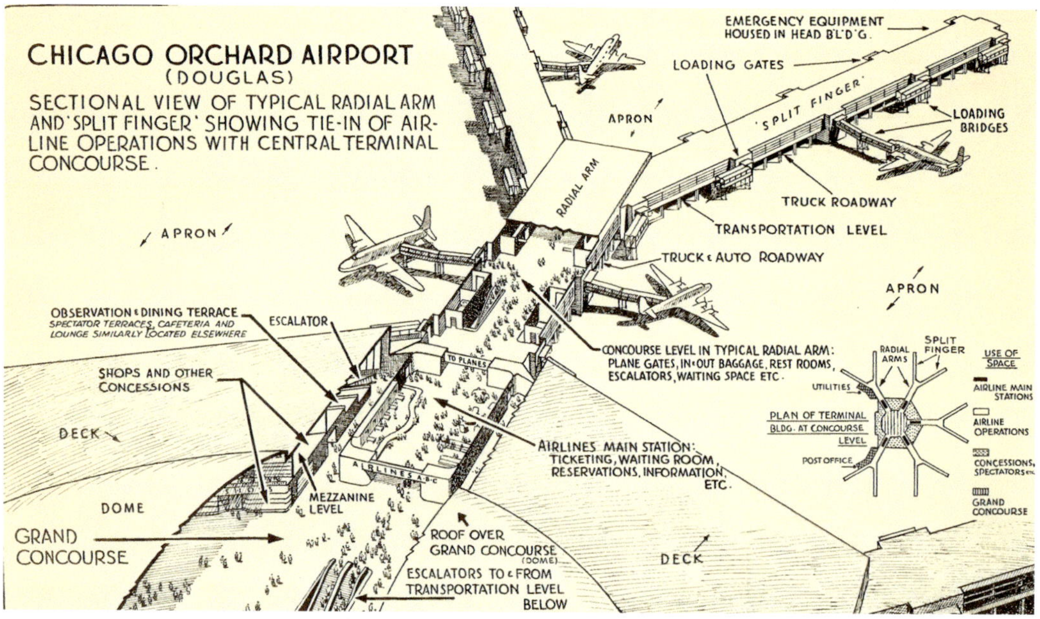

Figure 4.6 Finger pier with passenger boarding bridges.

SOURCE: CHICAGO ORCHARD (DOUGLAS) 1948 MASTER PLAN

Figure 4.7 United Airlines DC7 using the prototype Jetway Aero Gangplank at Chicago O'Hare 1958.
SOURCE: UNITED AIRLINES VIA AIRPORT HISTORY

airbridge design in the early fifties, and an early prototype was unveiled by Jetway Aero Gangplank at Chicago O'Hare (1958) (Figure 4.7).

By the late 1960s, Chicago was already looking for additional capacity. To alleviate congestion, O'Hare added a new east-west runway, which gave O'Hare three sets of parallel runways (Figure 4.8). The Airline Deregulation Act of 1978 had a profound impact on operations at O'Hare. Domestic deregulation led to United Airlines and American

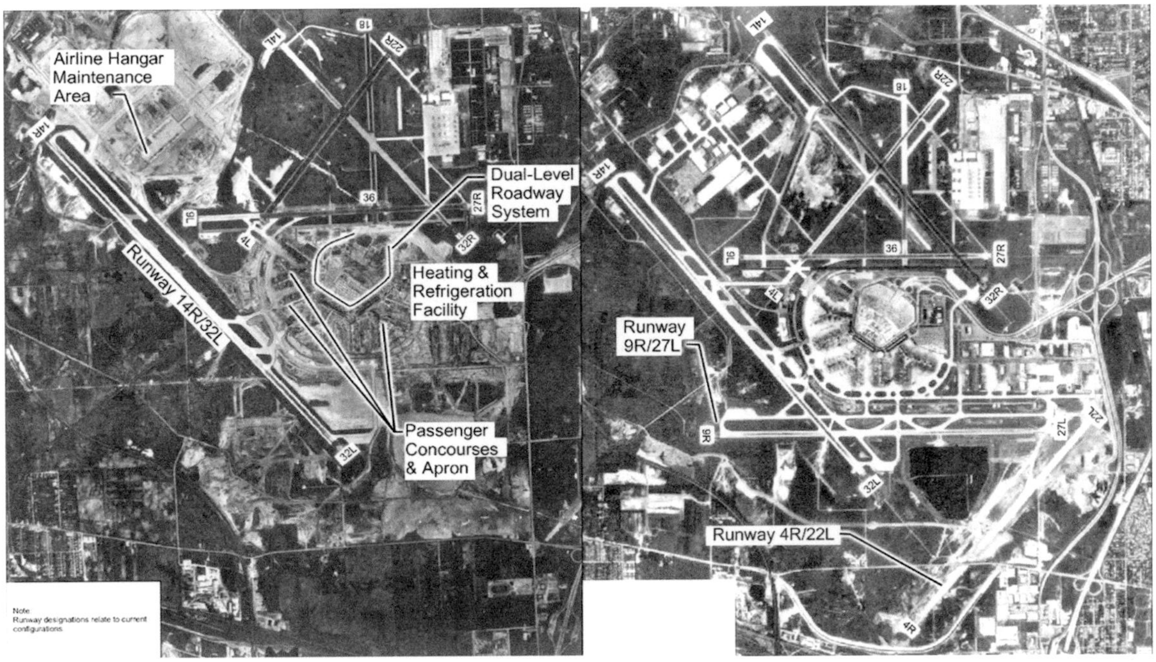

Figure 4.8 Chicago O'Hare, showing the 3 pairs of parallel runways to the right.
SOURCE: CITY OF CHICAGO VIA AIRWAYS MAGAZINE

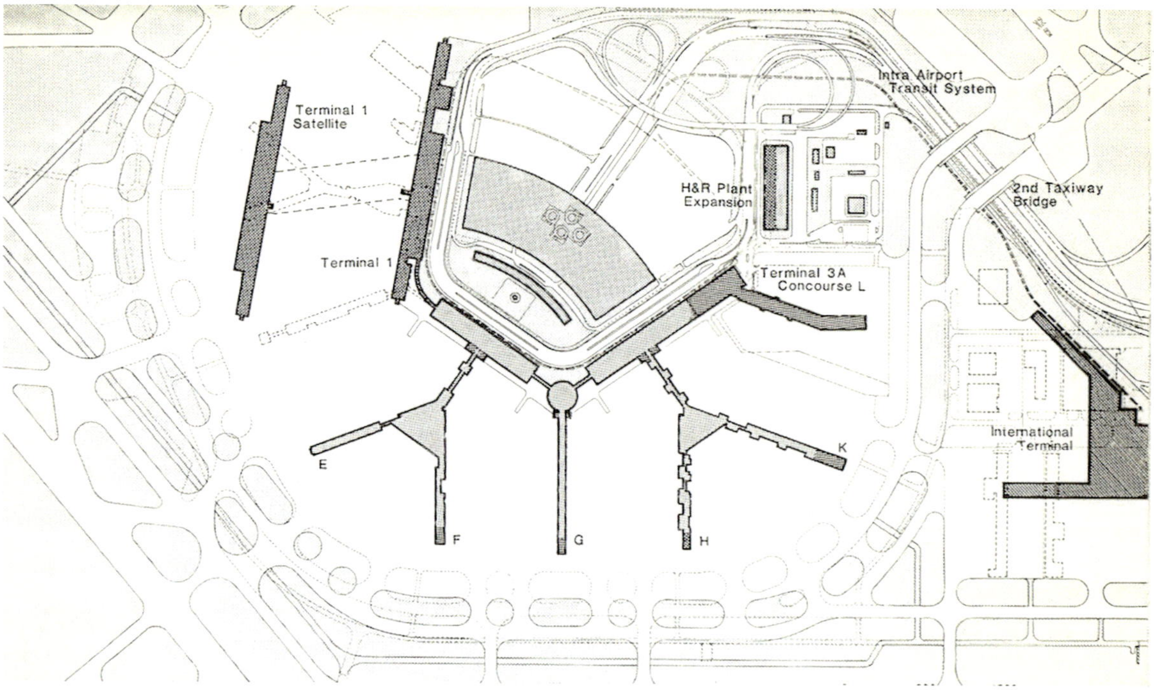

Figure 4.9 Chicago O'Hare, United Terminal International Development Program, showing the layout of the rebuilt Terminal 1, designed for United Airlines by Helmut Jahn, linked to a linear satellite by an underground walkway.
SOURCE: CITY OF CHICAGO

Airlines consolidating operations at the airport to form their hubs. O'Hare's terminals were reconfigured to serve the operations of the hubbing airlines as other airlines, such as TWA and Northwest Airlines, moved their hub operations elsewhere. As United and American's presence at O'Hare became more pronounced, plans were made for a new terminal complex.

In 1982, Chicago launched a new development plan designed to provide the airport with new or rehabilitated terminals and service areas by 1990. The plan called for a new International Terminal on the southeast side of the airfield, a commuter terminal and general aviation terminal located near the new International Terminal, construction of a new Terminal 1 – replacing the existing International Terminal – to be used by United and the expansion of Terminals 2 and 3. To accommodate the new larger jets, the master plan was modified with the original star-shaped piers becoming Y and I concourses.

In 1984, construction began on United's 'Terminal for Tomorrow', designed by Helmut Jahn (Figure 4.9). He described his design as drawing inspiration from the Kew Gardens glass houses and London Bridge and Liverpool Street railway stations in the UK, with open floor space and lots of natural light. The terminal was built next to the roadway, similar to the location of the other terminals, but a linear satellite was constructed in the middle of the airfield and connected to the terminal by an 850-ft tunnel (highlighted in yellow, (Figures 4.10–4.13)).

Future expansion plans are still evolving, and future master plans show the opportunities created for terminal expansion by simplification of the runway network while staying largely within the existing airport boundary (Figures 4.14 and 4.15).

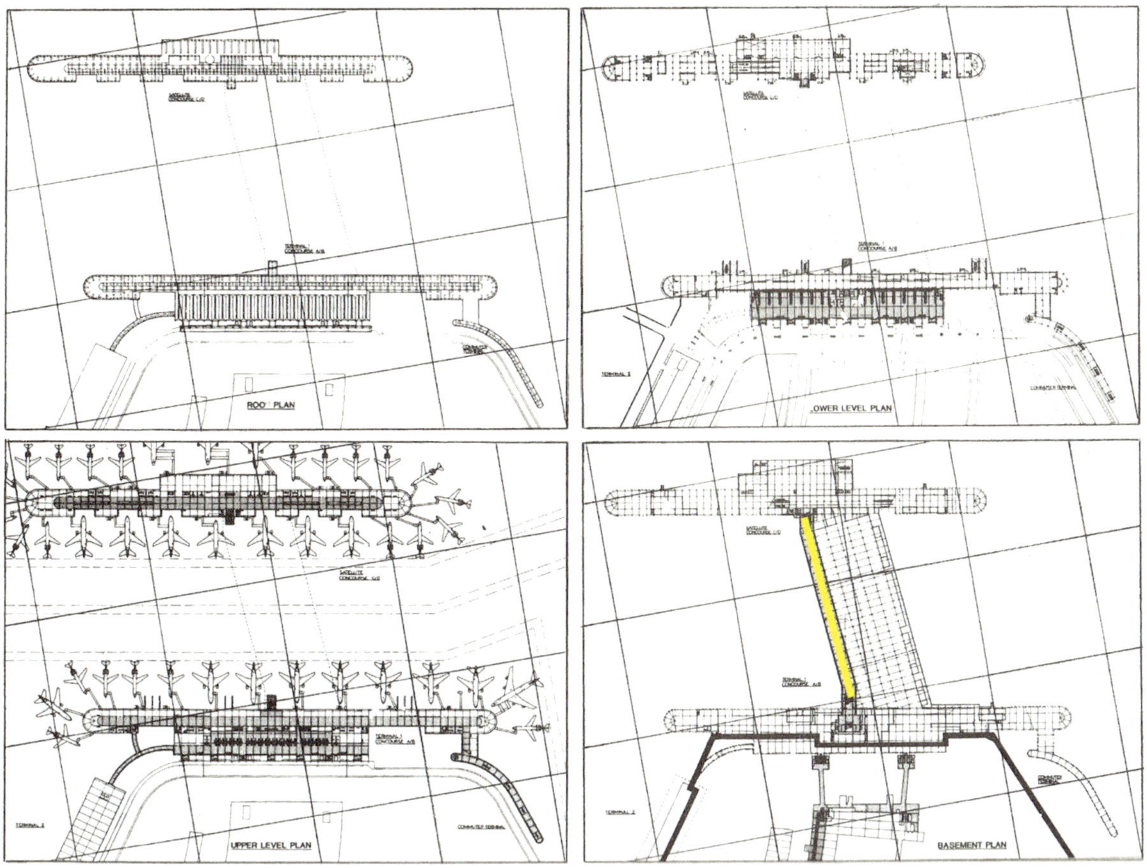

Figure 4.10 Chicago O'Hare, United Terminal, architect Helmut Jahn, with underground walkway highlighted in yellow.
SOURCE: CITY OF CHICAGO VIA AIRPORT TERMINALS

Figure 4.11 Underground walkway linking terminal concourse A & B.
SOURCE: MURPHY/JAHN VIA REDDIT

Figure 4.12 Satellite concourse B, central passenger aisle.
SOURCE: WIKIPEDIA

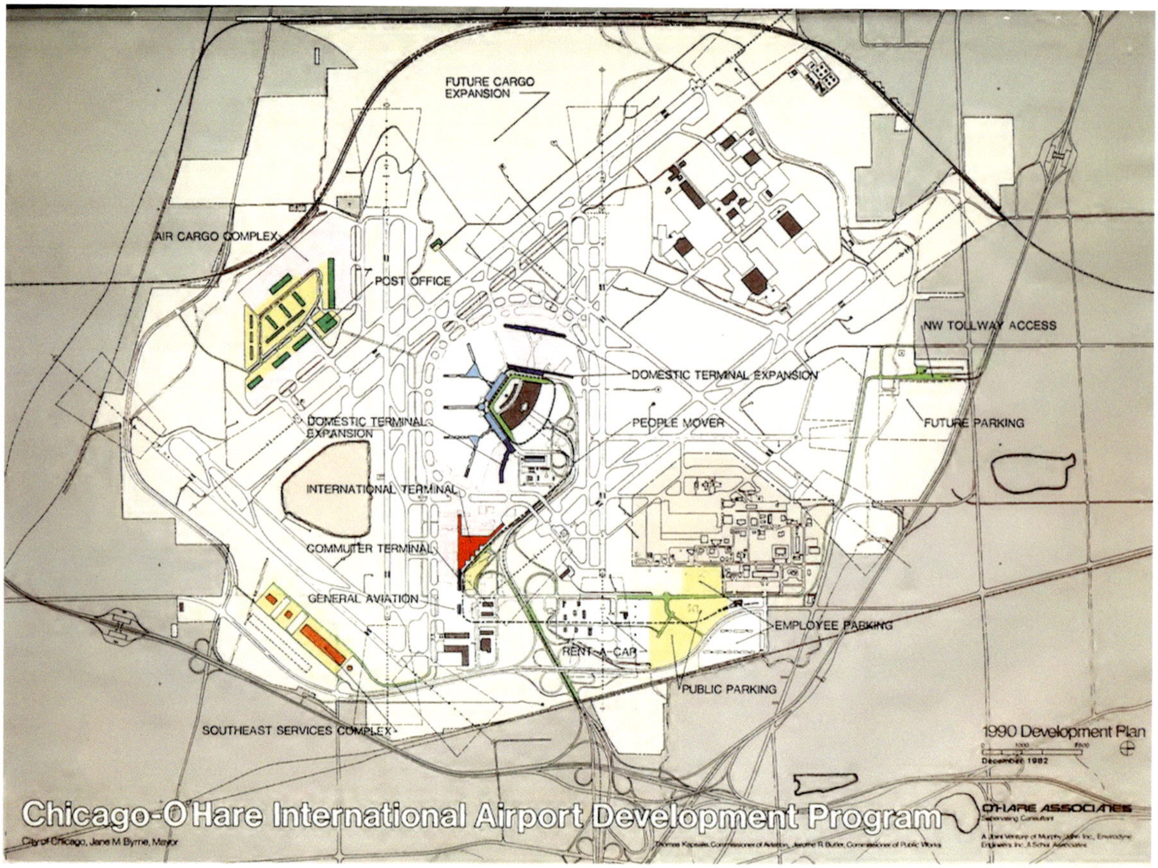

Figure 4.13 Chicago O'Hare, development plan 1990.
SOURCE: CITY OF CHICAGO VIA AIRWAYS MAGAZINE

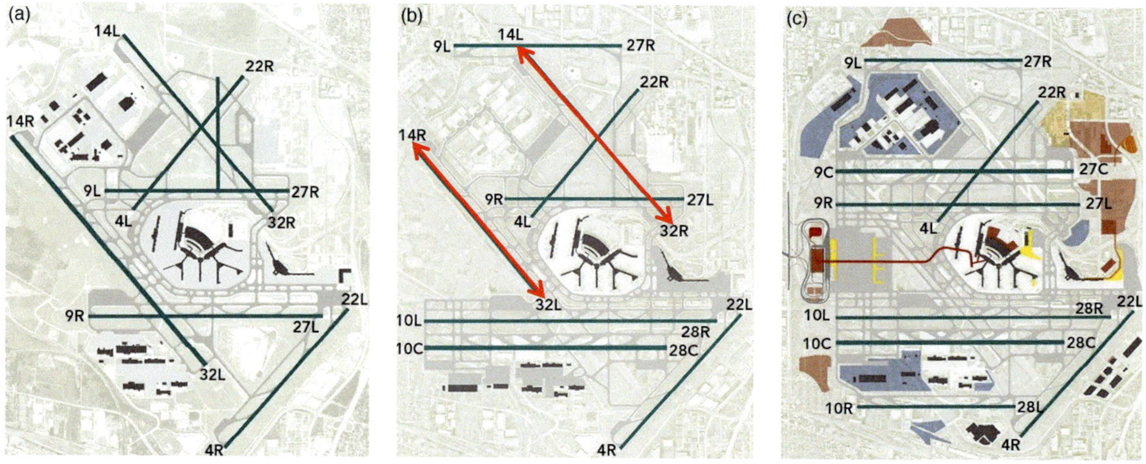

Figure 4.14 Chicago O'Hare, master plan evolution. (a) Airfield Layout 2003 (b) Airfield Layout 2015 (c) Future Airfield Layout.
SOURCE: CITY OF CHICAGO AVIATION DEPARTMENT

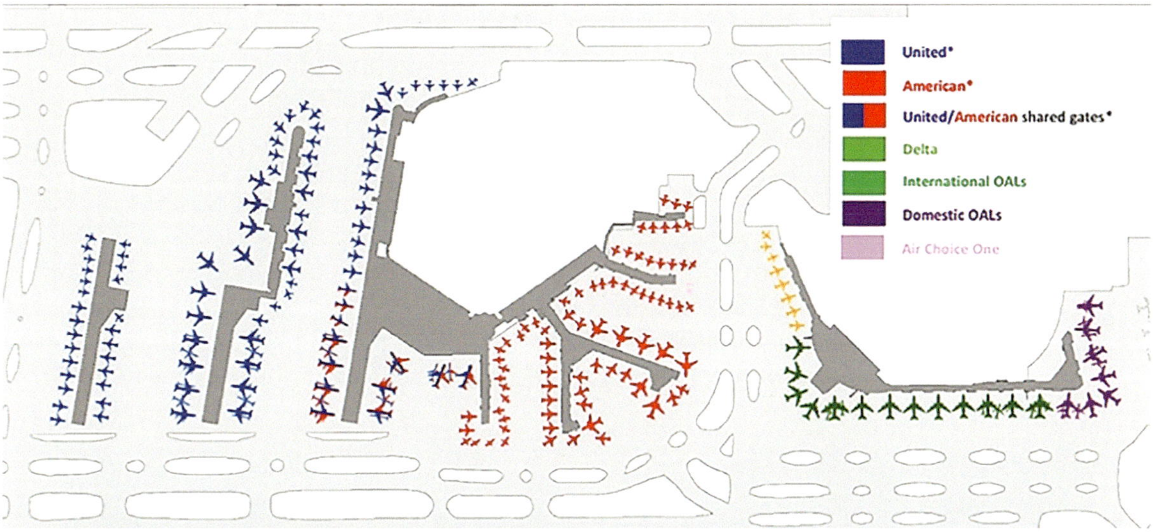

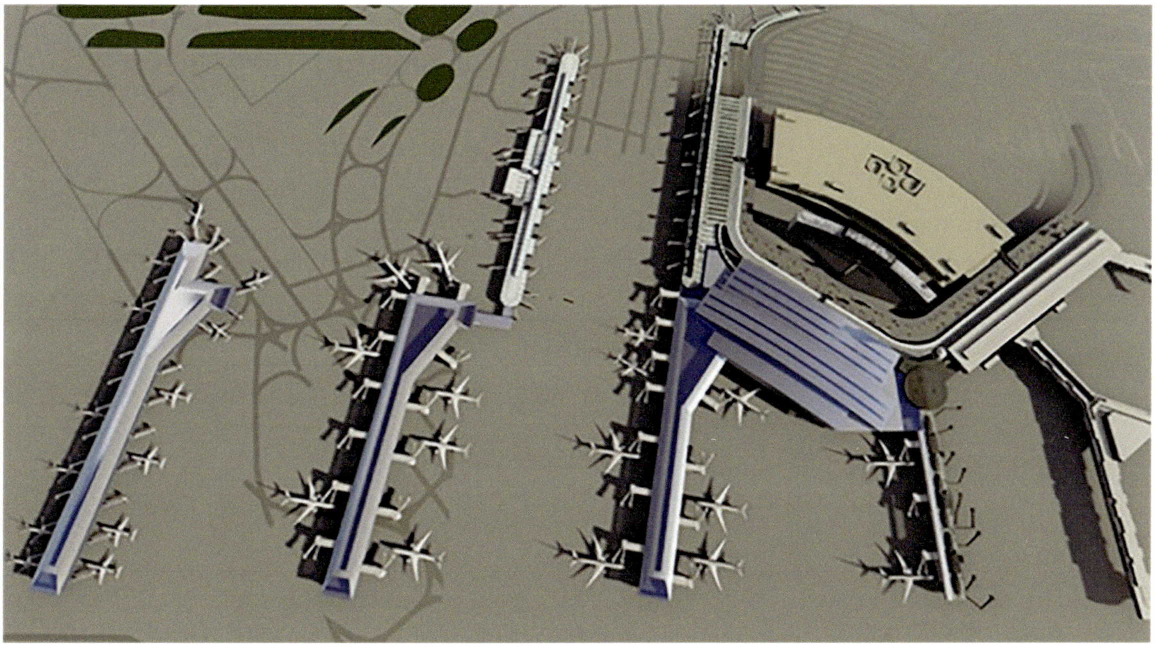

Figure 4.15 Chicago O'Hare, future master plan. First, two satellite terminals would be built to the west of Terminal 1, connected to the main terminals via walkways and tunnels. Second, the existing Terminal 2 would be demolished and replaced by an expanded international terminal. Third, doing both would allow United and American and their international partners to shift operations out of the existing international terminal, Terminal 5.

SOURCE: CITY OF CHICAGO AVIATION DEPARTMENT VIA WWW.CHICAGOBUSINESS.COM

11. Washington Friendship (Baltimore) (BWI) 1950

In 1946, Washington Friendship was one of the first airports to publish a master plan after the Second World War. It was also the first airport terminal designed with a finger pier, architect James R. Edmunds, to become operational when it was opened by President Truman on June 24, 1950, the day the Korean War started. The original masterplan had the terminal projecting into the airfield, with an integrated control tower as the focal point (Figure 4.16).

The airfield was laid out with three intersecting runways, with a claimed capacity of 60 movements an hour, though later one was decommissioned and replaced with a parallel runway, leaving the terminal in a mid-field location in the current masterplan which has allowed it to grow to meet increasing traffic demands (Figures 4.17 and 4.18).

Figure 4.16 Washington Friendship (Baltimore) postcard.
SOURCE: POSTCARD-MAGSGEN.COM

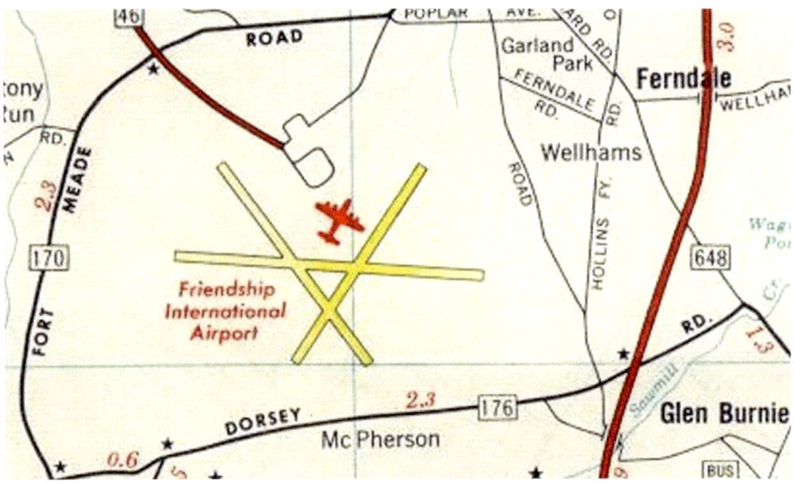

Figure 4.17 Washington Friendship, original airfield layout.
SOURCE: KILDUFFS.COM

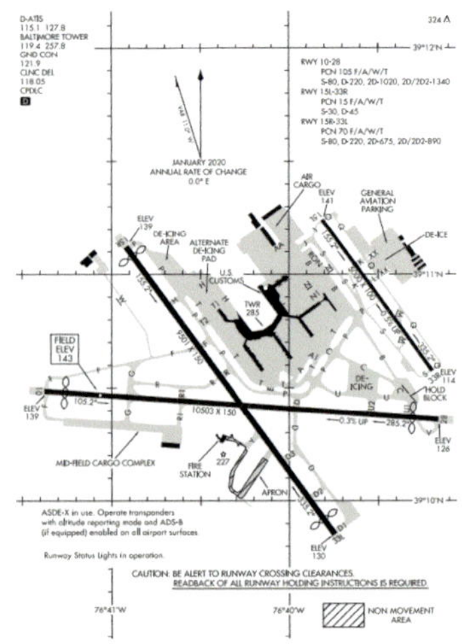

Figure 4.18 Masterplan.
SOURCE: WIKIPEDIA

Figure 4.19 Washington Friendship Landside view of two-level forecourt.
SOURCE: WTOPNEWS.COM

When built Friendship was one of the largest in the United States with an area of 3,200 acres, five times the area of Washington National Airport and four times the area of New York La Guardia. The terminal included a number of commercial facilities, including banks, a hairdresser, restaurants, meeting rooms and an integrated hotel with 22 small rooms. A viewing platform for spectators was provided at roof level over the finger pier. The control tower was integrated with the terminal and was the focus of the architectural composition, which was executed in a late Art Deco style (Figure 4.19).

Passengers were handled at an upper or lower floor level depending on whether they were outbound or incoming. With departures and check-in at an upper level, spiral baggage chutes delivered baggage down to a baggage loading hall at an apron level.

The airport was renamed Baltimore Washington International (BWI) in 1973. Since October 2005, the official name has been Baltimore Washington International Thurgood Marshall Airport. It has continued to expand as traffic has built up, and now incorporates multiple piers with passenger boarding bridges, but the original terminal building and control tower remain at the heart of the complex (Figures 4.20 and 4.21).

Figure 4.20 Washington Friendship (now BWI), showing the new parallel runway with the original terminal building in the centre.
SOURCE: GOOGLE EARTH

Figure 4.21 Expanded terminal area and rebuilt central pier. Note the original control tower is still the focal point.
SOURCE: BWI MARSHALL AIRPORT

12. San Francisco (SFO) 1954

Designed in the International Style, the seven-story building employed an innovative, dual-level forecourt layout that separated departures from arrivals. Dedicated roadways streamlined traffic to the terminal and panoramic views of the airfield and San Francisco Bay were provided from multiple spectators' terraces. It was one of the earliest airports to employ finger piers to access the aircraft (Figure 4.22).

Unlike many airports of this generation, the original terminal has been integrated into the subsequently revised redevelopment and continues to be at the heart of Terminal 2. The pier concept and the main access have been retained as they have evolved (Figures 4.23 and 4.24). The original runway layout has been retained and extended, but its configuration is a legacy of an earlier generation of airports when cross-wind runways were essential. The runway crossover is a limitation to the ultimate capacity of the airport when compared to Los Angeles Airport (Figure 4.25).

Figure 4.22 San Francisco 1954, aerial view of approach to the terminal.
SOURCE: FLYSFO

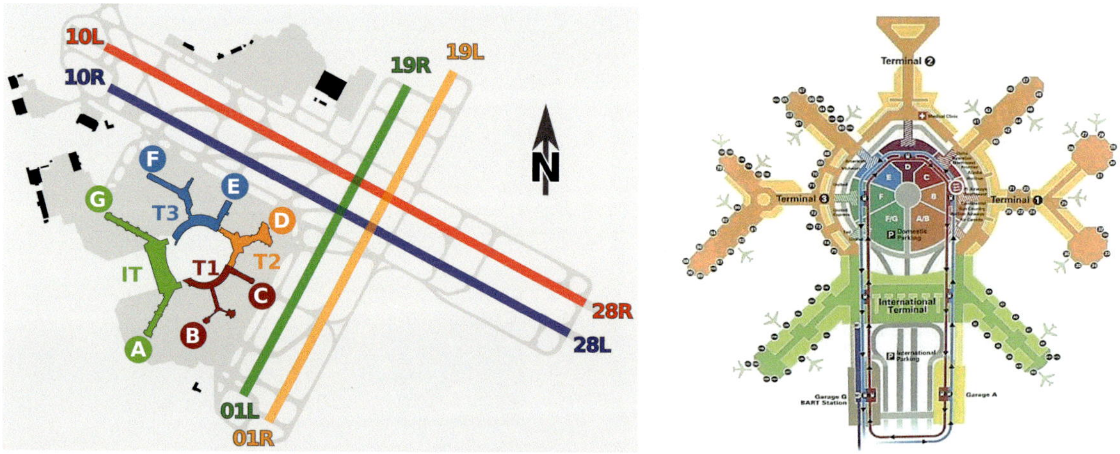

Figure 4.23 San Francisco current masterplan and terminal configuration.
SOURCE: WIKIPEDIA

Figure 4.24 San Francisco master plan 1967.
SOURCE: SFO MUSEUM

Figure 4.25 San Francisco, expanded terminal zone, with a new international terminal built across the approach roads.
SOURCE: WIKIPEDIA LICENSED UNDER THE CREATIVE COMMONSATTRIBUTION-SHARE ALIKE 4.0 INTERNATIONAL LICENSE

13. London Heathrow (LHR) 1955

Heathrow featured the first terminal area to be developed in the middle of the airfield. Unlike contemporary airport master plans for New York and Chicago, its unique 'Star of David' configuration required all the connecting surface access infrastructure to be routed underground to reach the central terminal area. The official story of Heathrow in 1955 recorded that 'The 1946 London Airport Layout Panel considered very seriously a tangential pattern of runways fanning out from the Central Area, but the runway plan decided upon concentrated on what was then described as a "parallel pattern" which was considered better suited operationally and to the topography of the site.'

The green arrow shows the underground road link from the A4, the red arrow the underground rail link (built later as the Heathrow Express and now forms part of the Elizabeth Line) and the blue arrow an extension to the Piccadilly underground line. While the underground road link was built at the outset for the opening of the first permanent terminal in 1955, it took until 1977 before the Piccadilly line extension to the Central Terminal Area was completed and 1998 before the Heathrow Express rail link opened. This was the first airport concept to envisage integrating both rail and underground metro links into the airport master plan. The underground surface access has over time proven to be one of the most intractable constraints to the evolution of the airport, around which subsequent upgrades have had to be planned (Figure 4.26).

A proposal for a new airport at Heathrow was developed under the cover of the Official Secrets Act during the Second World War, theoretically for long-range bomber aircraft, but in practice to become London's Post War International Airport to replace Croydon. By 1946, this had evolved into the familiar Star of David layout, doubling up the number of runways and adding a further three runways to the north that were not developed at the time. However, Heathrow opened with a much simplified phase 1 layout and temporary facilities to the north of Runway No.1 (Figures 4.27–4.30).

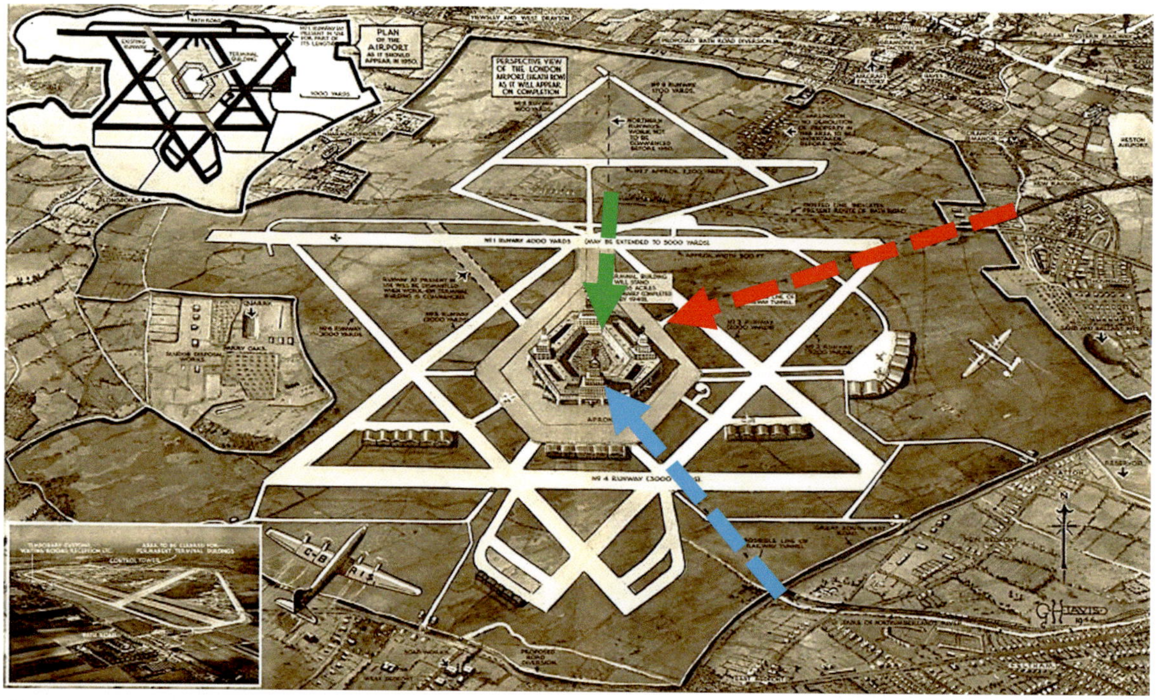

Figure 4.26 Heathrow master plan 1946.
SOURCE: ILLUSTRATED LONDON NEWS, SATURDAY 25 MAY 1946 WITH AUTHOR'S ADDED COLOUR

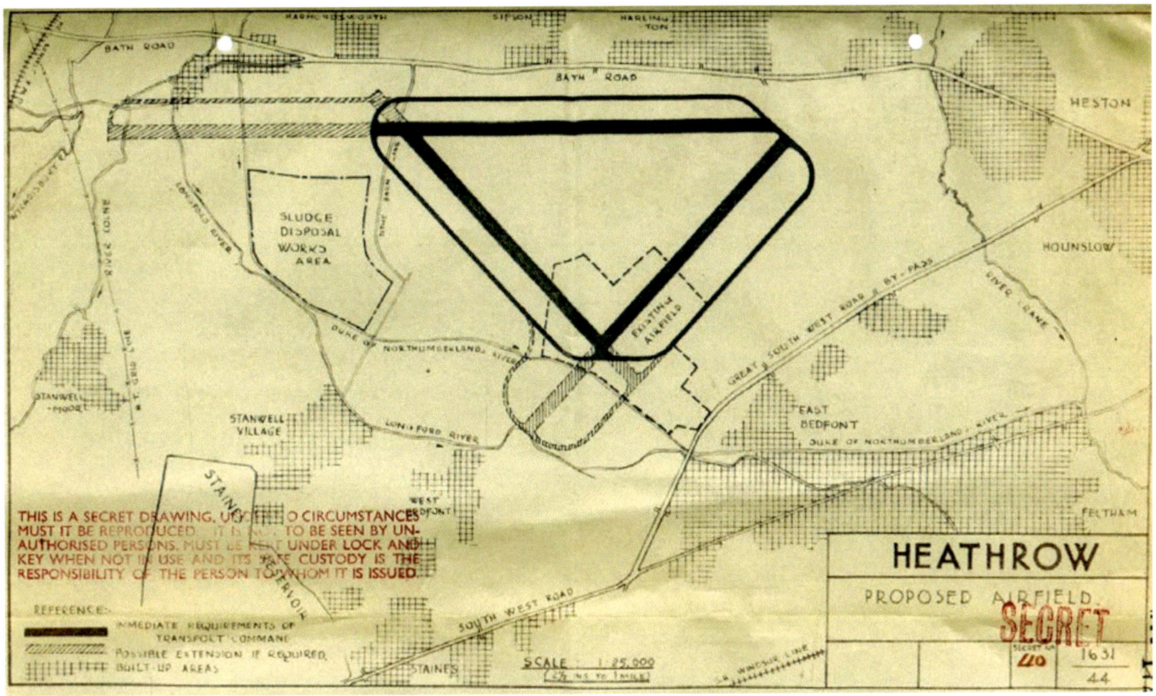

Figure 4.27 London Heathrow wartime master plan 1944.

SOURCE: HOUSE OF COMMONS BRIEFING PAPER 1136, 20TH JUNE 2018

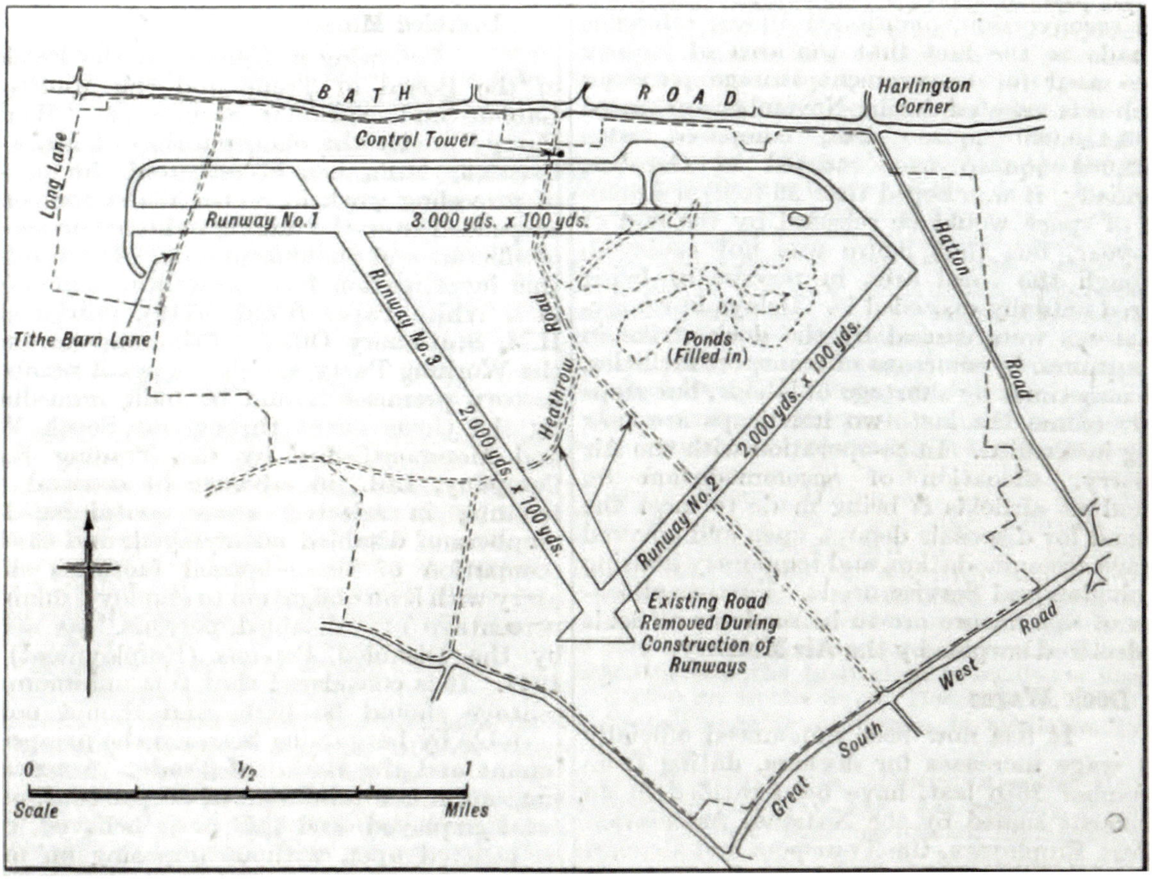

Figure 4.28 London Heathrow plan, showing phase 1 of the construction.

SOURCE: THE ENGINEER 11.01.1946

Figure 4.29 London Heathrow temporary tented terminal was located to the north of Runway No.1 and adjacent to the control tower 1946.
SOURCE: HEATHROW AIRPORT

Figure 4.30 Heathrow North 1946 – temporary tented terminal with W.H. Smith Bookshop, one of the earliest brand names to be introduced at an airport, in the background.
SOURCE: HEATHROW AIRPORT

Figure 4.31 Heathrow satellite concept 1950.
SOURCE: FLIGHT 04.05.1950

The linear satellite concept can first be seen in a 1950 concept for Heathrow Airport, by W.O. Biernacki, an architectural student at University College London. The design proposed a finger pier and two parallel satellites reached by underground walkways, serving aircraft via 83 telescopic loading bridges. Despite encouraging publicity in the architectural press, the idea did not progress (Figure 4.31).

The Heathrow Europa Terminal, later renamed T2, was the first completed terminal at Heathrow, opening in 1955, but has now been demolished (Figures 4.32 and 4.33). The architect was Frederick Gibberd. It was a unique example of a single-sided linear unit terminal in the UK, with a reversible passenger flow, with facilities capable of being operated in either departures or arrivals mode. It is a rare example of a terminal where time separation was employed to keep arriving and departing passengers separate with 10 reversible passenger lanes (Figure 4.34). It was subsequently converted to a conventional centralised layout to make the operation more efficient and use the processing facilities all the time.

Helicopter links to the city centre have often been short-lived. The Waterloo Air Terminal on the London South Bank was converted from the Waterloo entrance building to the 1951 Festival of Britain. Helicopters flew to London Airport Central (Heathrow) after BEA moved their services there from their post-war base at Northolt. The service opened on 25.07.1955 and lasted just ten months, closing on 31.05.1956 possibly because the helicopters could hold just five people and were uneconomical as well as being very noisy (Figure 4.35).

British European Airways (BEA), now the short-haul part of British Airways, built the West London Air Terminal in 1957 to provide town check-in facilities in London's West End. From there, passengers were transported to Heathrow by specially designed buses with a separate compartment for checked-in baggage. This was replaced by a new

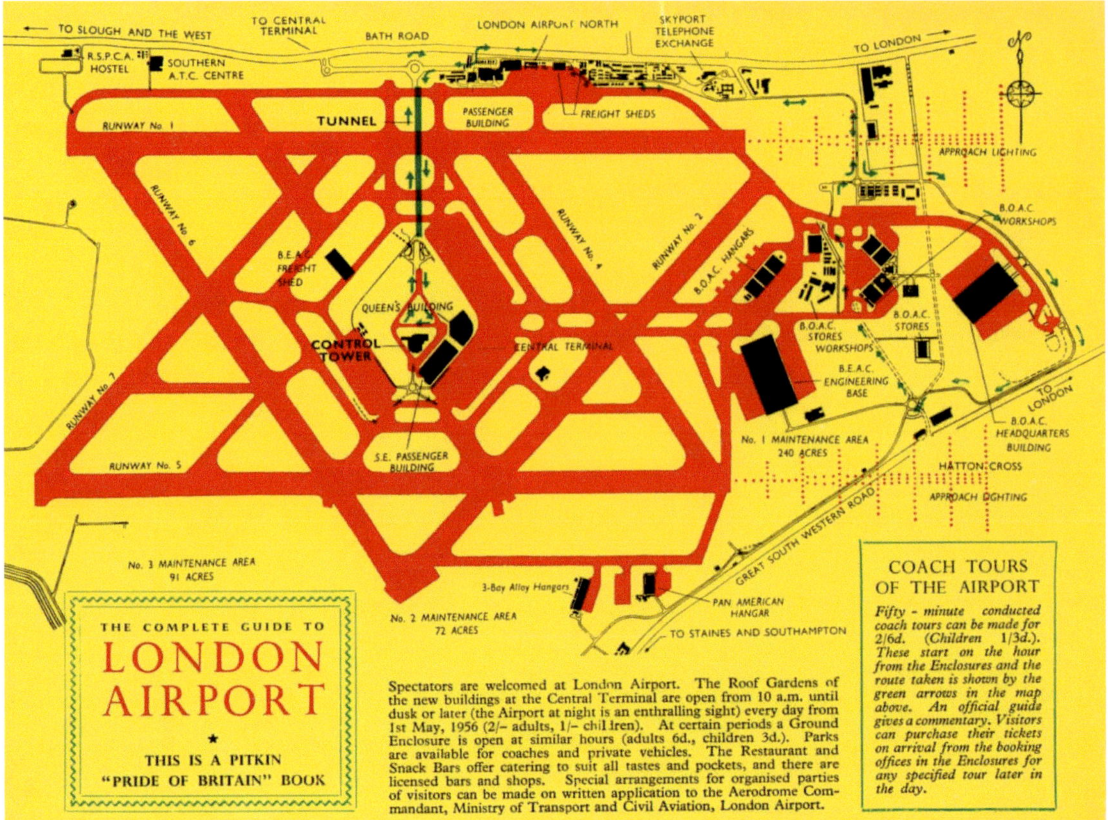

Figure 4.32 London Heathrow, 1955 showing the layout of the central terminal area and the developed runway and taxiway configuration in 1955 opening.

SOURCE: THE COMPLETE GUIDE TO LONDON AIRPORT, SIR MILES THOMAS

Figure 4.33 London Heathrow aerial view of Queens Building, Europa Terminal & Control Tower 1955.

SOURCE: HEATHROW AIRPORT VIA THE HISTORY PRESS

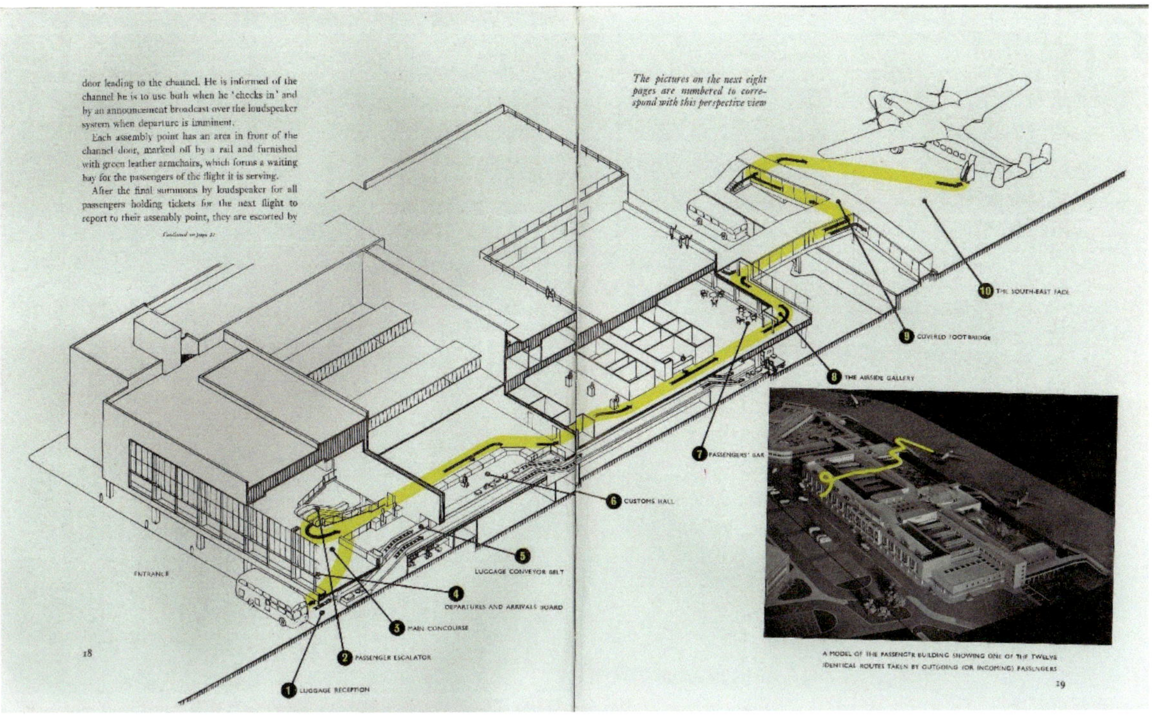

Figure 4.34 Heathrow Europa Terminal 1955 showing reversible passenger flow.

SOURCE: LONDON AIRPORT, THE OFFICIAL STORY OF THE NEW WORLD AIR CENTRE

Figure 4.35 Waterloo Air Terminal 1953.

SOURCE: THE UNDERGROUND MAP

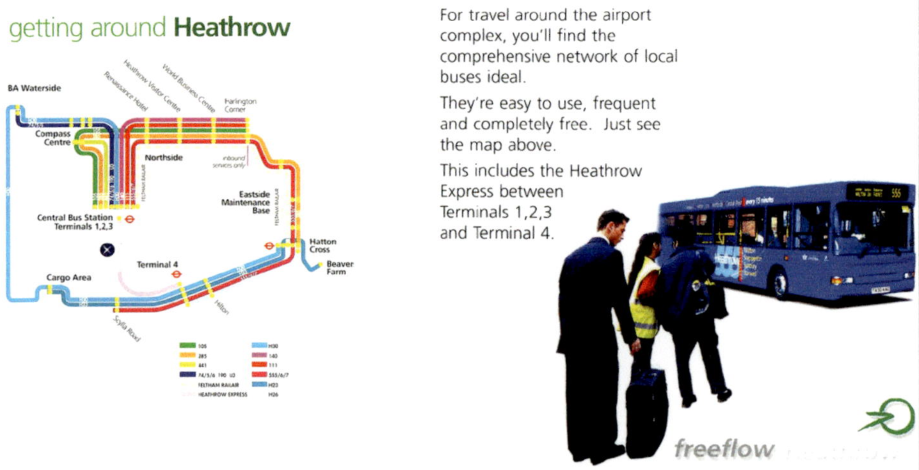

Figure 4.36 Heathrow central bus station.
SOURCE: HEATHROW AIRPORT LTD

building in 1963 built over Gloucester Road Tube Station and included a head office for BEA. It continued in use until 1973 when the check-in facilities were closed but the coach service continued. The demise of the coach service was confirmed, following the extension of the Piccadilly Underground line to Heathrow Central in 1977 and finally ceasing operation at the end of 1978.

Coach and bus operators have taken advantage of the volume of custom that airports generate to develop coach stations at airports that serve not only passengers flying from airports but also passengers who never leave the ground and are simply transferring from one coach service to another. Heathrow houses one of the busiest bus stations in the UK, second only to Victoria (Figure 4.36).

As Heathrow continued to expand with the construction of Terminal 3 (originally the Oceanic Terminal 1961) and Terminal 1 (1969) the constraints of the Star of David runway layout can be clearly seen (Figure 4.37).

Figure 4.37 Heathrow aerial view, showing Terminals 1, 2 &3 in the CTA 1977. As Heathrow continued to expand with the construction of Terminal 3 (originally the Oceanic Terminal 1961) and Terminal 1 (1969) the constraints of the Star of David runway layout can be clearly seen.
SOURCE: BRITISH AIRPORTS AUTHORITY

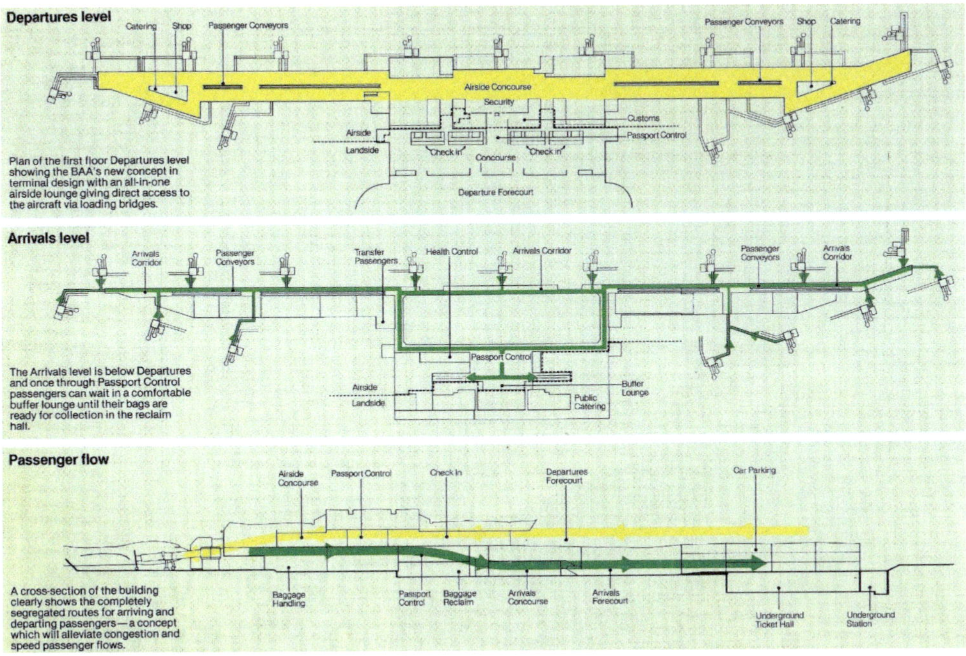

Figure 4.38 Heathrow Terminal 4 in 1987.

SOURCE: BRITISH AIRPORTS AUTHORITY

The next step was the construction of T4, a linear terminal, in the south which opened in 1986, Architects Scott, Brownrigg & Turner, and incorporated double-sided piers with open-planned gate lounges (1986). It introduced level separation for arriving (green) and departing (yellow) passengers. It was also the first terminal to be designed with security search from the outset (Figures 4.38 and 4.39).

Figure 4.39 Heathrow Terminal 4, architects Scott, Brownrigg & Turner.

SOURCE: BRITISH AIRPORTS AUTHORITY

14. St Louis Lambert (STL) 1956

St Louis Lambert International Airport is one of the most historic airports in the United States as the home of Charles Lindbergh, the first person to fly the Atlantic singlehanded. The terminal is notable for a number of innovations, including the separation of departing and arriving vehicles into a two-level forecourt and the early adoption of the finger pier concept which safeguarded space for the later introduction of aerobridges (Figure 4.40). Perhaps the most memorable feature is the signature vaulted copper-clad concrete roof of this innovative terminal (Figure 4.41).

The terminal has since evolved from a pier to a linear satellite layout (Figures 4.42 and 4.43). The future masterplan shows that Yamasaki's original terminal concept is planned

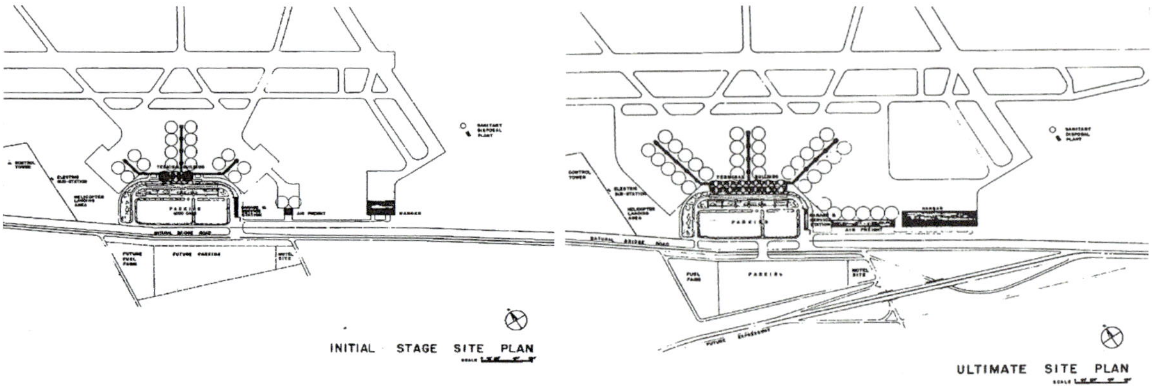

Figure 4.40 St Louis Lambert, initial and ultimate site plans c.1951.
SOURCE: ST LOUIS MUNICIPAL AIRPORT, ECONOMIC STUDIES

Figure 4.41 St Louis Lambert, aerial view showing extended 4th terminal module & finger pier.
SOURCE: CMJ

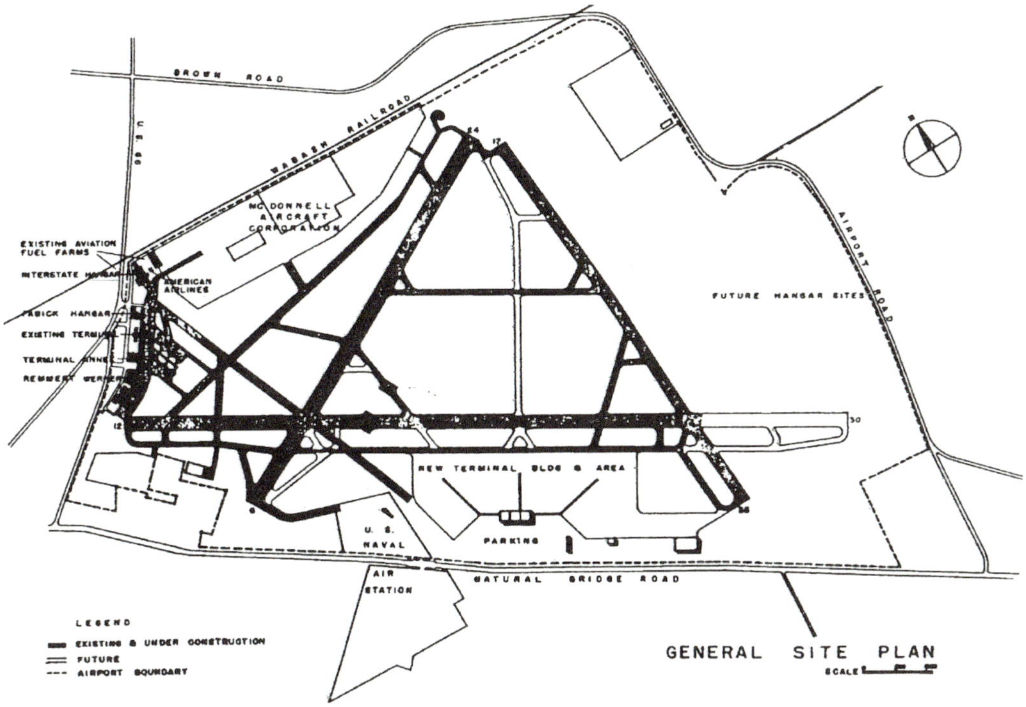

Figure 4.42 St Louis, Lambert, original master plan, c.1951.
SOURCE: ST LOUIS MUNICIPAL AIRPORT, ECONOMIC STUDIES

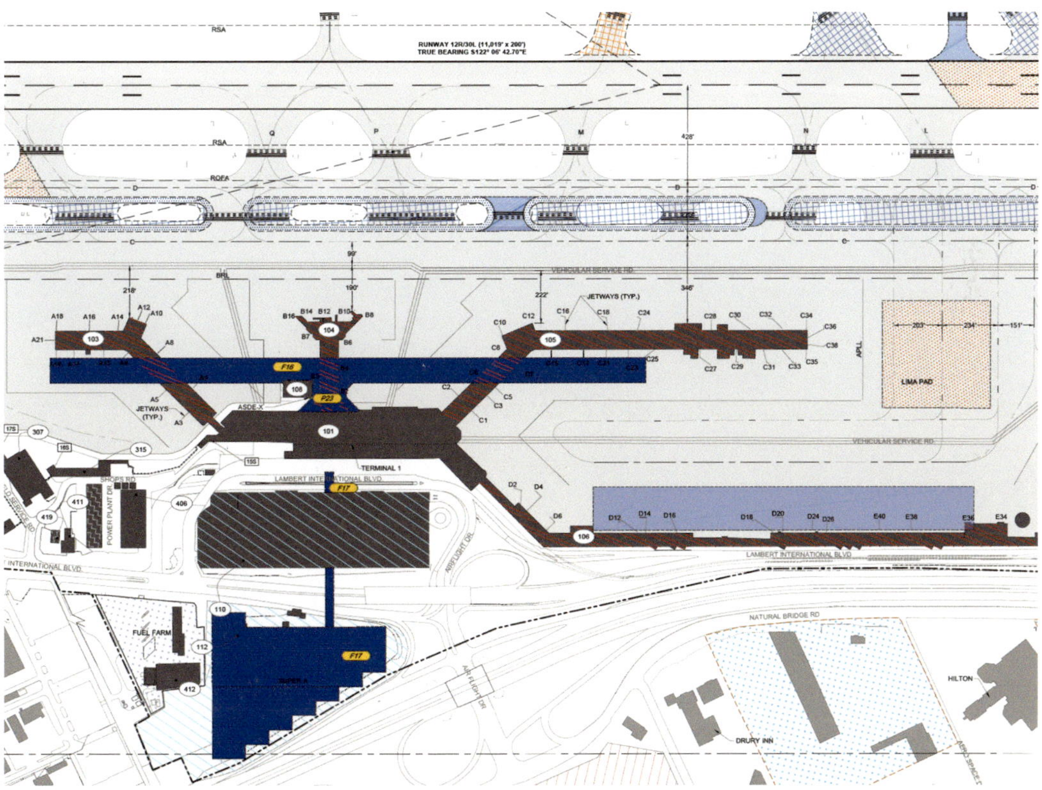

Figure 4.43 St Louis Lambert, future configuration. The future masterplan shows that Yamasaki's original terminal concept is planned to be retained but with the finger piers (cross-hatched in red) replaced with a double sided linear (dark blue) arrangement to accommodate increasing aircraft sizes.
SOURCE: FLYSTL.COM

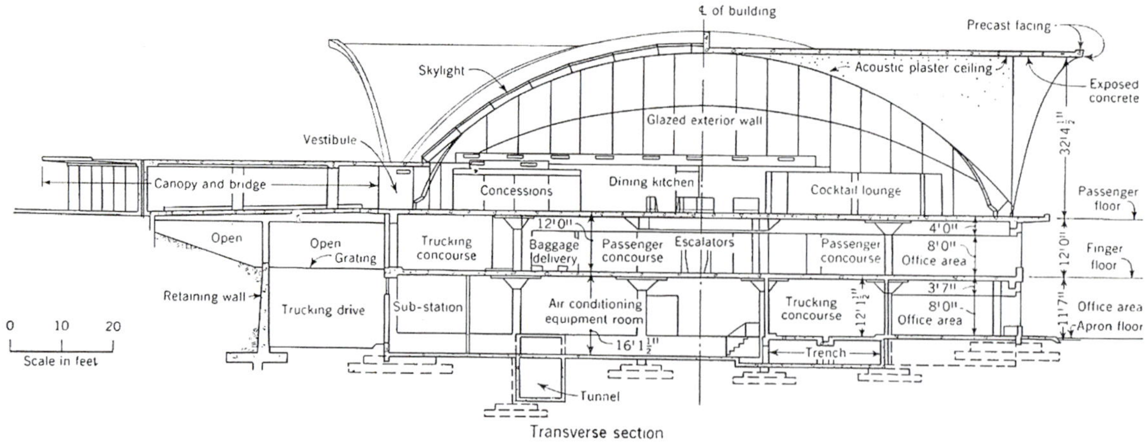

Figure 4.44 St Louis Lambert, 1956 section through terminal.
SOURCE: ARCHITECTURE USA, IAN MCCALLUM, ARCHITECTURAL PRESS 1959

to be retained but with the finger piers (cross-hatched in red) replaced with a double-sided linear satellite (dark blue) arrangement to accommodate increasing aircraft sizes. The section shows how it pioneered the use of escalators to allow passengers to change levels and incorporated, for the first time, a mechanical baggage handling system using conveyors (Figure 4.44).

Minoru Yamasaki of Hellmuth, Yamasaki and Leinweber introduced the shell concrete roof at St Louis Lambert (Figure 4.45). The terminal prefigured the glamour of jet travel, with its aerodynamic lines and a series of low-slung arches that celebrate the idea of flight. The original terminal building had three vaulted sections, each 120 ft square and 32 ft tall. An identical fourth section, architecturally in keeping with the

Figure 4.45 St Louis Lambert, view from forecourt.
SOURCE: LANDMARKS ASSOCIATION OF ST LOUIS

Figure 4.46 St Louis Lambert, departures concourse.

SOURCE: POSTCARD, THE LANDMARKS ASSOCIATION OF ST LOUIS, THE ST. LOUIS MERCANTILE LIBRARY AT THE UNIVERSITY OF MISSOURI, ST. LOUIS, THE LIBRARY OF CONGRESS' HARRIS AND EWING COLLECTION

rest, was added to the east in 1965. As well as the vaulted roof, the interior was also a trend setter by virtue of its freestanding cabins, designed like a self-supporting stage set, a feature that influenced Saarinen and later Foster and has become the norm in recent airport designs where a soaring roof is de-rigueur (Figures 4.46 and 4.47). The building and its vaulted copper roof have inspired the design of other terminals, such as the TWA terminal at John F. Kennedy International Airport in New York (Figure 4.48). It is regarded as an icon among airports.

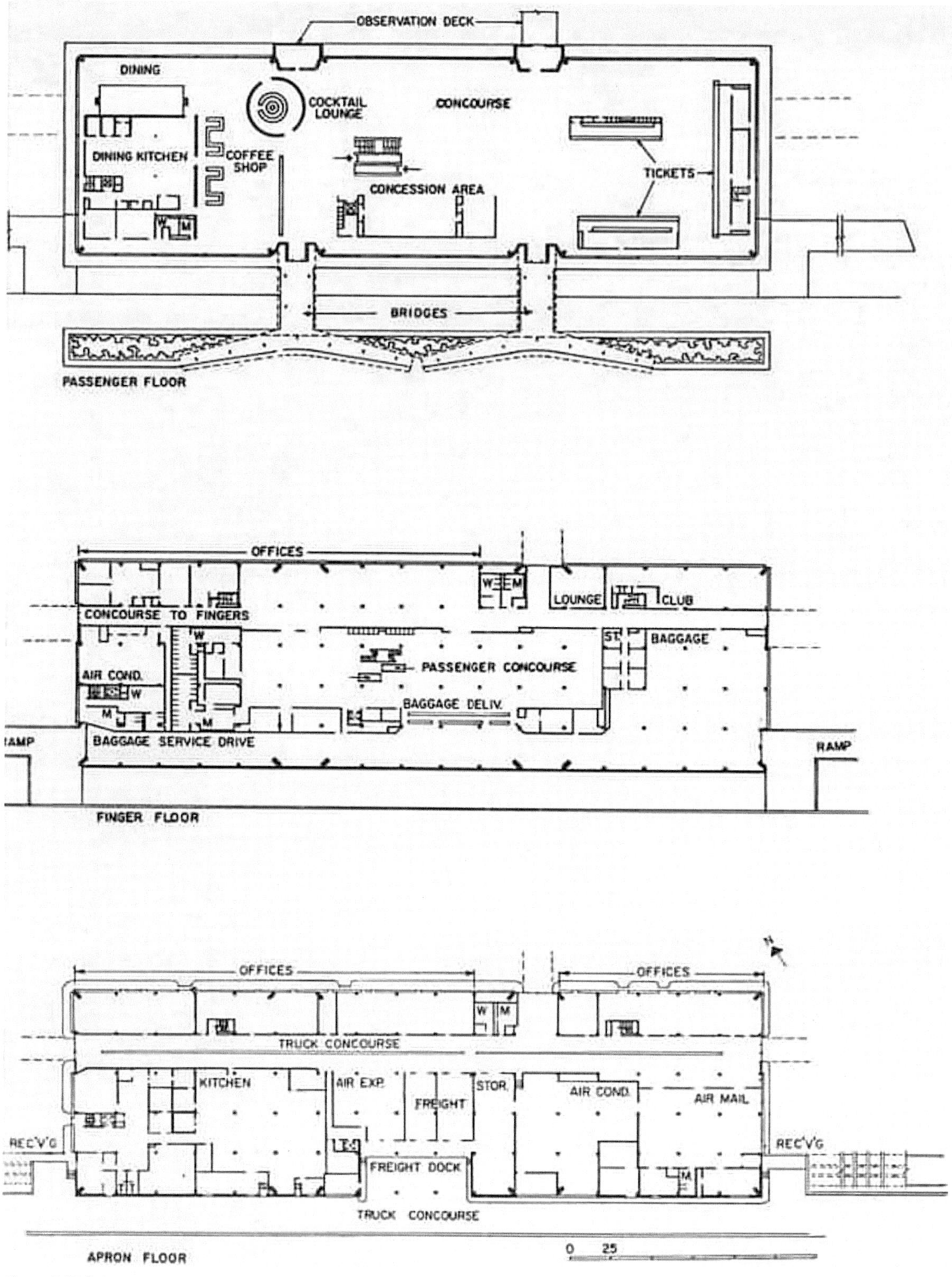

Figure 4.47 St Louis Lambert 1956, floor plans.
SOURCE: ARCHITECTURAL RECORD APRIL 1956

Figure 4.48 St Louis Lambert, departures concourse view looking back to the main entrance and showing the dramatic vaulted shell roof.
SOURCE: US LIBRARY OF CONGRESS

15. London Gatwick (LGW) 1958

The second Gatwick Airport of 1958 went further than the first airport in integrating rail travel by linking the rail station and air terminal under a single roof. It also introduced the first pier in Europe. The terminal was designed by Yorke, Rosenberg & Mardall (YRM) in the International Style with modular interchangeable panels that have allowed it to accommodate growth and change while finding some of the lightness and dynamism that were sought by the Futurists (Figure 4.49).

The terminal was dramatically located directly over and spanning the diverted A23 and was integrated with a new railway station (Figures 4.50). The Gatwick,

Figure 4.49 London Gatwick 1958.
SOURCE: BRITISH CALEDONIAN

Figure 4.50 London Gatwick 1958.
SOURCE: YORKE, ROSENBERG AND MARDALL (REPRODUCED WITH PERMISSION FROM RMJM)

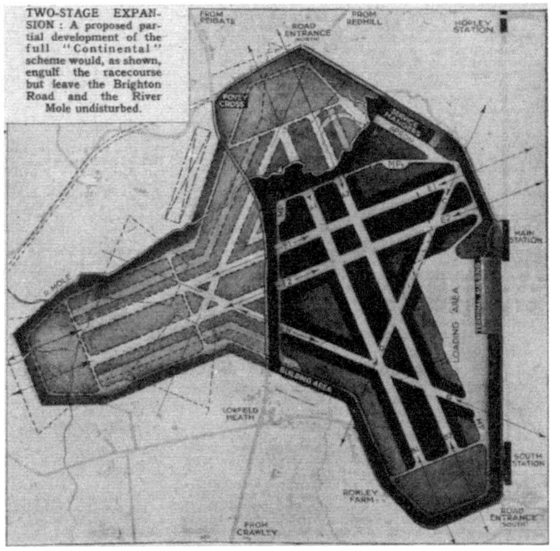

Figure 4.51 Gatwick continental master plan.

SOURCE: FLIGHT 25TH NOVEMBER 1943, P 578–580

continental master plan 1943, by Norman and Dawbarn was the antecedent of the new Gatwick airport with complex multidirectional runways but a simple small terminal linked directly to the railway (Figure 4.51).

Gatwick Airport's original 'Beehive' terminal had been closed in the mid-fifties to make way for a new, much more ambitious airport that was originally primarily aimed at handling diversionary traffic from Heathrow, but soon became a fast-expanding base for the burgeoning inclusive tour charter traffic. The new airport was located just to the north of the original airport on the site of the racecourse that had hosted the Grand National for several years during the First World War. It was the first airport to have a railway station integrated under one roof. As a result, Gatwick has always had a high percentage of passengers arriving by rail; in 2017, it was 34% of all passengers, against 36% arriving by car (Figure 4.52).

Figure 4.52 1958 aerial view with elevated forecourt over the A23 and railway station in the foreground.

SOURCE: ©BRITISH EUROPEAN AIRWAYS (NOW BRITISH AIRWAYS)

Figure 4.53 Gatwick central pier 1958 – architects Yorke, Rosenberg & Mardall (YRM).
SOURCE: AIRPORT HISTORY

In 1961, a few years after the new airport's opening, G.E. Kidder Smith wrote:

> *The cleverest feature of this south of London airport is the coalescing of three forms of transport – air, rail and road under one roof. Rarely has this been accomplished so effectively....... Another admirable feature is its quality of 'airportness': as soon as one arrives one feels that this is an airport and no other transport service. One is architecturally, indeed physically, projected into the field and made a part of the excitement, for no solid wall ever rises between the passenger and his aerial transport (Figures 4.53, 4.54 & 4.55).*
>
> (Kidder Smith, 1961)

Gatwick has been expanded greatly over the years, the first major extension being the doubling of the size of the terminal and the introduction of the north and south piers in

Figure 4.54 Gatwick 1958 aerial view.
SOURCE: BRITISH EUROPEAN AIRWAYS (NOW BA) VIA AV FRY POSTCARD & FAMGUS.SE

Figure 4.55 Gatwick Check-In concourse 1958.
SOURCE: YORKE, ROSENBERG AND MARDALL (REPRODUCED WITH PERMISSION FROM RMJM)

Figure 4.56 London Gatwick 1970 extended with three piers.
SOURCE: AIRPORT HISTORY

1964 (Figure 4.56). In 1970, the central pier was rebuilt on a larger scale with enclosed gate rooms and passenger boarding bridges to provide access to the aircraft. The existing landside forecourt, accessed from the A23, was demolished and the road access was moved to the east of the railway to provide access to the new M23 motorway (Figure 4.57). In 1983, a circular satellite designed to handle wide-bodied aircraft replaced the northern

Figure 4.57 London Gatwick, South Terminal 1977 with rebuilt central pier.
SOURCE: PPRUNE

Figure 4.58 Gatwick South Terminal with the satellite and APM in the foreground, 1983 and rebuilt central pier and northern extension beyond.
SOURCE: GATWICK AIRPORT

pier and was served by the first automated people mover (APM) in Europe (Figure 4.58).

A scheduled helicopter service, known as Airlink, flew between Heathrow and Gatwick from 1978 to 1986, but when the M25 opened the Government cancelled the licence. It was a fully commercial operation by British Caledonian Airways, supported by the British Airports Authority. Both of these organisations had a vested interest in introducing a high-speed connection between Gatwick and Heathrow airports. Since neither had helicopter expertise, they turned to another company resident at Gatwick, British Airways Helicopters (BAH), who could provide the expertise necessary for the CAA to issue an Air Operators Certificate for the service. With ten flights a day each way, it achieved an annual capacity of 50,000 to 60,000 people (Figure 4.59).

Figure 4.59 Airlink helicopter over Gatwick c1980.
SOURCE: GATWICK AIRPORT VIA ASHPOLE.ORG.UK

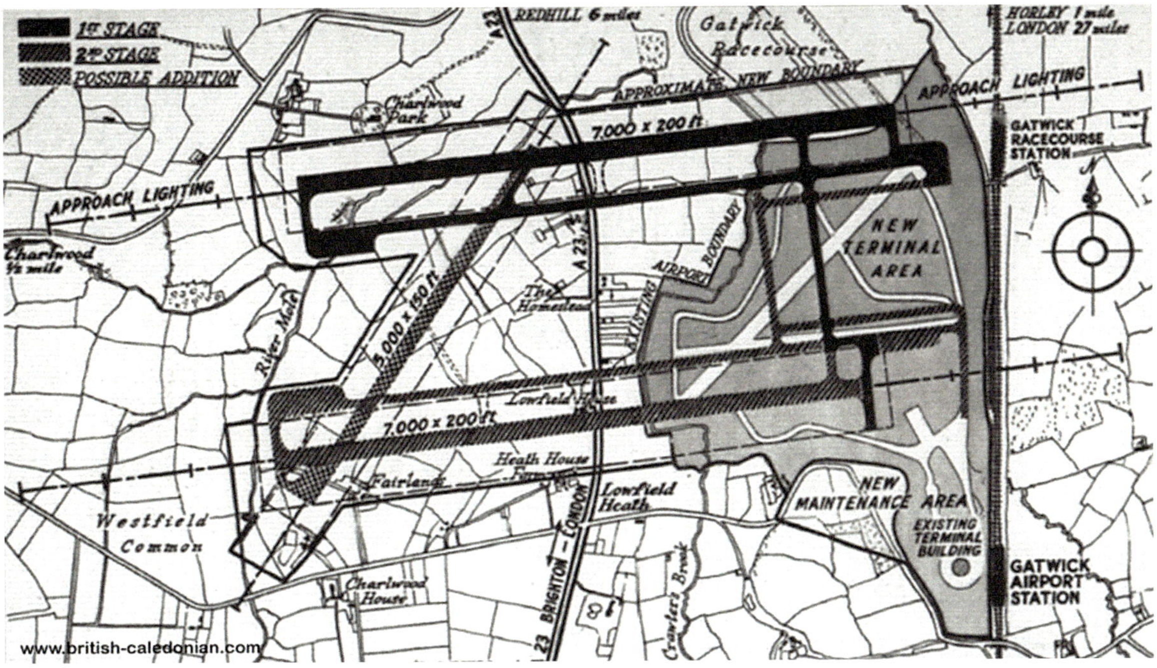

Figure 4.60 Gatwick 1952 land use plan showing the originally proposed second runway overlaid on the existing pre-war layout (hard-surface runways had been added during the war).
SOURCE: BRITISH CALEDONIAN

Gatwick had planned, until the 1970s, to build a second parallel runway to the north of the terminal, but later decided that terminal expansion took priority and in 1979 signed an agreement with the local planning authority not to seek a second runway for 40 years in exchange for their support for a second terminal (Figure 4.60).

The second terminal, now known as North Terminal, was then designed to occupy the site that had previously been safeguarded for the new runway. The evolution of the terminal facilities between 1958 and 1988 at Gatwick is ilustrated below (Figure 4.61).

The building was designed on three levels as a loose fit, almost entirely in dry prefabricated construction, with interchangeable cladding panels inside and out, developed from the system first used by YRM for IBM at Warwick University. That flexibility has allowed the terminal to respond to new trends, with major internal re-organisations, accommodating the growth of retail, stricter security, the introduction of automated passport control and self-service check-in (Figures 4.62 and 4.63).

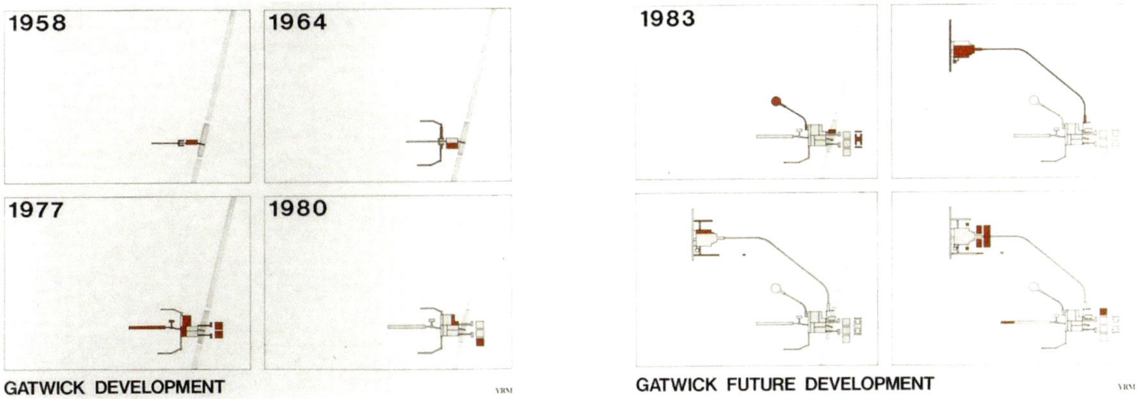

Figure 4.61 Evolution of Gatwick from 1958 to 1988.
SOURCE: YRM (REPRODUCED WITH PERMISSION FROM RMJM)

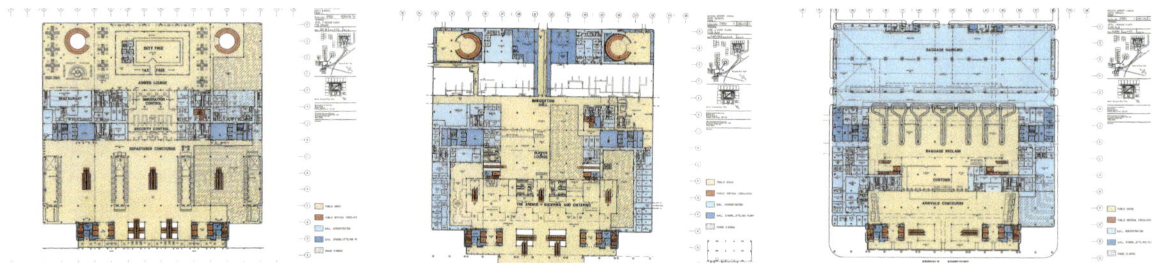

Figure 4.62 Gatwick North Terminal, departures, interchange and arrivals floor plans, 1988.
SOURCE: YRM (REPRODUCED WITH PERMISSION FROM RMJM)

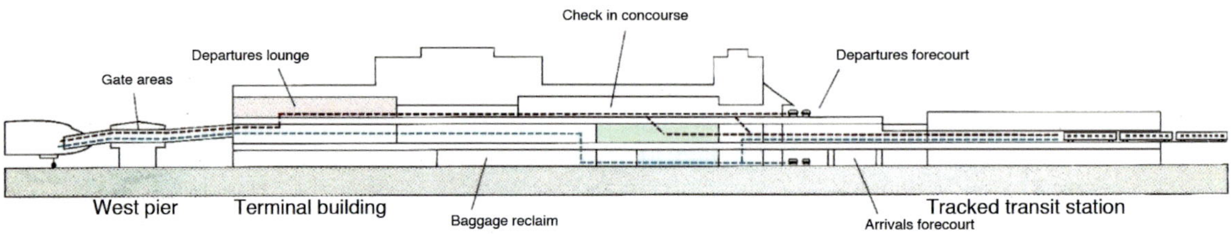

Figure 4.63 Gatwick North Terminal 1988, airside retail in pink and landside retail in light green at mezzanine level between departures and arrivals – architect YRM.
SOURCE: AUTHORS OWN COLLECTION

YRM's second terminal, Gatwick North, is reached from the rail interchange by another transit and by an extension of the road network from the South Terminal (Figures 4.64 and 4.65). YRM's publicity brochure of the time describes the terminal as follows:

> *'the impression of relaxing simplicity, with a total absence of unnecessary detail and varying quality and consistency of design that mars so many airports.'*

It was only after privatisation in 1986 that BAA regularly recruited retail experts who brought some of their learning from designing retail malls to airport commercial design. The North Terminal introduced to the airport familiar shopping mall concepts such as maximising retail footfall, including anchor tenants and destination outlets that would help draw passengers past other shops. Prior to that, convenience shopping had continued at airports for a very long time, as can be seen from the 1988 Gatwick South Terminal 'Skyshop' (Figure 4.66). The following plans show how retail design has evolved at Gatwick Airport's North Terminal from its opening in 1988 to the present day. The lounge

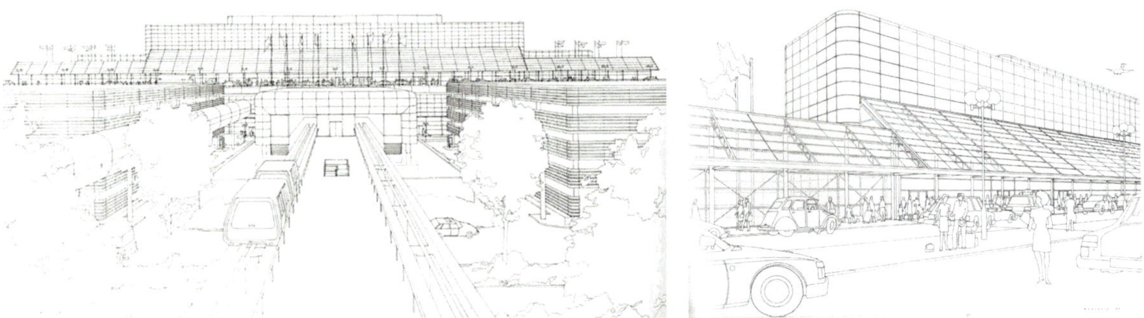

Figure 4.64 Gatwick, North Terminal landside interchange, transit station and forecourt design sketches.
SOURCE: YRM (REPRODUCED WITH PERMISSION FROM RMJM) – GRAPHICS: BO MAHADDIE

Figure 4.65 Gatwick with rail station, interchange and transit link to the North Terminal C 2000.
SOURCE: BAA

was originally planned as an outward-looking space with views over the airfield with an island-duty and tax-free shop. It has progressively morphed into an inward-looking space with the seating area surrounded by shops (Figures 4.67 and 4.68).

The most recent development is to introduce a walk through duty free shop that meanders from the exit of security screening to the lounge, but introduces additional walking distance as designers contrive to lead passengers past as many offers as possible.

Figure 4.66 Skyshop in the South Terminal Gatwick Village 1988.
SOURCE: AUTHOR'S OWN COLLECTION

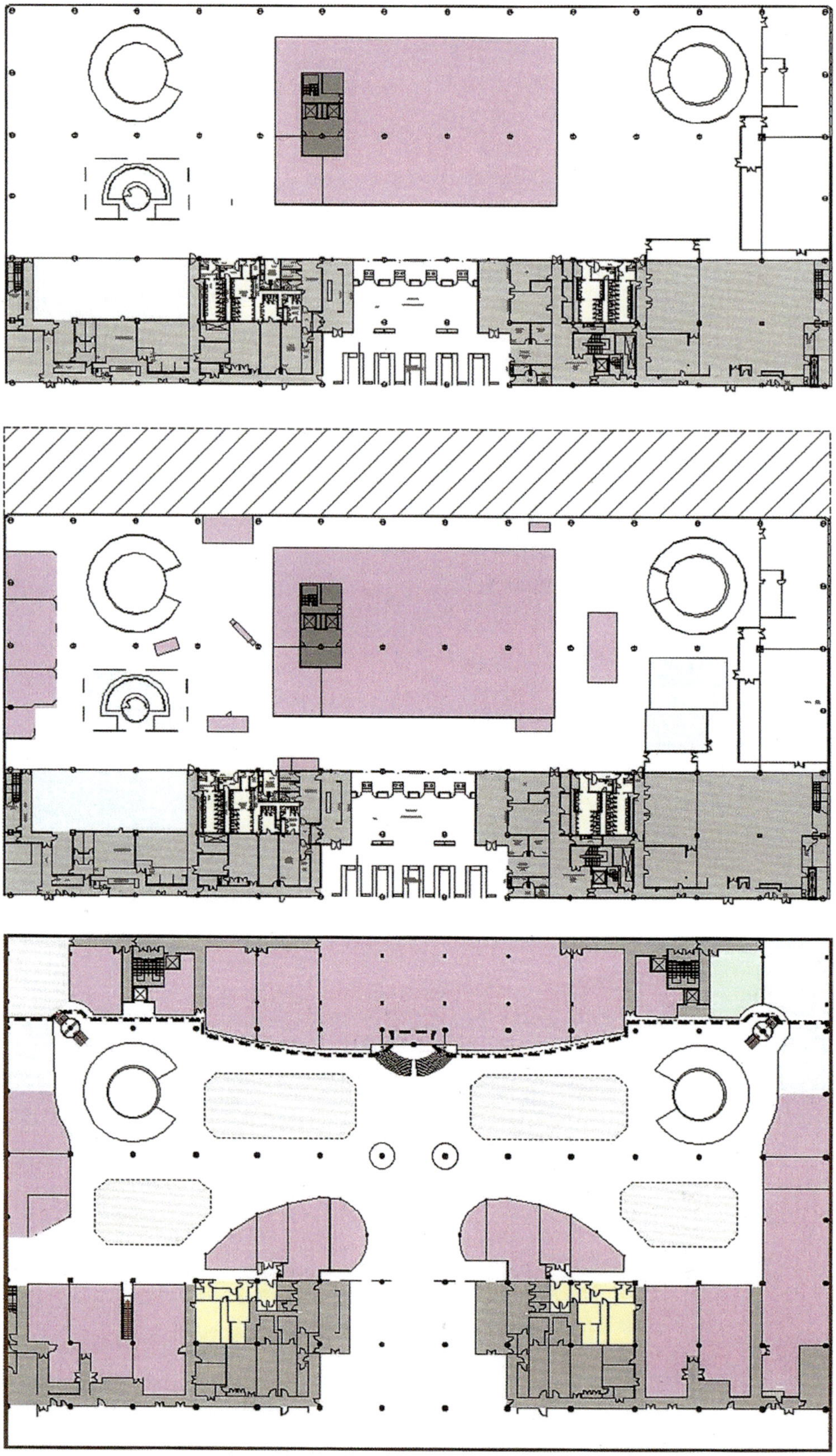

Figure 4.67 Gatwick North Airside Lounge Retail Evolution 1988/1991/1994.
SOURCE: GATWICK AIRPORT VIA AIRPORT INTERIORS

(a)

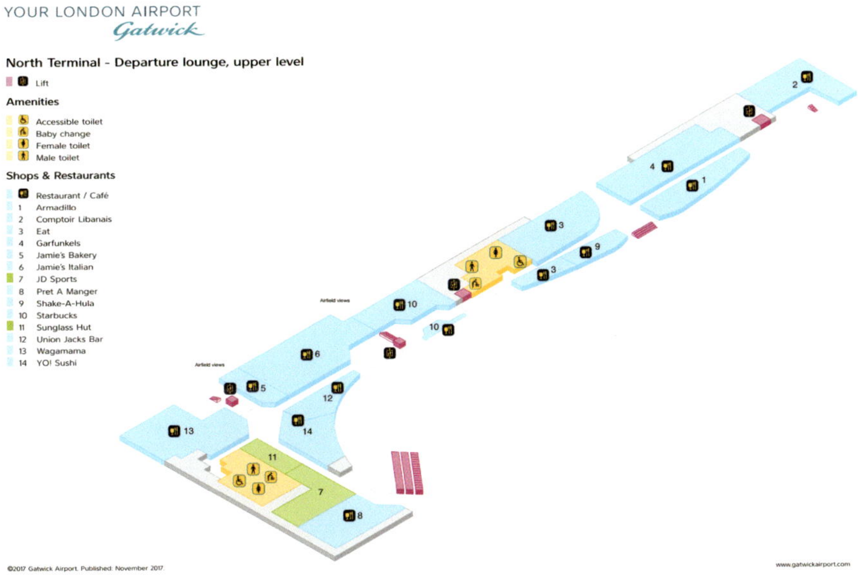

(b)

Figure 4.68 Airside departures lounge map 2017 (a) lower and (b) upper levels.
SOURCE: GATWICK AIRPORT

Colour plays a major role, with the blue of the vitreous enamel panels to the terminal building contrasting with the silver of the pier and white of the link bridges. In writing about YRM in 1992, to accompany an exhibition of their work at the RIBA Heinz Gallery, Alan Powers noted that (Figures 4.69, 4.70 & 4.71):

> *The advantage of YRM's consistency of style and gradual evolution is seen at Gatwick, where the turbulent zeitgeist has been calmed into submission and the architecture benefits from a continuous development of technical understanding.*
>
> (Powers, 1992)

Figure 4.69 Gatwick North Terminal, aerial view 1988.
SOURCE: BRITISH AIRWAYS

Figure 4.70 Gatwick North Terminal airside view.
SOURCE: YRM (REPRODUCED WITH PERMISSION FROM RMJM)

Figure 4.71 Gatwick North Terminal landside view from the transit linking the North and South terminals.
SOURCE: YRM (REPRODUCED WITH PERMISSION FROM RMJM)

Figure 4.72 Gatwick Bridge over the taxiway from the North Terminal to pier 6
SOURCE: UKAIRFIELDGUIDE.NET

The later passenger bridge was designed to link a remote satellite, known as pier 6, with the North Terminal. Early consideration had been given to a tunnel alternative, but following a visit to Denver, which had pioneered a bridge link to its first satellite, BAA was convinced that a bridge was both more economical and offered a better passenger experience. Unlike Denver where the bridge was only designed to span over short-haul B737 aircraft, the North Terminal Bridge was designed to allow B747 Jumbo aircraft to pass under (Figure 4.72). Other airports, such as Brussels and New York, LaGuardia have now followed this example.

While much of the simplicity and clarity have been lost as the airport has expanded, full advantage has been taken of the flexibility of the cladding and other components. Many components have been successfully moved from one location and re-used in another as the airport has expanded. Gatwick today is the busiest single-runway airport in the world.

Looking to the future, the draft Master Plan sets out Gatwick Airport Limited's ambitious vision for the future and looks at how the airport could grow across three scenarios, looking ahead to the early 2030s (Figure 4.73):

- *Main runway* – using new technology to increase capacity.
- *Standby runway* – bringing an existing emergency "standby" runway into routine use.
- *Additional runway* – safeguarding for the future – While Gatwick is not currently actively pursuing the option of building a brand-new runway to the south of the airport — Gatwick believes it is in the national interest to continue to safeguard this land for the future as part of its draft master plan.

Figure 4.73 Gatwick master plan with second runway and third terminal.
SOURCE: GATWICK AIRPORT LTD

CHAPTER 5

JET AGE 1960–1979

There was a false start to the jet age in 1952 with the introduction of the structurally flawed De Havilland Comet on a route between London and Johannesburg and later on other Commonwealth routes. It immediately ended the reign of flying boats on BOAC's long-haul routes, but services were suspended in 1954 after a series of crashes. It was relaunched in late 1958 as the Comet 4 shortly before the Boeing 707 was introduced. The 707 radically transformed commercial aviation with its large seating capacity and, with the 320 version introduced in 1962, the ability to cross the Atlantic non-stop with a full payload.

Jet flights brought with them a requirement for much longer runways and a radical increase in noise nuisance, causing the rise of the environmental protest movement as New York Idlewild (JFK) and Chicago O'Hare were adapted to the new larger jets with longer runways. A series of new terminals were designed specifically to support jet operations, including the Pan Am and TWA terminals at Idlewild (JFK) and new jet-age airports at Los Angeles International (LAX) and Washington Dulles were constructed.

The first of these terminals, the International Arrivals Building (IAB) at New York Idlewild (JFK), was designed to receive the new generation of large jet aircraft and came into operation in 1957, but at the time of opening, it still relied on passengers walking across the open apron to board the aircraft from the piers. Pan American's Worldport at Idlewild (JFK), which opened in 1960, can be seen as a transitional design with its overhanging canopy conceptually reminiscent of Berlin Tempelhof and its open aircraft access walkways, an anachronism when apron-drive airbridges had already been introduced. Its oval form was inherently difficult to expand, and its elegance was lost with the extensions that overwhelmed the simple original design. Its major legacy, however, was the introduction of nose-in aircraft parking, which was much more space-efficient than self-manoeuvring parallel parking and is now universally adopted at new large airports.

The Iconic TWA terminal that opened in 1962 captured perfectly the spirit of fight at the dawn of the jet age, but while its sculptural form is rightly admired, its planning was so bespoke that it proved difficult to adapt to larger aircraft, increased volumes of traffic and new operational practices. An innovation that has been widely copied is the satellite concept with an above-ground link.

Los Angeles International Airport (LAX) (1962) was originally designed in the late fifties with a circular terminal and satellites but evolved into lozenge-shaped satellites accessed via underground pedestrian links with passenger conveyors. Over time, the satellites have been progressively linked back to the entrance buildings to become conventional piers. The LAX Theme Building also produced one of the most futuristic and enduring airport images, evoking the space age, but like the TWA terminal, it has become redundant as airport processes have changed.

DOI: 10.4324/9781003323624-5

Eero Saarinen's second airport design for Washington Dulles (1962) has had the most profound influence on subsequent airport design, with an impact greater than any other airport before or since. It boasted the simplification of runway layout made possible by the improved performance of jet aircraft in cross-wind conditions, a mid-field terminal layout with a central control tower and a gyratory landside road layout that has become the norm for all recent airport multi-runway plans. Its surface access included a dedicated road link and safeguarded metro link that was installed later. The airfield layout was the first to adopt the 'toast-rack' layout which optimised aircraft movement between the runways.

The terminal building with its dramatic upswept clear-span roof, although constructed in concrete can be seen as the forerunner of the modern long-span steel structures that are an integral part of contemporary airport design. The interior incorporated freestanding cabins, reminiscent of exhibition stands, accepting that the interior had become a flexible stage set. The terminal also introduced the concept of an indeterminate architecture that looked complete and resolved, whether in its original size or when extended. These ideas predate Norman Foster's claims to have originated the ideas by thirty years.

The one feature that was to prove a wrong move was the proposal to use mobile lounges to access aircraft. Saarinen enlisted the help of Charles Eames to produce one of the most compelling cartoons promoting an airport concept of operation. For a time, mobile lounges were all the rage and several new airport designs were developed following these principles including the ill-fated and short-lived Montreal, Mirabel airport. Mobile Lounges proved to be slow, unreliable and expensive and ultimately were not able to carry sufficient numbers of passengers, as aircraft sizes grew and the volume of passengers increased. Washington Dulles' greatest virtue has been its ability to accommodate growth and change without losing either its simplicity of layout or integrity of form, because that was planned for from the outset, a feature that many more recent designs have failed to address.

The 1970s saw the introduction of the Concorde, the Boeing 747 and other wide-body jets, aircraft that were heavier and required even longer runways and forced many of the major airports to strengthen and lengthen their runways. As terminal sizes increased to handle these larger volumes of traffic and multiple terminals proliferated, the alternative to either walking or riding on a bus or in a mobile lounge was to travel on an automated people mover (APM).

The APM was originally introduced as a ride to reach various attractions at Disneyland, but was first adapted to airport use at Tampa Airport. The most complex system was installed at Dallas Fort Worth (DFW) (1973) to link together the multiple unit terminals and allowed a concept that was particularly suited to originating and destinating (O&D) flights to work effectively for transfer passengers. Transfer passengers were provided with unrivalled panoramic views of the airfield as they rode between the multiple semi-circular terminals.

Dallas Fort Worth had, at the time of opening, the largest land take of any in the world, and its layout with a central access spine and multiple terminals was much copied, including the stillborn 3rd London airport at Maplin Sands and later the original master plans for Hamburg Kaltenkirchen, Paris Charles de Gaulle and numerous subsequent airport designs by Aeroport de Paris (ADPI).

The airports selected to illustrate the key characteristics of the era are:

- New York Idlewild (JFK) 1962
- Washington Dulles (IAD) 1962
- Los Angeles (LAX) 1962
- Amsterdam Schiphol (AMS) 1967
- Dallas Fort Worth (DFW) 1973
- Paris Charles de Gaulle (CDG) 1974

16. New York Idlewild (JFK) 1962

New York Idlewild (JFK) introduced the Terminal City concept, a glamorous name for a mixture of individual airline terminals. It was a compromise that was presented as a new vision of the future but was in practice an uncoordinated free for all, with individual airlines developing their own terminals (Figure 5.1). The airlines were, however, the originators of some of the most advanced terminal concepts and most evocative images of the jet age, the most memorable of which were by Pan American and TWA (Figure 5.2).

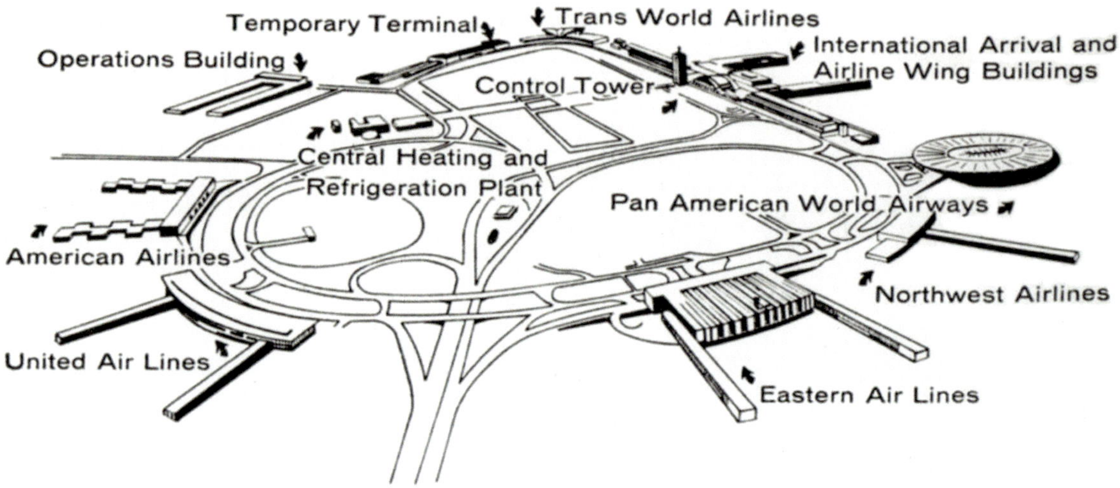

Figure 5.1 New York Idlewild (JFK) the terminal city concept with multiple airline terminals.
SOURCE: AIRPORTHISTORY.ORG

Figure 5.2 New York Idlewild (JFK) TWA Terminal, JFK 1962.
SOURCE: RARE HISTORICAL PHOTOGRAPHS

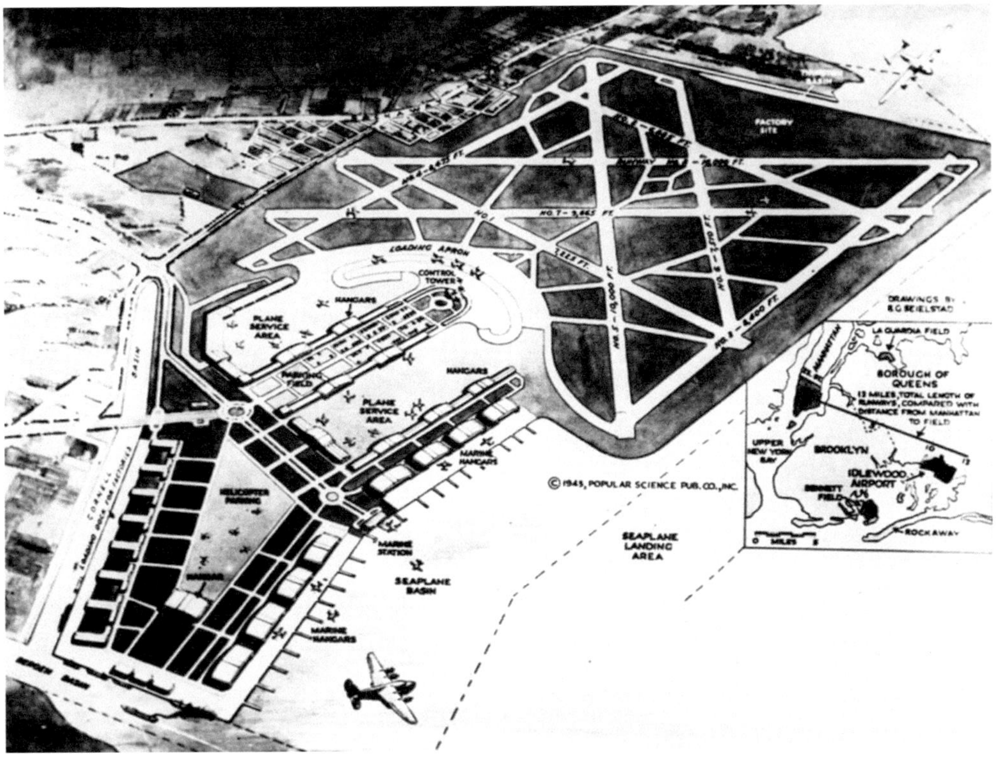

Figure 5.3 New York, Idlewood (later Idlewild) Airport 1943.
SOURCE: POPULAR SCIENCE PUB. CO. INC.

The history of JFK airport shows that the airport has been able to evolve to suit changing circumstance, but there can be little doubt that if there had been less of a free for all and more overall master planning the airport could, today, have successfully handled more passengers. The planned way forward is with fewer but larger terminals. As it is, the airport which was once in the vanguard of innovation and pre-eminent in the USA, has been overtaken in volume of traffic and quality of service by many other airports. The TWA building, now converted to a hotel, reminds us of the glamour of the jet age and with its soaring, bird-like roof, it still perfectly captures the spirit of flight.

The airport was proposed in the early 1940s when it became apparent that La Guardia would not have enough capacity in the long term. An early master plan prepared by Delano and Aldrich and published in Popular Science in 1943 looked backwards to the third prize proposal in the Lehigh Airports Competition and the Chicago Midway layout with its cruciform runways and terminal on the periphery. The original layout also had some similarities with La Guardia, with a large seaplane base and aircraft maintenance complex. It did, however, foresee much larger aircraft and incorporated two runways of 10,000 ft (3,048 metres) (Figure 5.3).

Feedback from the airlines was that with the development of long-range, land-based aircraft facilities for seaplanes were no longer needed, so subsequent master plans focused on land aircraft only. However, a series of more ambitious master plans soon evolved with mid-field terminal concepts and multiple radiating runways. One of the early 1945 centralised concepts promoted by American Airlines is shown below. Many concepts were considered and dismissed after that; the initial centralised perimeter development gave way to a unit terminal concept (1947) and a centralised scheme with finger piers (1948) (Figures 5.4–5.6). The airport originally opened in 1948 as New York International Airport

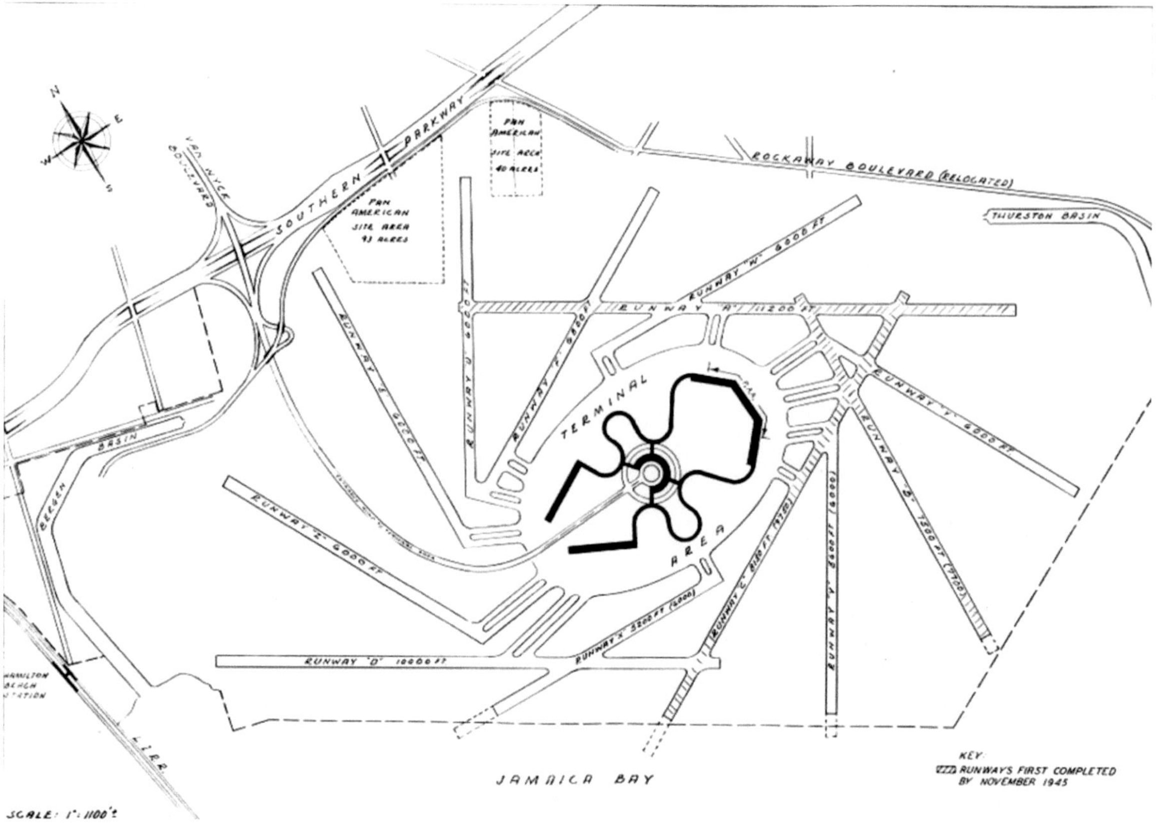

Figure 5.4 New York Idlewild (JFK) 1945 Master Plan.
SOURCE: CRADLE OF AVIATION MUSEUM VIA AIRPORTHISTORY.ORG

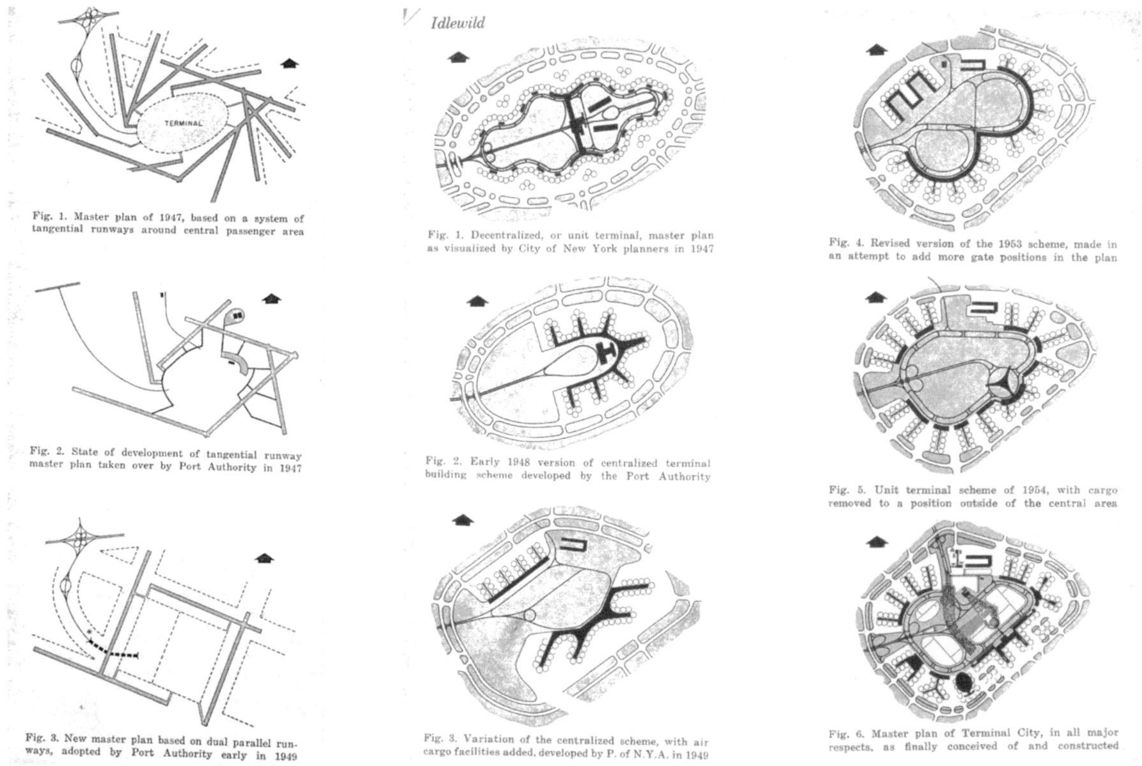

Figure 5.5 New York Idlewild (JFK) Master Plan Evolution 1947–1962.
SOURCE: PANYNJ VIA ARCHITECTURAL RECORD

Figure 5.6 New York Idlewild (JFK) early centralised terminal concept.
SOURCE: AIRPORTHISTORY.ORG

Figure 5.7 New York Idlewild (JFK) temporary terminal, 1947, prior to opening.
SOURCE: AIRPORTHISTORY.ORG

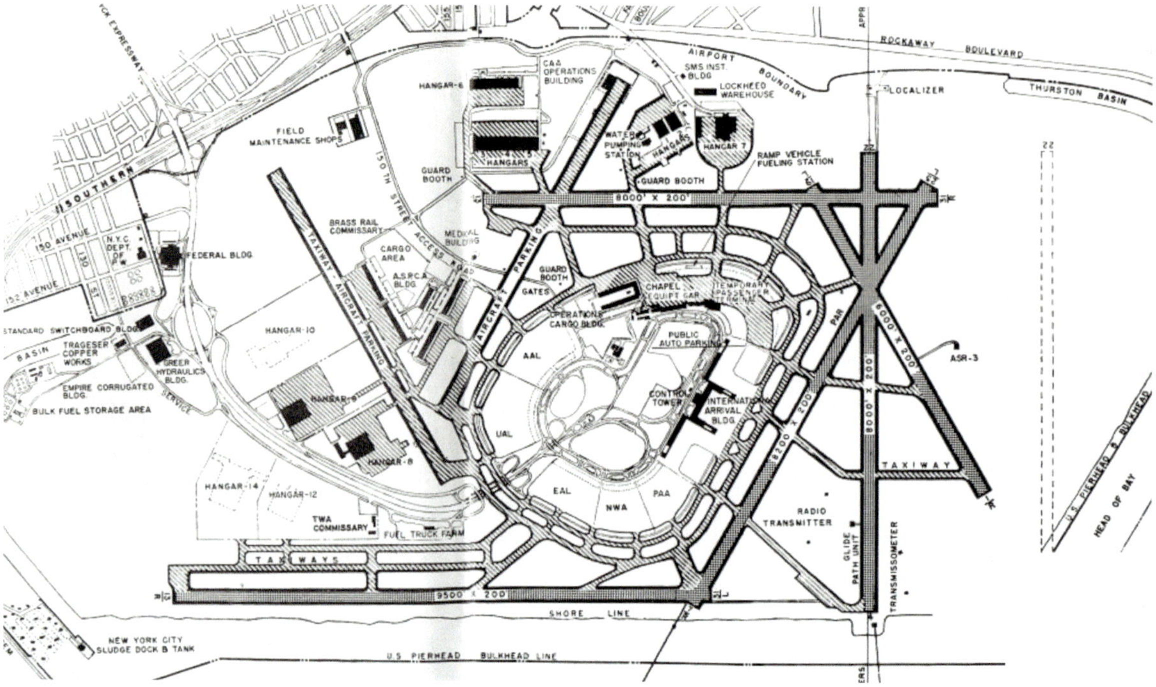

Figure 5.8 New York Idlewild (JFK) master plan showing the location of the International Arrivals Building and the simplified runway layout.
SOURCE: AIRPORTHISTORY.ORG

but was commonly known as Idlewild Airport. In the early years, it featured a series of temporary terminals (Figure 5.7).

In parallel with the evolution of the terminals, the runway system was simplified with the improvement in aircraft performance in cross winds, resulting in fewer but longer runways (Figure 5.8). Critically, this evolution was centrally planned by the Port Authority and the FAA rather than by the airlines who were responsible for the uncoordinated individual terminal designs.

The International Arrivals Building, or IAB, was the first new permanent terminal at the airport, opening in December 1957 and was the first to be planned to receive jet aircraft. The building was designed by Skidmore, Owings and Merrill. The terminal stretched nearly 700 metres (2,300 ft) and was parallel to runway 7R. The terminal had 'finger' piers at right-angles to the main building, allowing more aircraft to park (Figures 5.9 and 5.10).

Figure 5.9 New York Idlewild (JFK) Artists Impression of The International Arrivals Building. The Building was expanded in 1970 to accommodate passenger boarding bridges. However, by the 1990s the overcrowded building was showing its age and it did not provide adequate space for security checkpoints. It was demolished in 2000 and replaced with Terminal 4.
SOURCE: AIRPORTHISTORY.ORG

Figure 5.10 New York Idlewild (JFK) 1958, International Arrivals Building
SOURCE: JON PROCTOR

One final iteration of the cantilever roof concept can be found at the oval Pan American Airways Terminal *'Worldport'* at New York JFK (1960). Early versions of the design in the mid-fifties provided for propeller-driven aircraft to park parallel to the terminal and be accessed by moveable gangways under a giant cantilever canopy. This evolved during design development to accommodate the new generation of jet aircraft that had been ordered by Pan Am and revised to a nose-in parking configuration (Figure 5.11). The gangways can be seen as an interim step towards the passenger boarding bridge, but the terminal design was completed when the latter was only in the early prototype stage (Figure 5.12). Later, conventional telescopic apron-drive passenger boarding bridges were retrofitted to the terminal.

The terminal was designed as a showcase for Pan American and to help them promote international jet travel and is particularly famous for its *'flying saucer'* roof suspended far from the outside columns of the terminal by 32 sets of pre-stressed steel posts and cables (Figures 5.13–5.15).

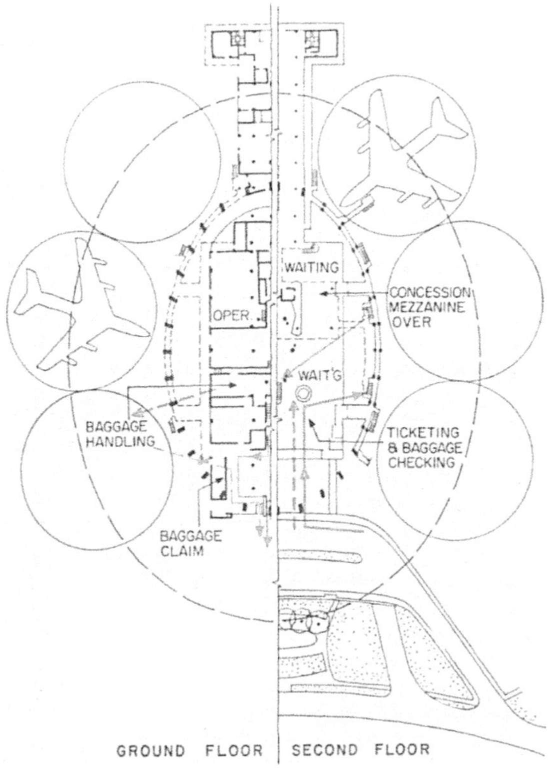

Figure 5.11 Pan American Terminal, New York Idlewild (JFK) 1960. Floor plan.
SOURCE: ARCHITECTURAL RECORD

Figure 5.12 Pan American Terminal, New York Idlewild (JFK) 1960. Models and photograph of the gangway.
SOURCE: THE PAN AM HISTORICAL FOUNDATION

Figure 5.13 New York Idlewild (JFK) 1960 photo c2000.
SOURCE: WIKIPEDIA HTTPS://WWW.JETPHOTOS.COM/PHOTO/6867311 GFDL 1.2 HTTP://WWW.GNU.ORG/LICENSES/OLD-LICENSES/FDL-1.2.HTML

Figure 5.14 New York Idlewild (JFK) Pan American Airways Terminal view.
SOURCE: LIFE

Figure 5.15 New York Idlewild (JFK) Pan American Airways Terminal aerial view, showing the cable suspended cantilever roof.
SOURCE: LIFE

Figure 5.16 New York Idlewild (JFK) aerial view of the Pan Am terminal after extension.
SOURCE: PAN AMERICAN AIRWAYS

The architect, Walter Prokosch, adopted this obsolescent concept for the Pan American Terminal by persevering with the cantilever canopy, although he had previously invented the apron-drive passenger boarding bridge as well as the finger pier, in conjunction with his engineering colleague Charles Froesch some 15 years earlier. His earlier idea had been to develop a powered enclosed telescopic bridge, which could be raised and lowered to align with the aircraft door and also adjusted for any slope inherent in tail-dragging aircraft such as the Douglas DC3. While at the prototype stage, the passenger boarding bridge had not evolved sufficiently for adoption at the time of the terminal design, though it was subsequently retrofitted to the terminal.

The limitations of the elliptical shape are clearly seen in the following image which shows the extended terminal designed to accommodate the larger Boeing 747 aircraft. The original terminal, originally designed to serve the first generation of jets, the Boeing 707, was then reconfigured to handle the new generation of short-haul jets, the Boeing 727 and 737 (Figure 5.16).

Trans World Airlines opened the TWA Flight Center in 1962, designed by Eero Saarinen with a distinctive winged-bird shape and flowing futuristic curvilinear interior. He articulated TWA's desire for one space to flow into another, *'We wanted to reveal the terminal, not as a static, enclosed space, but as a place of movement and transition'.* It was described by TWA as the *'Winged Gateway to the World of Flight.'* The following early sketch, terminal plan and images illustrate the design (Figures 5.17–5.26).

The satellite concept was also introduced by Eero Saarinen at the New York Idlewild (JFK) TWA Terminal (1962). The terminal was linked by over-ground walkways to the satellites designed to accommodate the new generation of jet aircraft, a concept taken forward by a number of subsequent terminal designs (Figure 5.27). The planning and

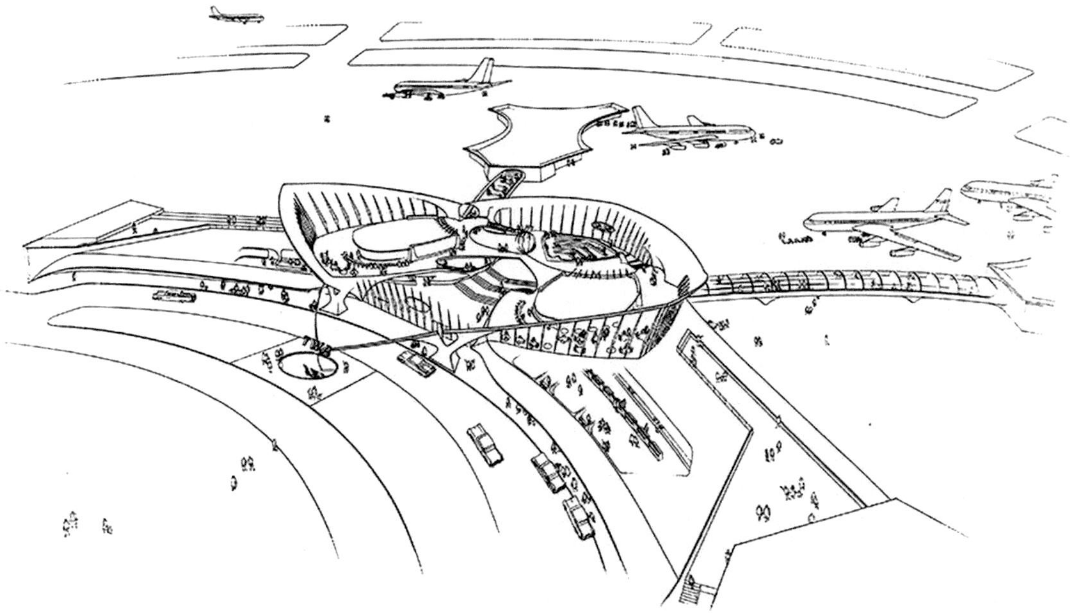

Figure 5.17 Preliminary sketch of the TWA Flight Center – note the shape of the satellite to the left which was in the form of the eventual shape of the second satellite while the first satellite to the right was simplified to a hexagon.

SOURCE: PHOTO: © EERO SAARINEN COLLECTION (MS 593)/MANUSCRIPTS AND ARCHIVES, YALE UNIVERSITY LIBRARY

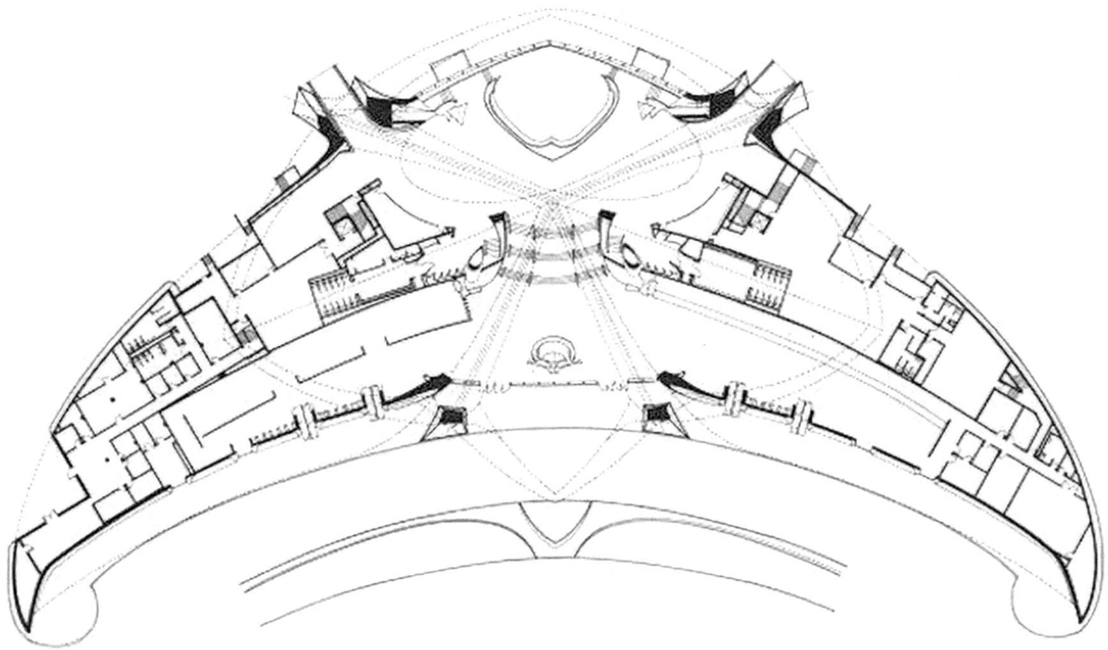

Figure 5.18 TWA Flight Center floor plan.

SOURCE: PANYNJ VIA WIKIARQUITECTURA.COM

Figure 5.19 TWA Flight Center.
SOURCE: PANYNJ

Figure 5.20 TWA Flight Center.
SOURCE: PANYNJ COURTESY/JOHN BARTELSTONE PHOTOGRAPHY

Figure 5.21 TWA Flight Center.
SOURCE: LOUIS VUITTON WOMEN COLLECTION CRUISE 2020 – LOUIS VUITTON MALLETIER

Figure 5.22 TWA interior 1962.
SOURCE: DANICA O. KUS

Figure 5.23 TWA lounge area.
SOURCE: WIKIMEDIA

Figure 5.24 TWA, Flight Information.
SOURCE: WIKIPEDIA PHOTO: WARREN LEMAY

Figure 5.25 TWA, Seating Area.
SOURCE: DANICA O. KUS

Figure 5.26 TWA Flight Center, open day in 2012.
SOURCE: WIKIPEDIA

Figure 5.27 New York Idlewild (JFK) TWA Terminal 1962.
SOURCE: PANYNJ

design of the terminal were so specifically geared to the TWA operation of the day, and although a second much larger satellite was added, the terminal complex proved inflexible and incapable of long-term adaptation to changed circumstances.

The high point of development at JFK was reached in 1962 when most of the airlines had completed their own terminals, but traffic hadn't grown beyond their capabilities (Figures 5.28 and 5.29). Table 5.1 summarises the development of the terminal city at New York JFK. In December 1963, Idlewild airport was renamed John Fitzgerald Kennedy Airport (JFK) in memory of the late president who had been

Figure 5.28 New York Idlewild (JFK) Aerial view of the original Airport City 1962.
SOURCE: AIRPORTHISTORY.ORG

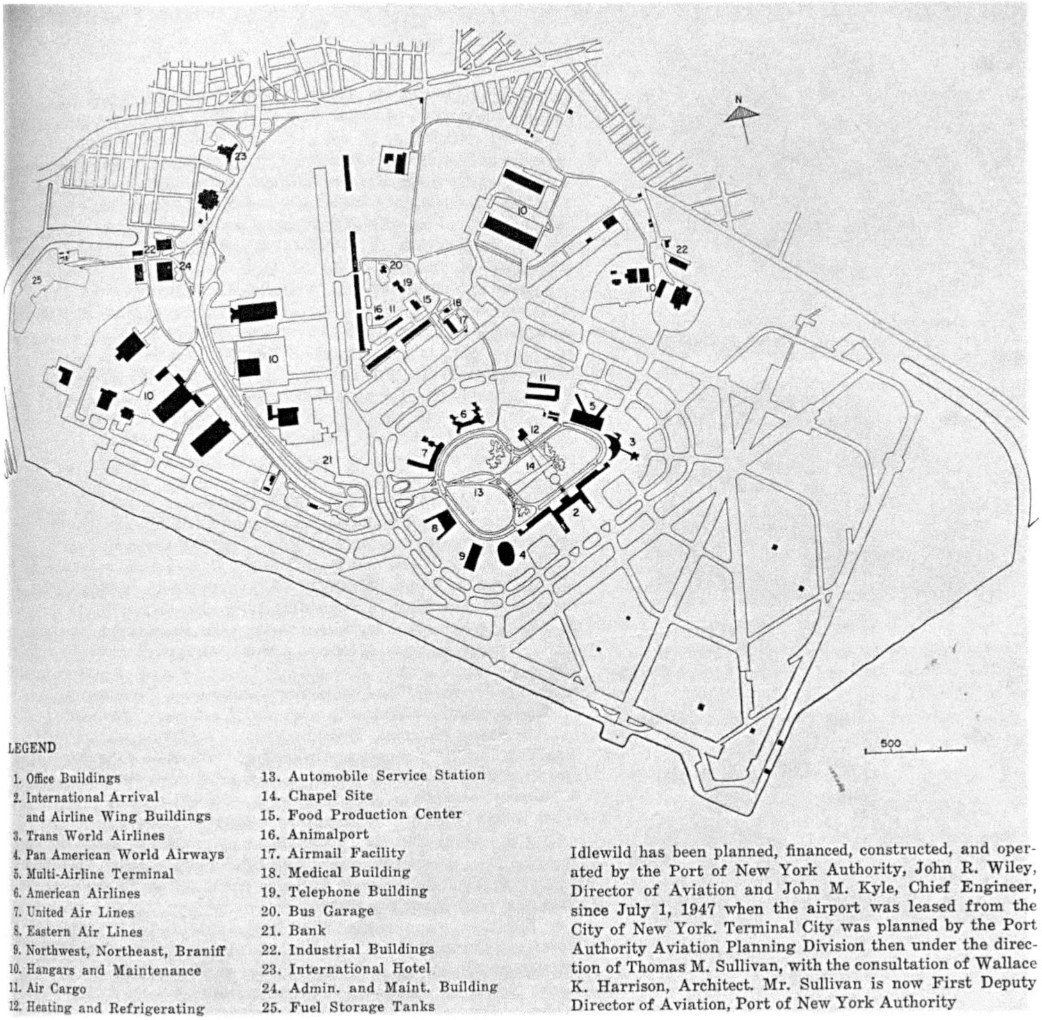

Figure 5.29 New York Idlewild (JFK) Master Plan 1961.
SOURCE PANYNJ VIA ARCHITECTURAL RECORD, SEPTEMBER 1961

Table 5.1 Summary of JFK terminals

Open	*Terminal*	*No*	*Close*	*Size*	*Cost*	*Stands*	*Architect*	*Key characteristics*
1957	IAB	T4	2001	55,742 m^2	\$30M	24	SOM	• All international arrivals • First finger pier at JFK
1959	Eastern	T1	1995	33,564 m^2	\$20M	16	Chester L Churchill	• Ramps instead of stairs, Two single level piers
1959	United	T9	2000	23,226 m^2	\$14.5M	16	SOM	• First boarding bridges • International style
1960	American	T8	2000	17,326 m^2	\$14M	16	Kahn & Jacobs	• First nose in stands • Stained glass facade
1960	Pan Am + Worldport	T3	2013	9,446 m^2	\$12M	8 +8 in ph2	TAMS	• Oval Cantilever Roof Extension for B747s
1962	Braniff, NE/NW	T2	2023	23,226 m^2	\$10M	10	White & Mariani	• Simplest & longest lived • Gates around terminal
1962	TWA	T5	2001	13,995 m^2	\$15.2M	14	Eero Saarinen	• Seminal expression of flight, only satellites
1970	National Sundrome	T6	2008			12	I M Pei	• All glass mullions • Diagrid ceiling
1970	BOAC	T7	2026*				GMW	• Brutalist architecture • Rooftop helipad
1983	Tower		2000			3		• Converted Hangar

Source: AirportHistory.org *Projected closure.

Figure 5.30 Walter Gropius (centre) with a model of the Pan Am Building, showing its juxtaposition with Grand Central Station the foreground.
SOURCE: OTRO

Figure 5.31 Aerial view of the Pan Am Building, a NY Airways helicopter about to land.
SOURCE: THE PAN AM HISTORICAL FOUNDATION

assassinated earlier in the year. The Airport City Master Plan of 1961 and aerial image of the Airport in 1962 can be seen overleaf.

Perhaps the best-known and most dramatic helicopter route was operated by Pan American Airways. It linked New York JFK airport, with Central Manhattan with the landing platform located on the roof of the Pan Am Headquarters which was in turn located directly above Grand Central Station, surely the ultimate interchange. The Pan Am building was designed by Walter Gropius and his practice, The Architects' Collaborative or TAC (Figure 5.30). It is the only aviation-related building designed by one of the founders of the modern architectural movement to have been realised in practice and demonstrates his preoccupation with the aerial view, a concept he introduced to architectural theory (Figures 5.31 and 5.32). This service operated between 1965 and 1968, briefly resuming in 1977, but was terminated in May after a fatal accident (Figure 5.33). All three of New York's heliports are now located at the river edge rather than on building roofs.

With the demise of TWA in 2001, the terminal remained vacant until 2005 when JetBlue Airways and the Port Authority of New York and New Jersey (PANYNJ) financed the construction of a new 26-gate terminal partly encircling the Saarinen building. Called Terminal 5 (Now T5), the new terminal opened on October 22, 2008. T5 is connected to the Saarinen central building through the original passenger departure-arrival tubes that connected the building to the outlying gates. The TWA Flight Center is now preserved on the National Register of Historic Places and has since been converted into the TWA Hotel (Figures 5.34–5.36).

Figure 5.32 Grand Central Terminal.
SOURCE: RAREHISTORICALPHOTOS.COM

Figure 5.33 Pan Am heliport 1965–1968.
SOURCE: THE PAN AM HISTORICAL FOUNDATION

An aerial view of the terminal zone before the demolition of the Pan Am terminal (T3 seen in the bottom right foreground) is shown in the photograph on the next page, and the current terminal area configuration is shown in the terminal area plan (Figures 5.37 and 5.38). JFK is currently spilt across six separate terminals and connected by the AirTrain light rail that was retro-fitted along with a complex road network. Future plans involve the creation of fewer and larger unified terminals that would allow easier transfers for passengers with flight connections (Figures 5.39 and 5.40).

Figure 5.34 View of the TWA terminal after restoration as a hotel.
SOURCE: DANICA O KUS

Figure 5.35 The TWA terminal after conversion to a hotel.
SOURCE: WIKIPEDIA. PHOTO: ROLAND ARHELGER

Figure 5.36 The TWA terminal after conversion to a hotel.
SOURCE: WIKIPEDIA. PHOTO: JAG9889

Figure 5.37 New York, JFK, aerial view.
SOURCE ARCHITECTURE DAILY PHOTO: JEFFREY MILLSTEIN

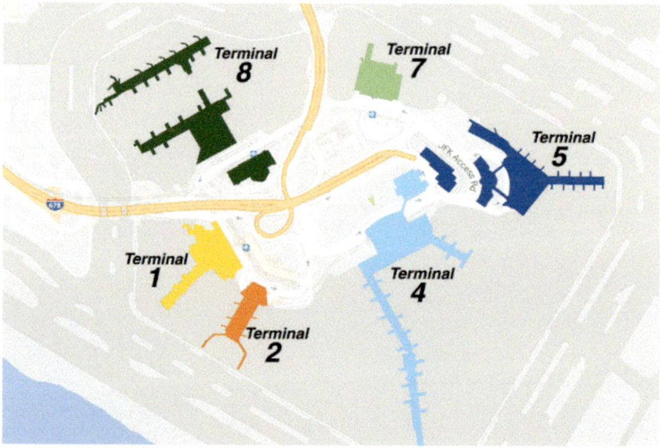

Figure 5.38 New York, JFK Terminal area plan.
SOURCE: JFK AIRPORT

Figure 5.39 New York, JFK Image of future Terminal 2.
SOURCE: NEW YORK GOVERNOR'S OFFICE

Figure 5.40 New York, JFK, aerial view of the redeveloped airport.
SOURCE: NEW YORK GOVERNOR'S OFFICE

17. Washington Dulles (IAD) 1962

Washington Dulles was the first new airport of the jet age, boasting a number of planning and design innovations that have become the norm for subsequent major international airport designs, including:

- a simplified runway configuration
- a toast rack apron layout
- a mid-field terminal location
- a central control tower
- a landside gyratory road system
- a clear-span terminal roof
- an indeterminate architecture
- a safeguarded metro link

Washington Dulles' greatest virtue has been its ability to accommodate growth and change without losing either its simplicity of layout or integrity of form because expansion was planned from the outset (Figure 5.41). The one feature that was to prove a wrong move was the proposal to use mobile lounges to access aircraft (Figure 5.42).

The terminal building, with dramatic upswept clear-span roof, although constructed in concrete can be seen as the forerunner of the modern long-span steel structures that are an integral part of contemporary airport design. The interior incorporated freestanding cabins, reminiscent of exhibition stands, accepting that the interior has become a flexible stage set. The terminal design also introduced the concept of an indeterminate architecture that

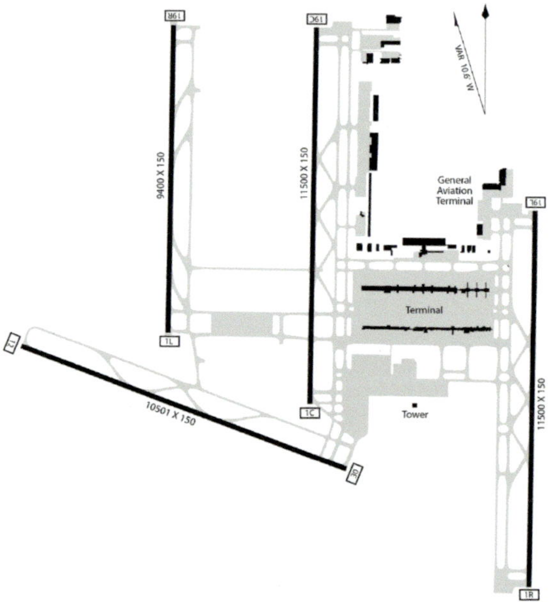

Figure 5.41 Washington Dulles, master plan.
SOURCE: FAA VIA AUTHOR'S COLLECTION

Figure 5.42 Washington Dulles interim update with later satellites served by mobile lounges.
SOURCE: LIBRARY OF CONGRESS

looked complete and resolved, whether in its original size or when extended. These ideas predate Norman Foster's claims to have originated the ideas by thirty years.

Washington Dulles was designed to allow growth that could not be accommodated at Washington National Airport. The airport was originally designed with an open apron, with passengers conveyed by cumbersome and slow mobile lounges between the terminal and the aircraft, which were parked in a series of parallel rows placed at right angles to the runways. This layout relied on the simplification of the runway layout into two primary parallel runways that was made possible by the introduction of jet aircraft that were less susceptible to cross winds.

The master plan and the indeterminate architecture of the mid-field terminal have facilitated major changes since opening. Changes include extending the terminal seamlessly, adding a landside metro to Washington, DC, introducing a linear pier and satellites accessed by an underground transit system and adding a third parallel runway. This has been achieved without damaging the character of Eero Saarinen's original terminal concept and has reinforced the airport's role as a major hub and gateway to the USA. Its master plan and mid-field terminal layout continue to influence the design of large airports to this day.

The terminal, designed by Eero Saarinen, was located between the runways, with landside access to one face of the terminal and airside access to the other. The Washington Dulles Airport master plan can therefore be described as the first of the jet age and a prototype for future mid-field hub airports (Figure 5.43). It has subsequently been extended and reconfigured with satellites linked by an underground automated people mover (APM) as the mobile lounge concept proved expensive to operate, unreliable and increasingly impractical as traffic grew.

Washington Dulles took surface access design a step further by providing a dedicated airport access road, known as the Dulles Airport Access Road (DAAR) and covering initially about 12 miles and opening with the airport in 1962. It was subsequently extended to intersect Interstate 66 before leading into Washington passing close to Washington National Airport. This is a four-lane freeway for airport traffic only and has no

Figure 5.43 Washington Dulles 1962.
SOURCE: MWAA/IAD VIA WASHINGTON POST

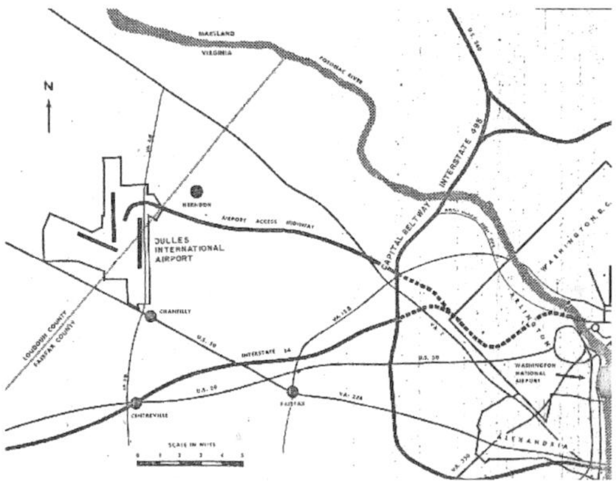

Figure 5.44 Washington Dulles Airport, access road.
SOURCE: FAA, HANDBOOK OF INFORMATION, IAD VIA AUTHOR'S COLLECTION

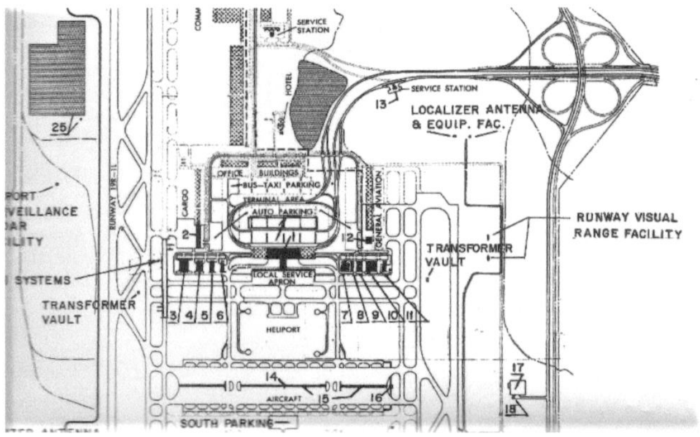

Figure 5.45 Washington Dulles Airport, landside zone.
SOURCE: FAA, HANDBOOK OF INFORMATION, IAD VIA AUTHOR'S COLLECTION

toll. There are sparse interchanges: on-ramps only for traffic inbound to Dulles and off-ramps only outbound traffic towards Washington, D.C. to prevent rat running. The road was built by the Federal Aviation Agency (FAA) and they obtained a wide enough right-of-way for parallel roadways to be eventually built on each side (Figure 5.44).

Later, the Dulles Toll Road (DTR) was built. It opened in 1984, with parallel roadways on either side of the DAAR. Cars, buses and trucks use the road without restriction, unlike La Guardia. The DTR has frequent full interchanges for commuter traffic. The DTR and DAAR are essentially freeways in a dual-dual configuration, although they are two separate highways. The recirculating loop road layout at the entrance to the terminal and the segregation of arrivals and departures flows into two forecourt levels have become the model for highway access for all subsequent mid-field airports (Figures 5.45 and 5.46).

Washington Dulles' greatest virtue has been its ability to accommodate growth and change without losing either its simplicity of layout or integrity of form because expansion was planned from the outset (Figures 5.47–5.51).

The one feature that was to prove a wrong move was the proposal to use mobile lounges to access aircraft. Saarinen enlisted the help of Charles Eames to produce one of the most compelling cartoons promoting an airport concept of operation based on the use of mobile lounges. For a time, mobile lounges were popular and several new airport designs were developed following these principles, including the ill-fated Montreal, Mirabel airport. Mobile Lounges proved to be slow, unreliable and expensive and ultimately were not able to carry sufficient numbers of passengers as aircraft sizes grew and volumes of passengers increased (Figure 5.52).

However, this did not prevent the seamless extension of the terminal as seen in the images of the check-in concourse and forecourt (Figures 5.53–5.55).

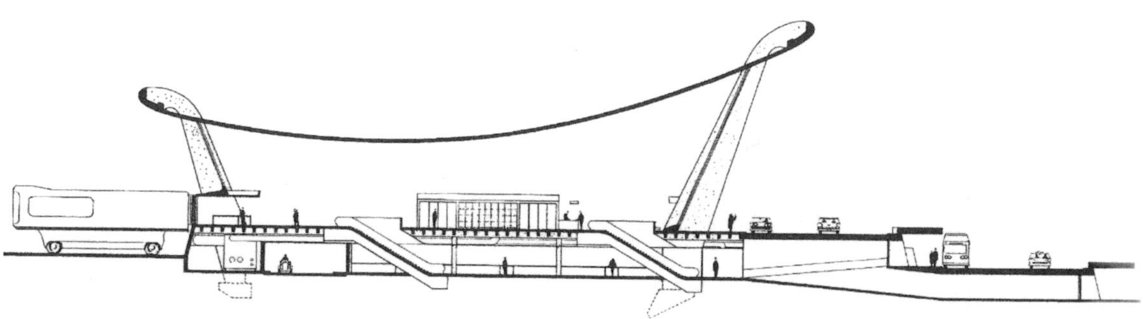

Figure 5.46 Washington Dulles 1962, showing two-level forecourt and pedestrian underpass in section.
SOURCE: FAA VIA AUTHOR'S COLLECTION

Figure 5.47 Washington Dulles 1958 – model of terminal and forecourts seen from the car park.
SOURCE: US LIBRARY OF CONGRESS, BALTHAZAR KORAB COLLECTION

Figure 5.48 Washington Dulles.
SOURCE: MWAA/IAD VIA EATER WASHINGTON DC

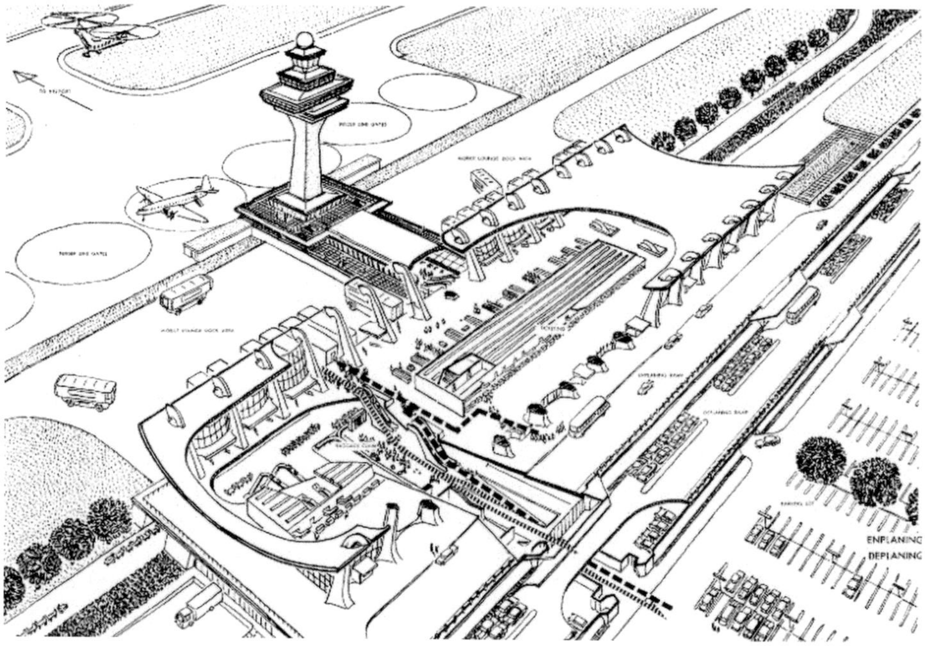

Figure 5.49 Washington Dulles, cut-away perspective of terminal.
SOURCE: FAA VIA AUTHOR'S COLLECTION

As Saarinen developed his architectural vision further at Washington Dulles Airport. He wrote:

> *The tradition of Federal architecture is static, but a jet airport should be essentially non static, expressing the movement and the excitement of travel. We thought that if we could bring these two things together into a unified design we could have a very interesting building. There was also the problem of the site – a beautiful flat plain. In a way, architecture is really placing something between earth and sky. We came to the conclusion that a strong form that seemed both to rise from the plain and to hover over it would look the best. The horizontal element, or roof, would be the highest element. It should be tilted forward so the building would be seen. The terminal should also have a monumental scale in this landscape and the vastness of this huge airfield.*

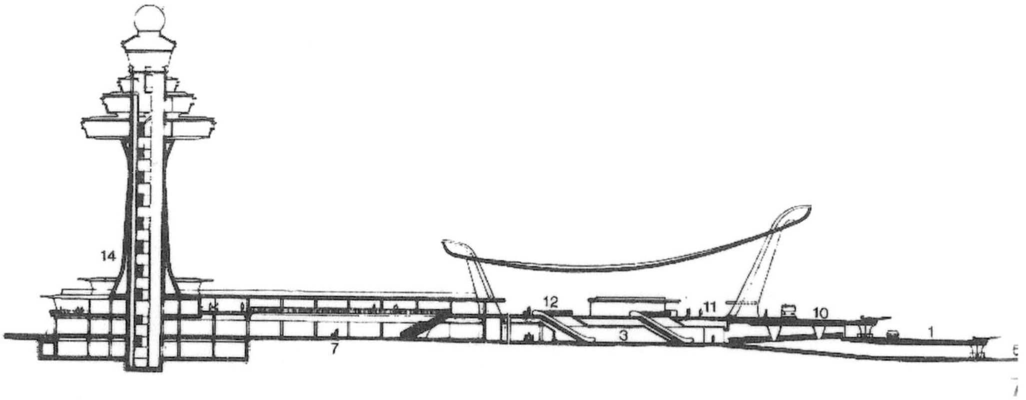

Figure 5.50 Washington Dulles, section through terminal and control tower.
SOURCE: FAA VIA AUTHOR'S COLLECTION

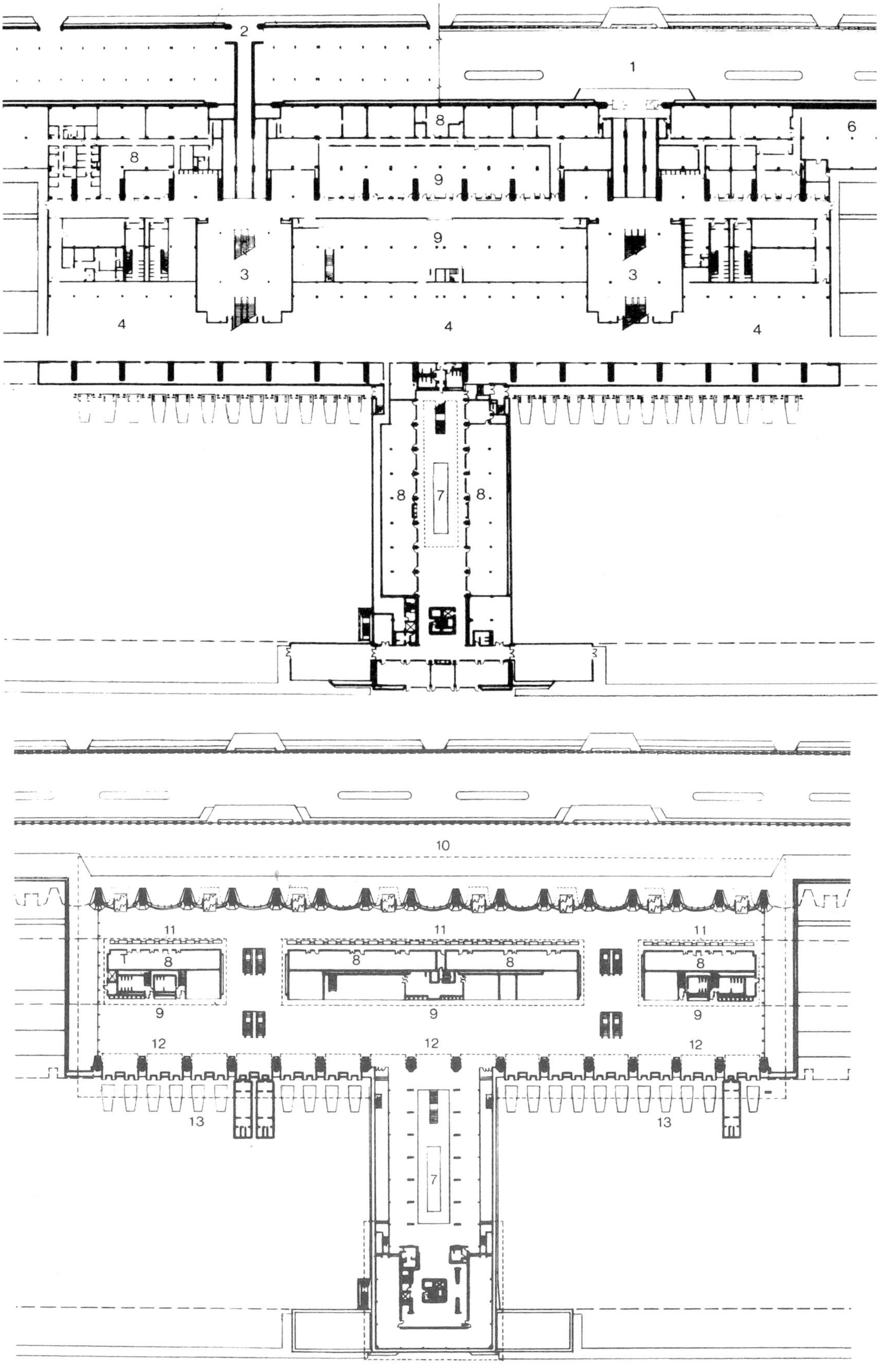

Figure 5.51 Washington Dulles, lower-level arrivals and upper-level departures plans of terminal.

SOURCE: FAA VIA AUTHOR'S COLLECTION

Figure 5.52 Washington Dulles, view of 'Planemate' mobile lounge loading aircraft.
SOURCE: MWAA/IAD VIA PINTEREST

Figure 5.53 Washington Dulles, view of check-in concourse at opening.
SOURCE: FAA VIA AUTHOR'S COLLECTION

Figure 5.54 Washington Dulles, view of check-in concourse after extension

SOURCE: WIKIPEDIA HTTP://CREATIVECOMMONS.ORG/LICENSES/BY-SA/3.0/ – PHOTO – DAVID BENBENNICK

The roof is supported by a row of columns forty feet apart on each side of the concourse, sixty-five feet high on the approach side, forty feet high on the field site. It is like a huge, continuous hammock suspended between concrete trees. It is made of light suspension-bridge cables between which concrete panels of the roof deck fit. The concrete piers are sloped outward to counteract the pull of the cables. But we exaggerated and dramatized this outward slope as well as the wide compressive flange at the rear of the columns to give the colonnade a dynamic soaring look as well as a stately and dignified one.

(Federal Aviation Administration, 1962)

Figure 5.55 Washington Dulles after 1996 Extension.

SOURCE: WIKIPEDIA -PHOTO: JOE RAVI LICENSE CC-BY-SA3.0

Figure 5.56 Washington Dulles 1962 Control Tower and Control Tower Restaurant Tulip Collection.
SOURCES: WTOP/DAVE DILDINE & IAD VIA PINTEREST

Washington Dulles made a major feature of its main restaurant by incorporating it at high level within the control tower with spectacular views. Unlike the LAX theme, restaurant it is integrated within the terminal complex (Figures 5.56).

The final aerial view shows that the masterplan has retained its simplicity and clarity as the airport has evolved and linear satellites have been added (Figure 5.57).

Figure 5.57 Washington Dulles, aerial view.
SOURCE: MWAA/IAD VIA FLYDULLES.COM

18. Los Angeles (LAX) 1962

Los Angeles (LAX) pioneered the satellite concept, accessed by underground walkways and created one of the most exciting images of the jet age, the Theme Building (Figures 5.58 and 5.59).

Later, to increase the number of stands available for aircraft parking, the satellites were reconnected with the head house terminal to create a series of finger piers. The resultant cul-de-sacs have resulted in less convenient aircraft circulation but increased capacity.

Figure 5.58 Los Angeles International aerial view 1963.
SOURCE: LOS ANGELES WORLD AIRPORTS FLIGHT PATH LEARNING CENTER

Figure 5.59 Los Angeles (LAX) Theme Building.
SOURCE: LOS ANGELES WORLD AIRPORTS FLIGHTPATH LEARNING CENTER

Figure 5.60 Aerial view of Mines Field circa 1931.

SOURCE: LOS ANGELES TIMES PHOTOGRAPHIC ARCHIVE, YOUNG RESEARCH LIBRARY, UCLA, USED UNDER A CREATIVE COMMONS LICENSE

The Theme Building is the most memorable image of the jet age at LAX and was immensely popular at the time but has become obsolete as the introduction of security has moved the passenger dwell time from landside to airside. It is now just a memory of what the jet age once promised.

Los Angeles Airport was originally known as Mines Field and began operations in 1928. It was originally secondary to Burbank's Lockheed Field (today, Bob Hope Airport), becoming Los Angeles' principal airport in the 1940s (Figure 5.60).

By the end of 1946, the new runways, intermediate passenger terminals, hangars, control tower and maintenance sheds were ready and American, Trans World, United and Western moved there from Burbank, followed by Pan Am shortly after (Figure 5.61 and 5.62).

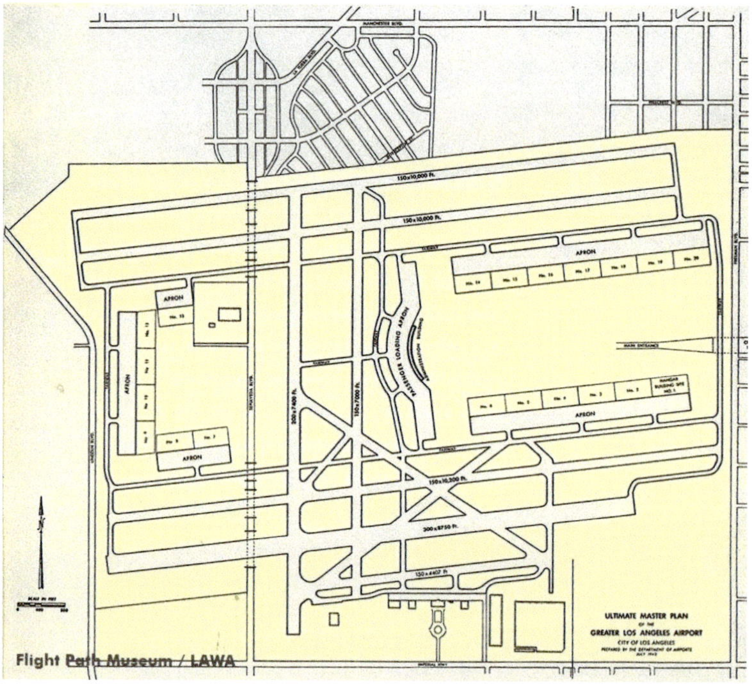

Figure 5.61 Master Plan 1944 for Greater Los Angeles Airport. The first stage would provide the airlines with intermediate air terminal facilities, and the second stage called for expansion and construction of massive new terminal facilities between two sets of runways as a mid-field complex.

SOURCE: FLIGHTPATH MUSEUM/LAWA

Figure 5.62 Aerial view of Los Angeles International Airport shortly after its post-war renovations showing the intermediate terminal facilities.

SOURCE: FLIGHTPATH MUSEUM LAWA

In the 1950s, Robert Pereira had the idea of creating Los Angeles Airport as a total leisure experience, based on his knowledge gained from working with Disney. He proposed a circular terminal at the heart of the airport serving a series of circular satellites. After that idea was rejected, he proposed a much more low-key strategy but with the futuristic Theme Building at its heart. His original concept below bears a striking resemblance to the Changi Jewel, Architect Moshe Safdie, which opened in 2019 (Figures 5.63–5.66).

This evolved into a new concept which had seven simple linear terminals linked by underground walkways to lozenge-shaped satellites, some parallel to the terminal and some at right angles. These surrounded a central zone with a large landside car park and

Figure 5.63 Los Angeles (LAX) Circular Terminal 1952.

SOURCE: FLIGHT PATH MUSEUM LAWA

Figure 5.64 Los Angeles, aerial view of circular terminal and satellites 1952.
SOURCE: LOS ANGELES WORLD AIRPORTS FLIGHT PATH LEARNING CENTER

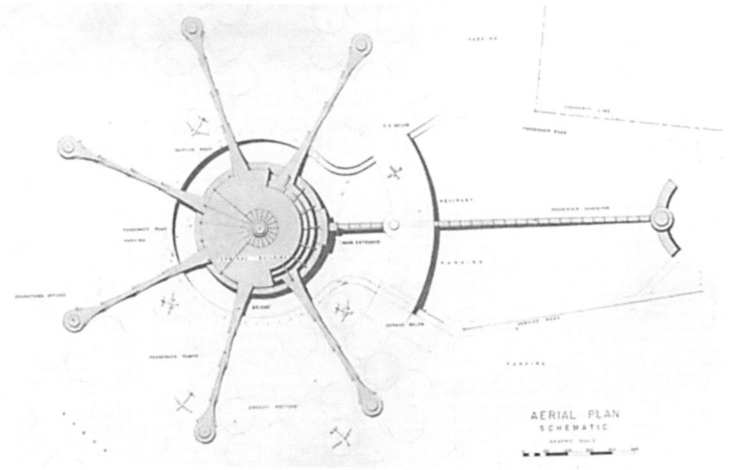

Figure 5.65 Los Angeles 1952, Terminal and Satellite Plan.
SOURCE: LOS ANGELES WORLD AIRPORTS FLIGHT PATH LEARNING CENTER

Figure 5.66 Los Angeles 1952, bird's-eye view of airfield.
SOURCE: LOS ANGELES WORLD AIRPORTS FLIGHT PATH LEARNING CENTER

commercial and leisure buildings. The landside element was eventually reduced and simplified to the iconic theme restaurant that can still be seen today. Work continued on updating the concept at LAX to make it suitable for the jet age. Robert Pereira's partner Charles Luckman summed up the challenge in his autobiography:

> *We were then, in 1955, planning an airport to be constructed in 1960, which was large enough for 1980. We had our work cut out for us. All architectural design anticipates future needs, but this was a different order: building to accommodate machines that existed only on drawing boards and which were expected to lead to large but hard-to-predict changes in global travel. The jets would of course need new facilities, such as longer, stronger runways.' But the aircraft themselves weren't the greatest test the airport faced. 'It is neither the number of airplanes, nor the number of operations which is significant,' Luckman wrote, 'but the number of people. Jets could carry more passengers than propeller planes; more significantly, they promised to dramatically reduce the cost of air travel. The prediction for 1980 was 18 million passengers a year. (It turned out to be 33 million.) How would this human torrent get to and from the airport and on and off their planes, without congestion and confusion?*
>
> (Luckman, 1988)

So they re-invented their 1950s satellite concept for the 1960s Los Angeles Airport, designing lozenge-shaped concourses linked by underground passages to a series of separate entrance buildings designed to meet the needs of individual airlines. Pereira & Luckman's plan separated arrivals and departures passenger movement into two levels, so each could flow in one direction only (Figure 5.67).

Figure 5.67 Los Angles aerial view of early master plan.
SOURCE: LOS ANGELES WORLD AIRPORTS FLIGHT PATH LEARNING CENTER

Figure 5.68 Los Angeles Satellite 1962.
SOURCE: JONPROCTOR.NET

More significantly, they broke away from having a single building; instead, a central complex was surrounded by elliptical satellite buildings. Luckman described the layout as looking like a gigantic horseshoe (Figures 5.68–5.71):

> *At the open end of the horseshoe we placed the control tower, giving the controllers total peripheral vision of field and sky. Along the outside of the horseshoe plan were the two-level satellite buildings for each airline, which looked like bubbles rising from a sea of concrete.*

Later, to increase even further the number of stands available for aircraft parking the satellites were reconnected with the head house terminal to create a series of finger piers. The resultant cul-de-sacs have resulted in less convenient aircraft circulation and increased the likelihood of delays to aircraft movement but the original satellite concept imposed unacceptable limits on the capacity of the airport (Figure 5.72).

Figure 5.69 Los Angeles, satellite with parallel parking and front and rear door boarding.
SOURCE: LOS ANGELES WORLD AIRPORTS FLIGHT PATH LEARNING CENTER VIA WWW.AIRPORTHISTORY.ORG

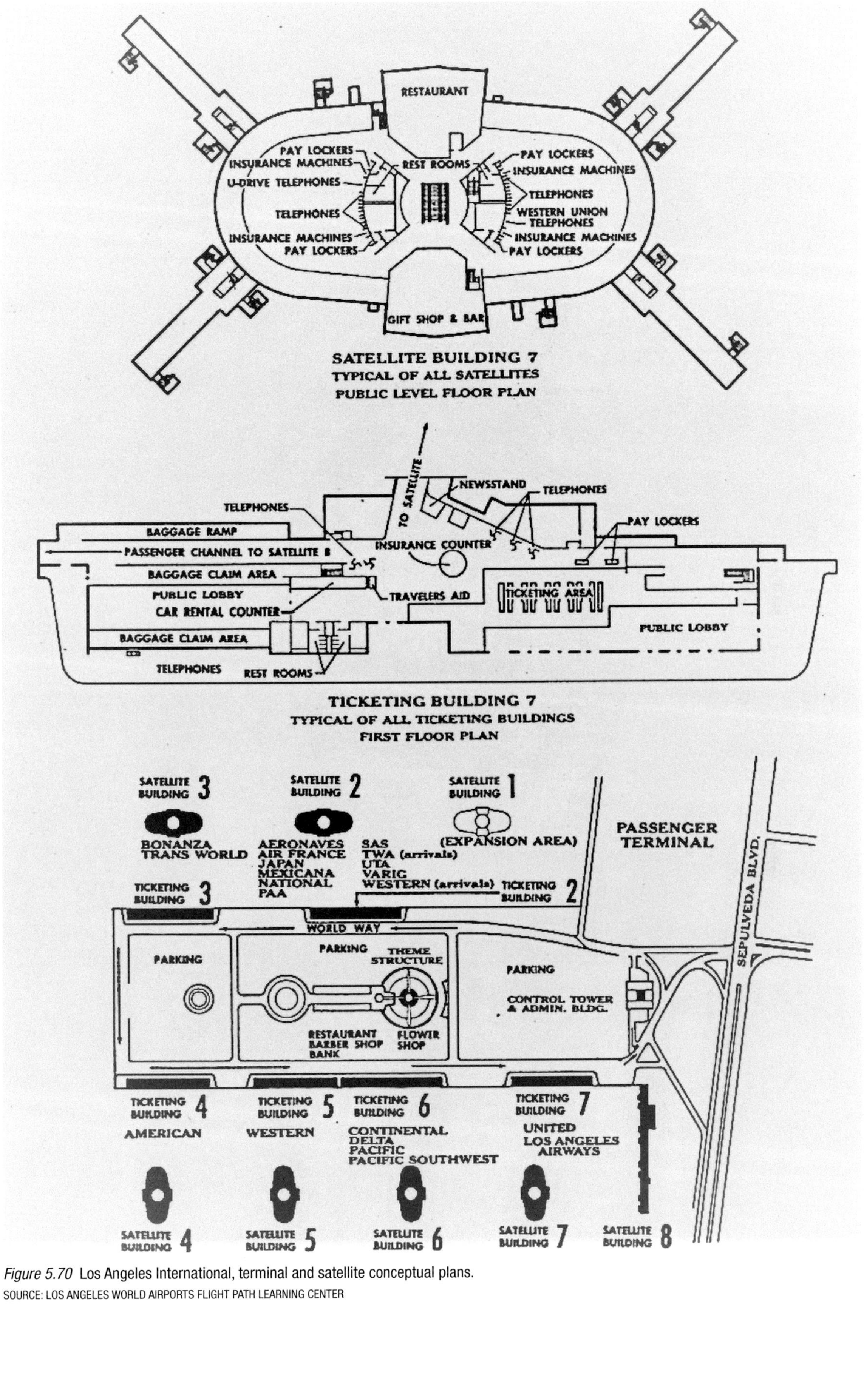

Figure 5.70 Los Angeles International, terminal and satellite conceptual plans.

SOURCE: LOS ANGELES WORLD AIRPORTS FLIGHT PATH LEARNING CENTER

Figure 5.71 Los Angeles, aerial view showing change to nose-in parking to increase stand capacity.
SOURCE: LOS ANGELES WORLD AIRPORTS FLIGHT PATH LEARNING CENTER

Figure 5.72 Los Angeles International, aerial view of satellites updated and reconnected with the terminal to form piers, now with nose-in aircraft parking and front door boarding only, to use the available site area more effectively.
SOURCE: LAWA VIA PINTEREST – FROM WIRED

Figure 5.73 Los Angeles, Theme Restaurant 1962.
SOURCE: TWITTER

The Theme Building was designed as the focal point of the airport. Robert Pereira adopted the aesthetic of space travel and echoed the form of a flying saucer for the rotating restaurant. However, it was later made stationary (Figure 5.73). Despite a major refurbishment by Disney, the restaurant has fallen into disuse. Los Angeles World Airports Executive Director Gina Marie Lindsey explained its eventual closure as follows:

> *It was built when bringing your family out to have dinner and watch the airplanes was a cool thing to do, but now passengers won't leave the terminals to have dinner there because it's unpredictable how long it'll take them to get through security again. Initially, the restaurant on top rotated slowly, giving the visitors a 360-degree dining experience and was very popular.*
>
> (Lindsey, 2013)

Imagery still plays a very important role at Los Angeles. The illuminated signs and obelisks erected at the entrance to the airport in 2000 are designed as a landmark and a gateway for those arriving and departing (Figure 5.74).

Figure 5.74 Welcome signs at the main entrance to Los Angeles.
SOURCE: LANDMARK SIGNS INC.

19. Amsterdam Schiphol (AMS) 1967

Schiphol is the oldest civil airport in the world, with over 100 years of operations. It has expanded and reinvented itself several times, starting with a simple grass field with hangars on one edge in 1920. It gained a modern terminal and control tower in 1929 and became only the second airport in Europe to have hard-surfaced runways in 1938. After the war, it expanded again and was one of the first to use airside coaches to take passengers to their aircraft. The radical expansion proposed in the new 1949 master plan was based on the Chicago O'Hare concept with a central mid-field terminal with piers and radiating tangential runways. A simplified version was later constructed and a new terminal opened in 1967 (Figure 5.75, Table 5.2).

The retention of the single terminal concept as Schiphol has expanded has helped the airport retain its position as one of Europe's premier airports. Its excellent connectivity and wide range of leisure facilities ensure it always achieves high marks in customer satisfaction surveys, making it an effective transfer hub for KLM and the Sky Team Alliance.

The adoption and consistent application of orthogonal geometry and International Style Architecture employing interchangeable modular components have helped Schiphol retain its simplicity, legibility and coherence as it has expanded, as can be seen in their future plans for Terminal A. Current expansion plans are geared towards the airport reaching 80 mppa+ in 2025.

The first military aircraft landed there on 19 September 1916, marking the start of Schiphol's service as a military airfield, on the simple grass field with hangars on one edge (Figure 5.76). On 17 May 1920, the first scheduled KLM aircraft, a De Havilland 16, touched down at Amsterdam Airport. There were two passengers on board, journalists from the British newspaper The Times.

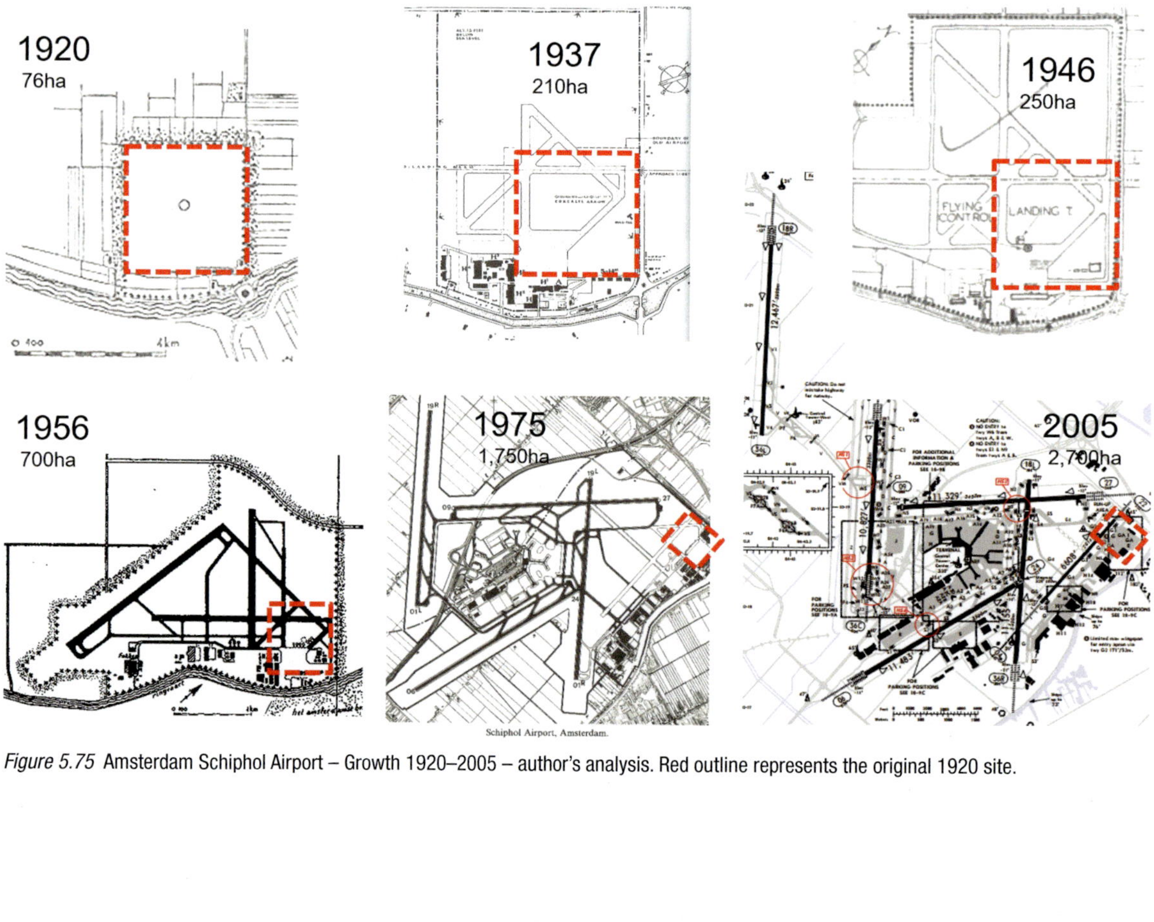

Figure 5.75 Amsterdam Schiphol Airport – Growth 1920–2005 – author's analysis. Red outline represents the original 1920 site.

Table 5.2 Growth of Schiphol Airport in size and traffic volumes

	1920	*1939*	*1960*	*1967*	*1970*	*1980*	*1990*	*1995*	*2000*	*2005*	*2019*
Area	76Ha	210Ha	700Ha	1250Ha	1480Ha	1700Ha	2000Ha	2400Ha	2678Ha	2787Ha	2787Ha
ATMs							246,504	290,689	432.483	420,736	497,000
PAX	440	16,290	1.4M		5.5M	9.7 M	16.5M	25.4M	39.3M	44.2M	71.7M

Source: Schiphol.

Schiphol was built on land reclaimed from the Haarlemermeer Lake and is some four metres below sea level and has required extensive pumped drainage infrastructure to keep it functioning, so planned intervention is ingrained in the DNA of the airport.

However, from the outset, there was a continuing problem with high groundwater. The problem became worse as aircraft became heavier. The soggy airfield attracted all kinds of nicknames: 'Schiphol Swamp' and 'Schiphol Mudport'. The French even referred to it as 'Schiphol-les-Bains': the Schiphol bathing resort.

From its earliest days, Schiphol has served as a transfer airport. As early as 1924, Imperial Airways, in its first year of operation, flew scheduled services on several European routes from London Croydon, including Berlin and Copenhagen, with transfers at Schiphol.

The evolution of the airport master plan is illustrated on the following page (Figures 5.77–5.84).

Figure 5.76 Schiphol as a military airfield 1916.
SOURCE: © PHOTO: AMSTERDAM CITY ARCHIVES

Figure 5.77 Location of Schiphol Airport with grass airfield 1930.
SOURCE: SCHIPHOL

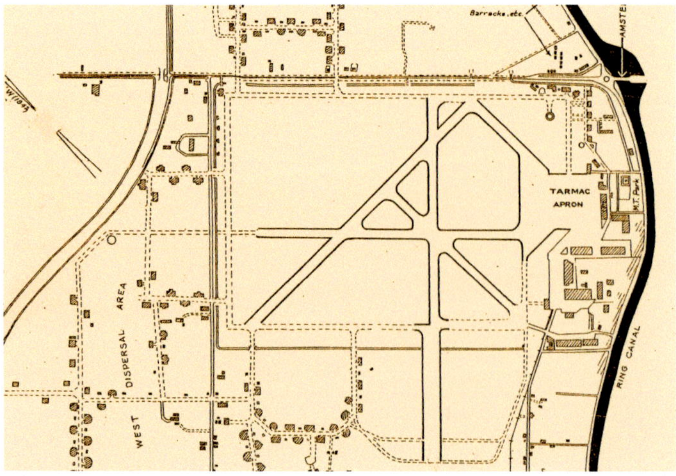

Figure 5.78 Schiphol, wartime airfield layout with concrete runways 1942.
SOURCE: WIKIPEDIA

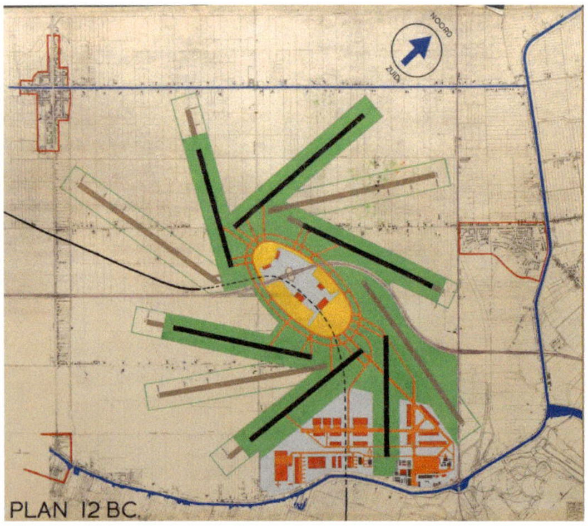

Figure 5.79 Schiphol, Master Plan 1949.
SOURCE: AMSTERDAM MUSEUM

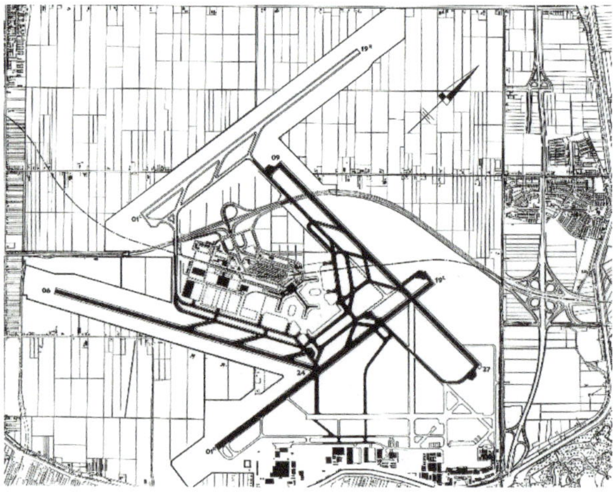

Figure 5.80 Schiphol 1967 on opening of new airport and terminal.
SOURCE: SCHIPHOL

Figure 5.81 Schiphol, Master Plan 2003.
SOURCE: SCHIPHOL

Figure 5.82 Schiphol, Future Master Plan.
SOURCE: SCHIPHOL

Figure 5.83 Schiphol, aerial view 1928.
SOURCE: SCHIPHOL VIA EBAY POSTER

Figure 5.84 Schiphol with new asphalt runways 1938.
SOURCE: SCHIPHOL

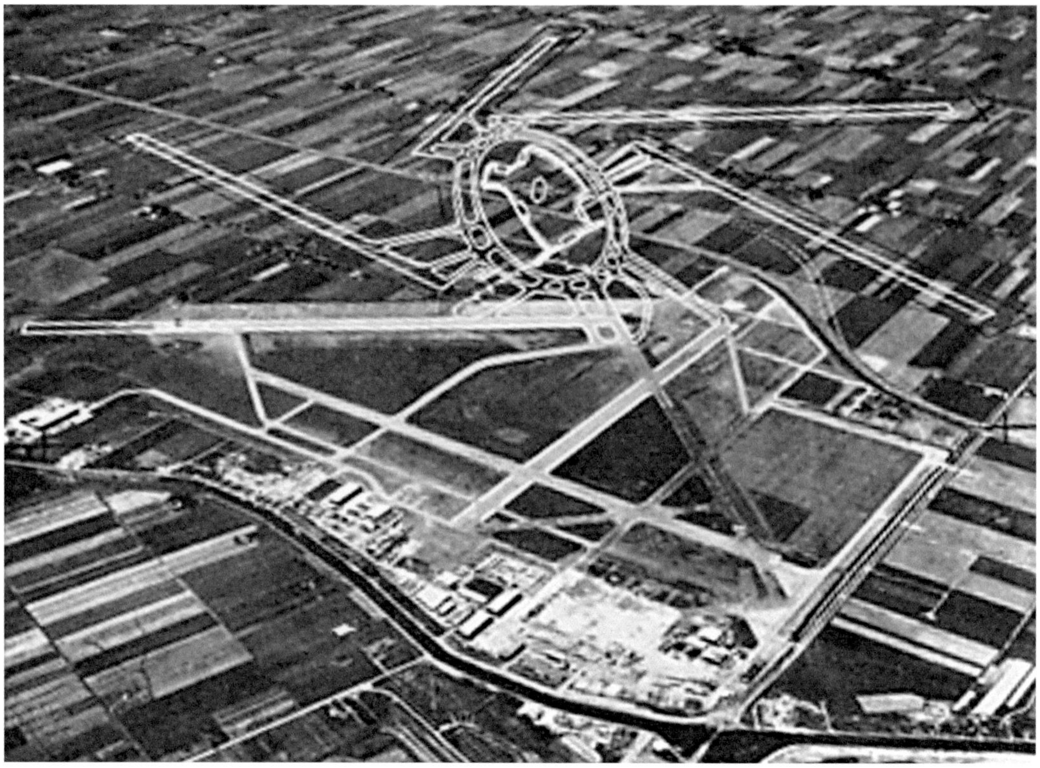

Figure 5.85 Schiphol with 1949 master plan superimposed over aerial photograph.
SOURCE: SCHIPHOL

Jan Dellaert's 1940s master plan placed the new terminal building, with its characteristic finger piers at the heart of the airport (Figure 5.85). It was eventually officially opened in 1967 (Figure 5.86). The master plan followed precedents set in the USA at Chicago O'Hare and New York Idlewild.

Figure 5.86 Schiphol in 1966 shortly before the official opening in 1967.
SOURCE: SCHIPHOL

Figure 5.87 Stillborn concept to rebuild Schiphol as an artificial island in the North Sea in 1994 by Rem Koolhaas.
SOURCE: OMA

There have been many plans to move the main airport in the Netherlands to new sites, primarily to reduce noise nuisance. The most radical plan for an airport at sea came from the architectural firm OMA of Rem Koolhaas, who is a grandson of Dirk Roosenburg, the pre-war in-house designer for KLM and architect of the first terminal. In 1999, OMA presented, on behalf of Noord-Holland, Amsterdam and Haarlemmermeer, a plan that opted for the complete transfer of the airport to an artificial island at a considerable distance from the coast. In Koolhaas's vision, the island should develop into a hypermodern 24-hour aerotropolis with a race track, hotels and shopping centres and would have been accessible from the mainland through a tunnel (Figure 5.87).

The aerial view shows Schiphol, as it has expanded, with the third parallel runway to the left, Amsterdam in the background and airport city in the foreground 2019 (Figure 5.88).

Figure 5.88 Aerial view of Schiphol.
SOURCE: GOOGLE EARTH

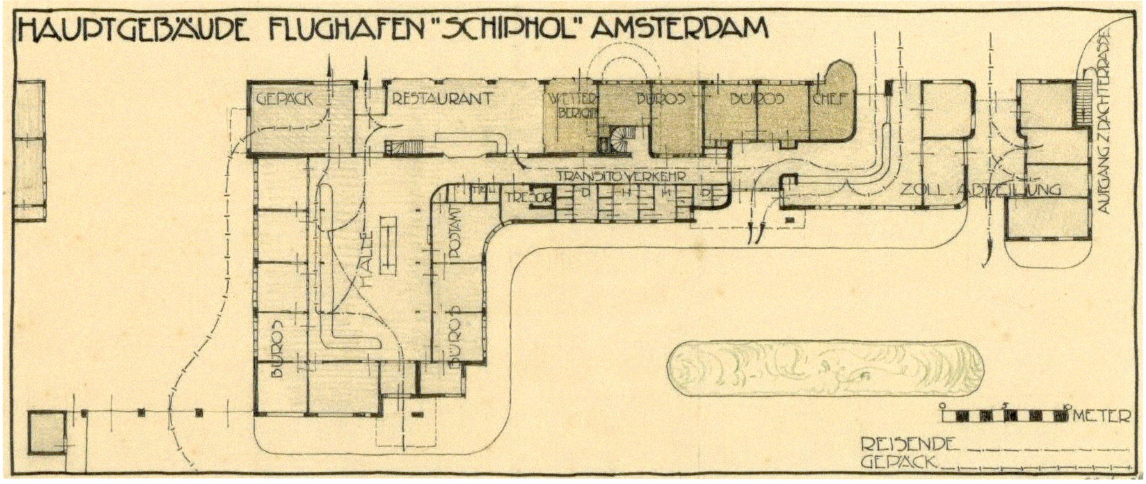

Figure 5.89 Schiphol layout plan of first terminal 1928.
SOURCE: COLLECTION HET NIEUWE INSTITUUT/BOEK, T63.4

The first terminal was designed by architect Dirk Roosenburg, who had earlier worked in the office of Hendrick Berlage. In 1919, he designed the original KLM logo, featuring an orange-white-blue flag with a black hexagon and KLM letters in black inside with wings on the left and right and a royal crown on top. The terminal opened in 1928 for the Amsterdam Olympics. It is unusual among airports in being constructed in brick and being asymmetrical in layout and form, with a semi-circular glazed control tower with an external viewing platform that resembled the railway signal boxes of the period. The interior of the control tower had a very modest capability compared with Croydon (Figures 5.89–5.94).

Figure 5.90 Schiphol airside view of first terminal 1928.
SOURCE: SCHIPHOL VIA EBAY.CO.UK/ITM/12400012629

Figure 5.91 Control Tower1928.
SOURCE: SCHIPHOL

Figure 5.92 Interior of the control tower.
SOURCE: DBNL.ORG

Figure 5.93 Amsterdam Schiphol 1929.
SOURCE: ANP FOUNDATION

Figure 5.94 Replica of the 1928 Terminal at the Aviodrome Museum at Lelystad Airport.
SOURCE: AVIODROME

Figure 5.95 Schiphol second terminal and new control tower showing large open apron.

SOURCE: KLM AEROCARTO NO 30658 Via famgus.se

The terminal and much of the airfield were destroyed in the Second World War. So the second terminal was constructed hastily using prefabricated components after the war while designs were being developed for Dellaert's ambitious master plan. A new and much larger control tower was added in 1950. The development resulted in a large open apron. To take passengers to the aircraft Schiphol pioneered the use of airside coaches (Figures 5.95–5.98).

Figure 5.96 Schiphol second control tower erected in 1950.

SOURCE: WIKIPEDIA AND SCHIPHOL

Figure 5.97 Schiphol second terminal and new control tower 1959 highlighting the major coaching operation.
SOURCE: NOORD-HOLLANDS ARCHIEF/COLLECTION *162, NR 36322*

Figure 5.98 Schiphol second terminal airside view 1959.
SOURCE: KLM AEROCARTO NO 609271 VIA FAMGUS.SE

Figure 5.99 Schiphol 1970s, airside view.
SOURCE: REMBRANDT, NOORD-HOLLANDS ARCHIEF/COLLECTION 7101, NR 15598

Construction of the current airport terminal, designed by architect Marius Duintjer, began in 1963. Inspired by the design for Orly, Duintjer came up with a *'layer cake concept'*, in which horizontal lines predominated for the new International Style Terminal. To get light and transparency into the terminal, glass became the most important facade element. The design aimed to be sober and sleek because the entire Schiphol complex had to fit into the rectilinear polder landscape of the Haarlemmermeer (Figure 5.99). The new airport, officially called Schiphol Centre, was opened on 28 April 1967 by HM Queen Juliana.

The new terminal, consisting of an arrival hall on the ground floor and a departure hall above it with three radiating piers, was located within a central zone of the four-runway system. The terminal was accessed from the adjacent autobahn that linked the commercial centre, Amsterdam, with the capital, The Hague. The layout was safeguarded for the later introduction of a rail link and airport station.

Amsterdam Schiphol was one of the first airports to open a main-line railway station that linked two major cities, the capital The Hague and the commercial centre, Amsterdam. Schiphol developed into what was referred to in the 1980s as a *'Mainport'* where various modalities would come together: air transport, road transport and rail transport (Figures 5.100–5.102).

In 1986, Schiphol opened its own railway station, which was located underground, and an associated public transport interchange and since then has invested heavily in the provision of rail services. In the 1990s, a direct rail

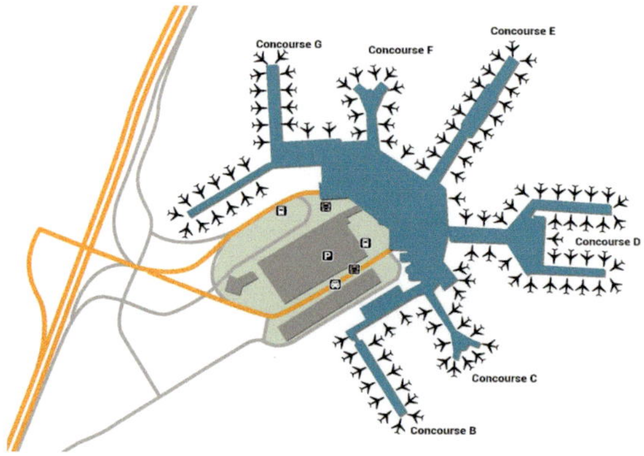

Figure 5.100 Schiphol extended with multiple piers but retaining the single terminal concept.
SOURCE: SCHIPHOL

Figure 5.101 Aerial photograph of Schiphol Airport.
SOURCE: GOOGLE EARTH

Figure 5.102 Amsterdam Schiphol 'Mainport' multi-modal interchange.
SOURCE: SCHIPHOL AIRPORT

connection to all major cities in Europe was realised and Schiphol now excels as a multi-modal transport interchange. The Schiphol Plaza is a landside shopping centre which incorporates the access to the railway station as the focus of the Airport City concept (Figure 5.103–5.106).

As it expanded, Schiphol persevered with a single terminal operation. With its development as a transfer hub, to support KLM, it invested heavily in a sophisticated baggage system to deliver excellent minimum connection times for transferring passengers that were vital to the airline's business model. Schiphol is now a highly regarded international transfer hub. It has even promoted itself as London's 3rd airport as it serves more UK regional airports than either Heathrow or Gatwick, making it an important transfer hub.

Schiphol Airport has recently carried out major changes to the terminal structure, and especially security control, by introducing transfer security checks at the root of each pier. In practical terms, this means that no one should have to pass through security control twice if they arrive from an EU country with the same security level as Schiphol. To achieve this, central security checks have been introduced for all who start their journey in Amsterdam and there will be a new transfer security control for passengers arriving from destinations outside the EU/Schengen. This has all been achieved while retaining the simplicity and consistency of design that characterises Schiphol.

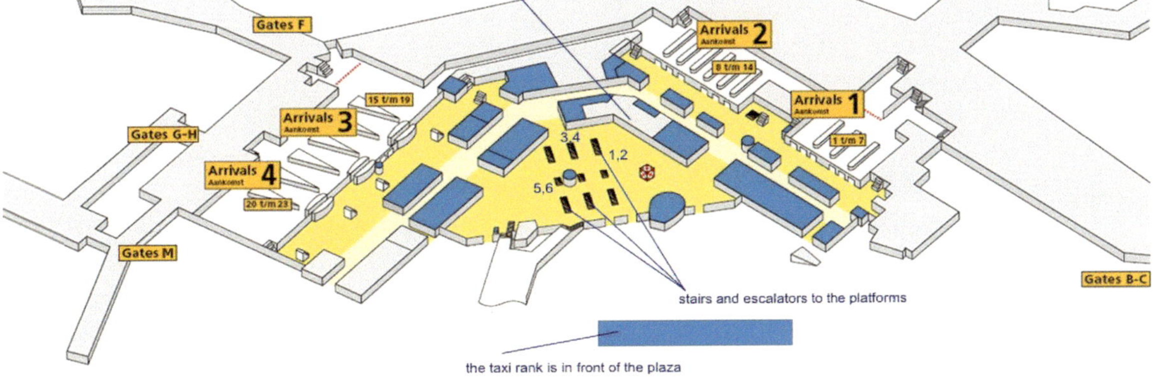

Figure 5.103 Axonometric of Schiphol Plaza and Railway Station.
SOURCE: SCHIPHOL

Figure 5.104 Aerial view of Schiphol Plaza triangular roof to the left and current control tower on the right. The 101-metre-tall tower opened in 1991.
SOURCE: AIRLINERS.NET PHOTOGRAPH BY NICK DE JONG

Figure 5.105 Schiphol Plaza with entry to the rail station on the left.
SOURCE: SCHIPHOL

Figure 5.106 Schiphol Railway Station platform.
SOURCE: SCHIPHOL

The first prototype of the Avio Bridge (passenger boarding bridge), using aluminium construction instead of the steel used in contemporary American experimental bridges, was designed by the Aviolanda Aircraft Construction Company and installed on the old Schiphol area. It was officially put into use on 7 August 1961, making Schiphol the first airport in Europe to employ airbridges (Figure 5.107).

A unique concept of an *'overwing'* airbridge was developed for the new generation of B747 Jumbo aircraft, but these faced a number of operating difficulties, and there were a number of minor clashes with aircraft due to very tight clearances as they passed over the wing. Although they demonstrated clear benefits from much faster aircraft boarding and alighting, they have since been removed and conventional bridges to doors 1 and 2 have been retrofitted.

The following photographs show the original overwing airbridges serving a single-level pier, Concourse F, where gate security was employed, allowing arriving and departing

Figure 5.107 Prototype 'Aviobridge' passenger boarding bridge.
SOURCE: SCHIPHOL

passengers to mix in the concourse. Below is a photograph of the conventional airbridges that were retrofitted while adding an upper storey to provide a segregated high-level arrivals route to respond to the later more onerous security requirement to segregate arriving and departing passengers that has become the norm in international airports (Figures 5.108 and 5.109).

Schiphol has seen the provision of leisure and recreation facilities as complementary to their retail offer while enhancing its reputation as an international

Figure 5.108 Schiphol, Concourse F with 'overwing' airbridges 2007.
SOURCE: SCHIPHOL

Figure 5.109 Schiphol, Concourse F with retrofitted conventional airbridges and segregated high-level arrivals route and transfer security facilities at the root of the pier.
SOURCE: SCHIPHOL VIA NACO

transfer hub. For a number of years, they had an airside casino. In 2002, the Rijksmuseum opened a satellite museum at Schiphol, making Schiphol the first airport in the world to house an art gallery. Airline passengers travelling through the airport can visit airside exhibitions about the Dutch masters; the exhibitions are updated regularly. Schiphol is also the first airport in the world to have its own library. The Airport Library makes it possible for passengers travelling with intercontinental flight services to discover Dutch culture while they wait for their connecting flight. The library offers a broad range of Dutch books in translation, photo albums, videos, mini exhibitions and music by Dutch artists (Figures 5.110–5.113).

Schiphol continues to expand its single terminal concept in the proposed new Terminal A (Figures 5.114–5.116).

Figure 5.110 Schiphol, Rijksmuseum.
SOURCES: RIJKSMUSEUM

Figure 5.111 Schiphol, Library.
SOURCE: SCHIPHOL

Figure 5.112 Schiphol, Rijksmuseum.
SOURCE: SCHIPHOL

Figure 5.113 Schiphol, Flower Shop.
SOURCE: SCHIPHOL

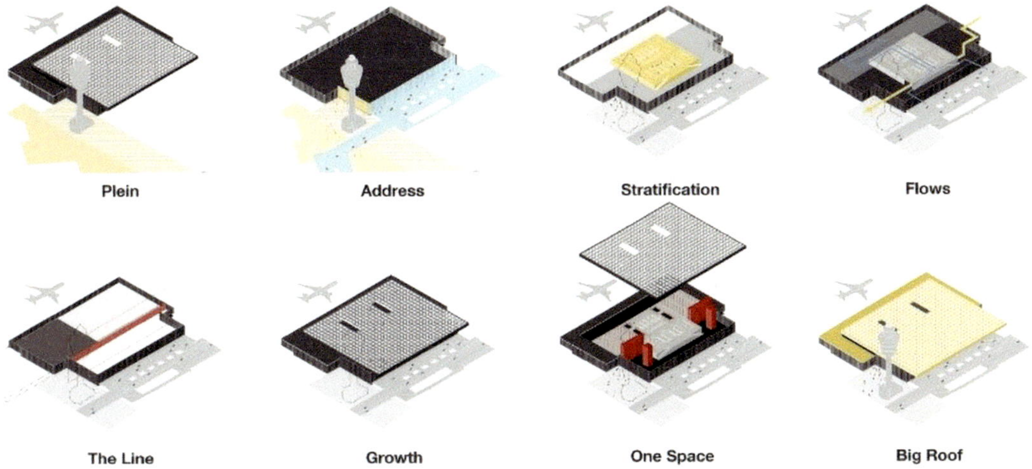

Figure 5.114 Schiphol, Terminal A, planning concepts.
SOURCE: ESTUDIO LAMELA & KAAN ARCHITECTEN

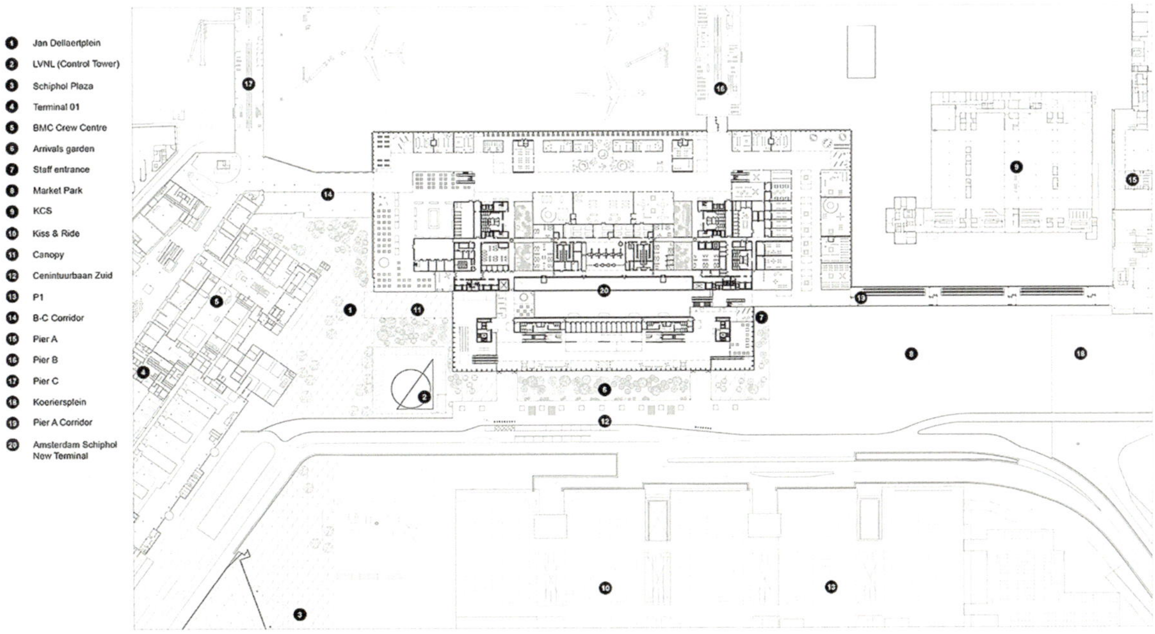

Figure 5.115 Schiphol, Terminal A Layout Plan.
SOURCE: ESTUDIO LAMELA & KAAN ARCHITECTEN

Figure 5.116 Schiphol, Terminal A, landside view.
SOURCE: ESTUDIO LAMELA & KAAN ARCHITECTEN

Schiphol also pioneered the Airport City concept that has ensured that the hinterland is consistently planned and supports rather than constrains the viability of the airport (Figures 5.117 and 5.118). The Schiphol Group sees its own airport as an 'Airport City' which they describe as *'A dynamic environment integrating and enhancing people and businesses, logistics and shopping, information and entertainment. This efficient, multi-modal hub for air, rail and road transport is a seamless link in the travel process that provides visitors a unique experience'*

Hubert-Jan Henket, supervisor of architecture and urban design, Schiphol 1996–2008 has observed, *'The average traveller these days is much less elite than the passenger almost half a century ago. They have become consumers looking for amusement and the airport responds by providing plenty of facilities. With the provision of an increasing number of facilities that could also be found in the city, the airport began to look more like an airport city. Shortly after I was appointed, this realisation became so strong that Schiphol even positioned itself as an Airport City and when this caught on, we began to market the concept.'*

Schiphol's future growth is currently restricted by the 500,000 ATM cap to control the environmental impact of noise nuisance which could result in limited future growth. It is for this reason Schiphol has recently adopted the strategy of decanting some of the smaller aircraft to Lelystad Airport to give room for growth of larger aircraft. Schiphol's future vision and proposed environmental strategy were published in July 2020.

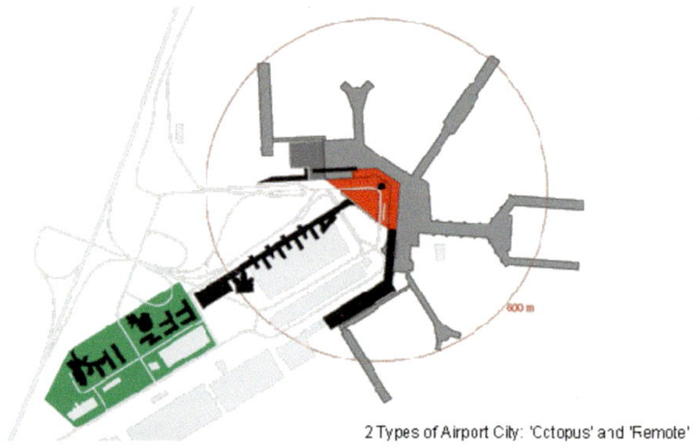

Figure 5.117 Schiphol, Airport City master plan.
SOURCE: © GÜLLER GÜLLER ARCHITECTURE URBANISM

Figure 5.118 Schiphol Airport City aerial view.
SOURCE: SCHIPHOL VIA KCAP.EU

20. Dallas Fort Worth (DFW) 1973

Dallas Fort Worth was designed to replace two separate airports, Dallas Love and Meacham Field, serving Fort Worth. When it opened in 1974, it was the largest airport in the world, with four semi-circular unit terminals that stretched out along a landside central spine linked by an elevated APM. The original master plan provided for expansion to thirteen terminal buildings along the central spine. The International Terminal was added later (Figure 5.119 and 5.120).

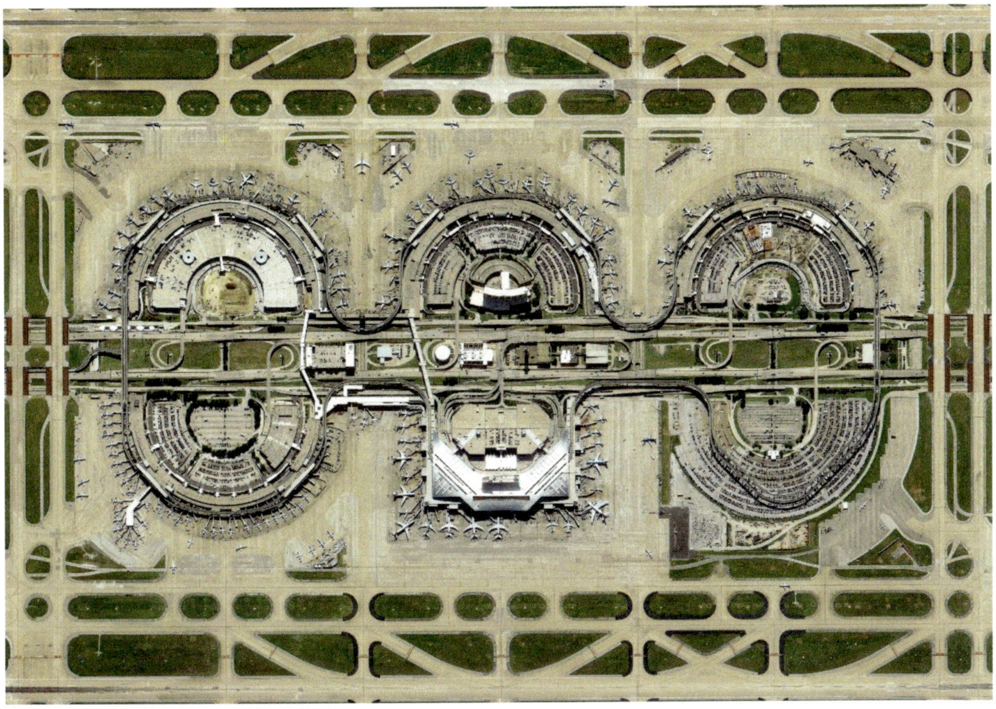

Figure 5.119 Dallas Fort Worth Terminal Zone – International Terminal at the centre bottom is a later centralised addition.
SOURCE: DFW VIA FROMABOVE.ALTAVISTA.ORG

Figure 5.120 Dallas Fort Worth International Terminal, Skylink Transit in the foreground.
SOURCE: WIKIPEDIA

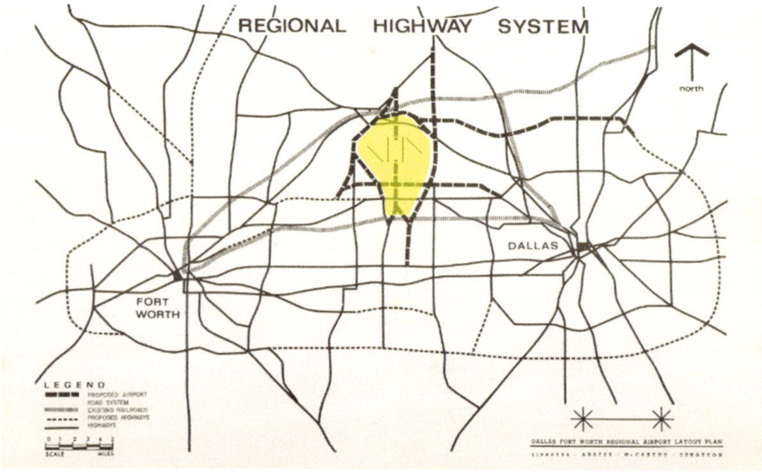

Figure 5.121 Dallas Fort Worth Regional Plan.
SOURCE: DFW VIA TIPPETTS ABBETT MCCARTHY STONE

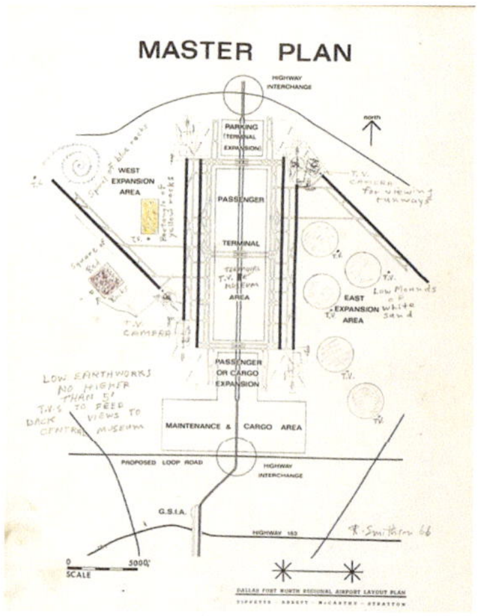

Figure 5.122 Master Plan.
SOURCE: DFW VIA TIPPETTS ABBETT MCCARTHY STONE

The central landside spine of the master plan and the unit terminal concept, once popular features, have largely been superseded by the centralised mid-field terminal concept seen at Washington Dulles. The complex transit system has also not been replicated. However, the idea of replacing individual airports close to the city centre with a single, larger and more remote airport serving a wider catchment has become the norm for new large hub airports (Figure 5.121).

The original 1966 plans show how the new airport was located half way between Dallas and Fort Worth, with a terminal zone between the two pairs of north/south parallel runways. Originally, a central spine of terminal buildings with finger piers was proposed by master planners, Tippetts-Abbett-McCarthy-Stone, with integrated multi-storey car parks similar to the design that had previously been developed for Houston Intercontinental (Figure 5.122).

This concept produced phasing challenges as both the elevated and basement roads would have had to be constructed the full length of the airport, even though there would not have been demand for the same amount of terminal space (Figures 5.123 and 5.124).

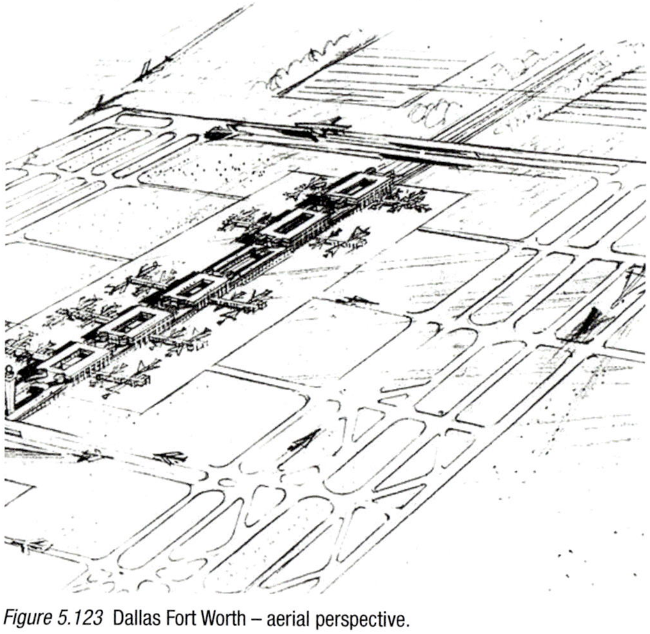

Figure 5.123 Dallas Fort Worth – aerial perspective.
SOURCE: DFW VIA TIPPETTS ABBETT MCCARTHY STONE

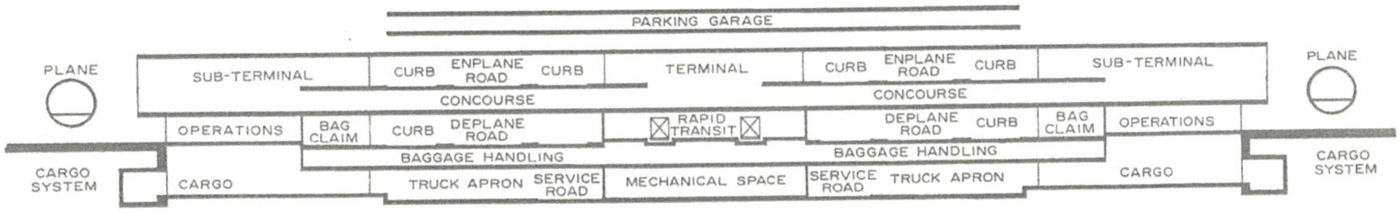

Figure 5.124 Dallas Fort Worth, cross section through original design terminal complex.
SOURCE: DFW VIA TIPPETTS ABBETT MCCARTHY STONE

A much simpler and more economical horseshoe-shaped series of unit terminals was then developed in conjunction with architects Hellmuth, Obata, Kassabaum and Brodsky (HOK), with semi-circular surface parking areas in front of each of the terminal zones. The ultimate design of thirteen terminals would have provided 234 aircraft-boarding gates (Figure 5.125).

The following section drawing, which is very simple compared with the original proposals and illustrates the Unit Terminal concept which focuses on short walking distances by decentralising all the processing. Each of the four initial terminals contained 790,000 ft^2 of covered space and could accommodate eighteen Boeing 747s (Figure 5.126).

Figure 5.125 Dallas Fort Worth – future aerial view of airport in 2001.
SOURCE: DFW VIA TIPPETTS ABBETT MCCARTHY STONE

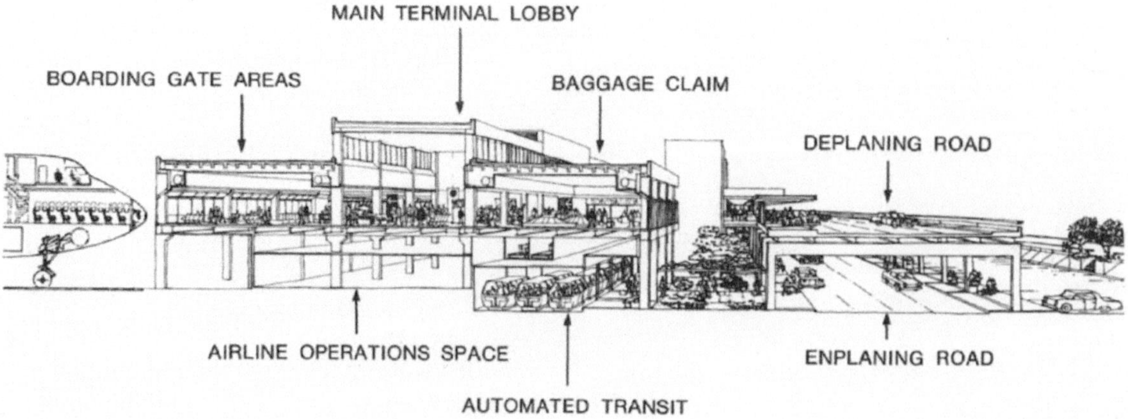

Figure 5.126 Dallas Fort Worth 1974 – Section through terminal.
SOURCE: IMAGES OF AVIATION DFW, BLEAKLEY, BRUCE A

Dallas Fort Worth was one of the key exponents of the Unit Terminal concept and morphed into a major transfer hub by virtue of its transit system that linked the terminals. The later international terminal abandoned the unit terminal concept in favour of a more conventional centralised terminal with the benefits of consolidated security and a compact retail layout that maximised footfall.

The original people mover train (*Airtrans)* APM was notoriously slow (17mph (27km/h)), uni-directional (running only in a counter-clockwise direction) and located outside the secured area, thus requiring travellers to go through security again. Uniquely, the system was also designed to carry goods to the terminals, but the latter was soon abandoned as it required double handling (Figure 5.127).

It was replaced by *'Skylink'* in April 2005 after serving approximately 250 million passengers. *'Skylink'* serves all five terminals at a considerably higher speed (up to 35mph (56km/h)), is bi-directional and is located airside providing the transfer passenger with panoramic views of the airfield.

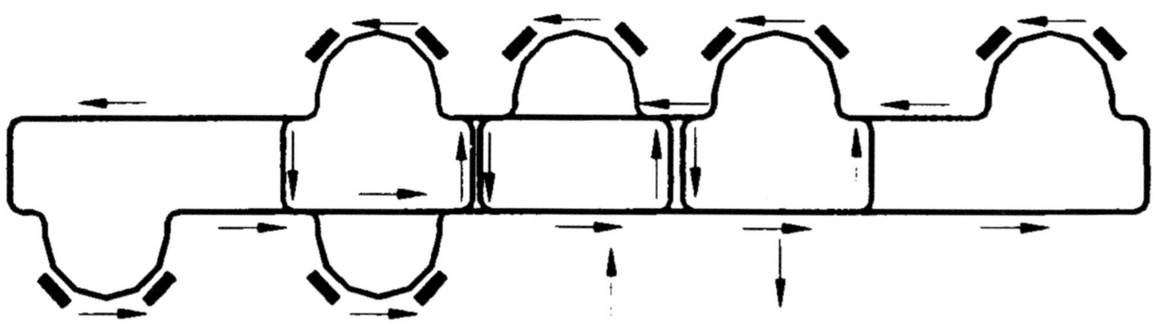

Figure 5.127 Dallas Fort Worth 1974 – Network APM system.
SOURCE: JOURNAL OF ADVANCED TRANSPORTATION, VOL. 33, NO.1

In 2018, Dallas Forth Worth handled 69mppa while Dallas Love handled 16mppa. The airfield has continued to expand, and its long-term master plan envisages additional widely spaced parallel runways accessed via end-around taxiways (EATs), as pioneered by Atlanta, to avoid aircraft having to cross the main runways (Figure 5.128).

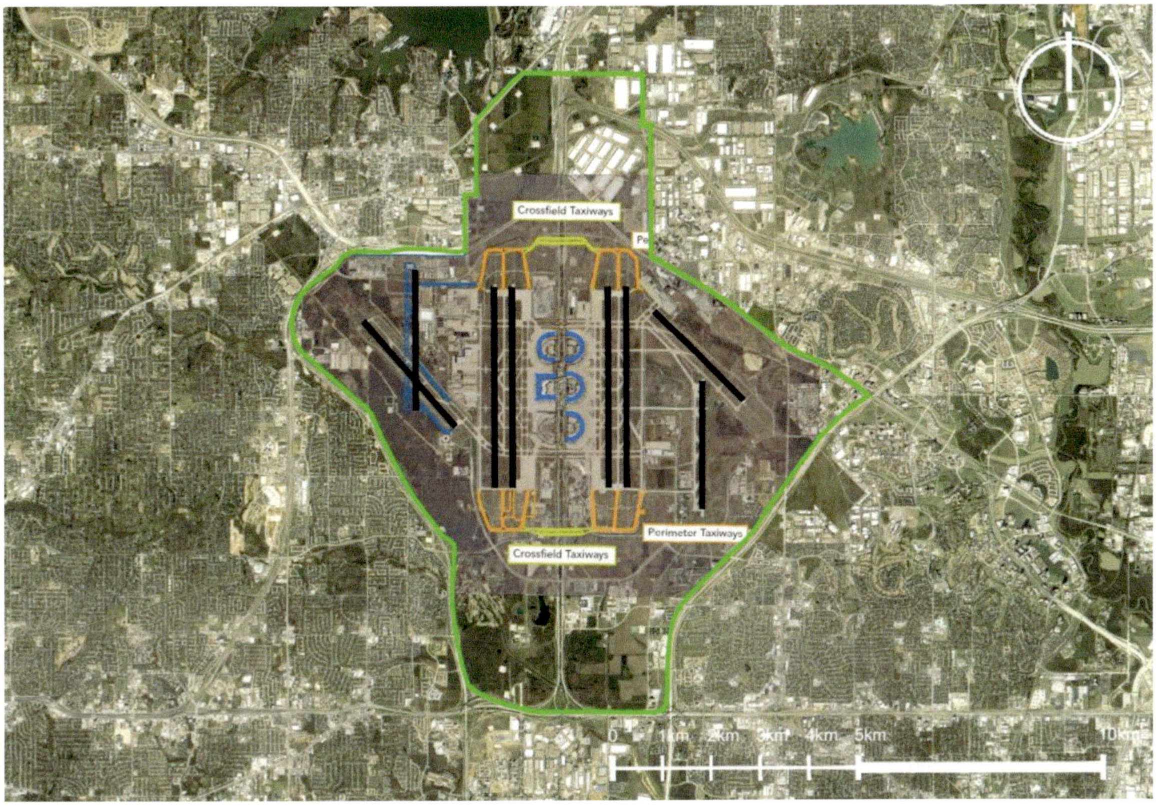

Figure 5.128 Dallas Fort Worth, End–Around Taxiways (in orange).
SOURCE: DFW VISION FOR THE FUTURE 2030

21. Paris Charles de Gaulle (CDG) 1974

The original master plan envisaged a series of 5 circular terminals supported by satellites. Fifty years later, the annual capacity is now almost eight times that figure, and the original master plan has evolved radically as new concepts (linear terminals for T2 and later supported by satellites) have been introduced to address lessons that have been learnt (Figures 5.129 and 5.130).

One area where the airport has excelled is in the provision of rail services and an associated public transport interchange. This facility is located between Terminal 2 A&B and Terminal 2 C &D. It links the airport directly with the French TGV international rail network.

The most recent design for Terminal 4 abandons the previous individualistic concepts and incorporates a more conventional single, large processing building with a series of piers under one continuous roof (Figure 5.131).

Could it be that the airport has finally learnt the lessons from its previous architecturally striking but functionally flawed designs and understands that operational efficiency, commercial opportunity and designing for growth and change underpin successful airports?

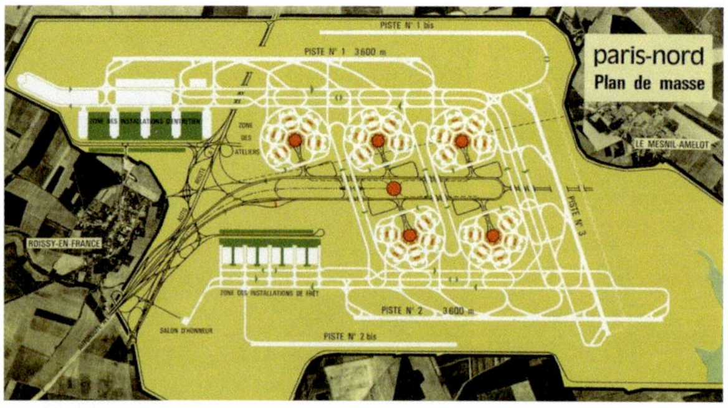

Figure 5.129 Charles de Gaulle 1965, master plan.
SOURCE: SIDESHARE.NET JEANMARIE-CHEVALLIER ADP

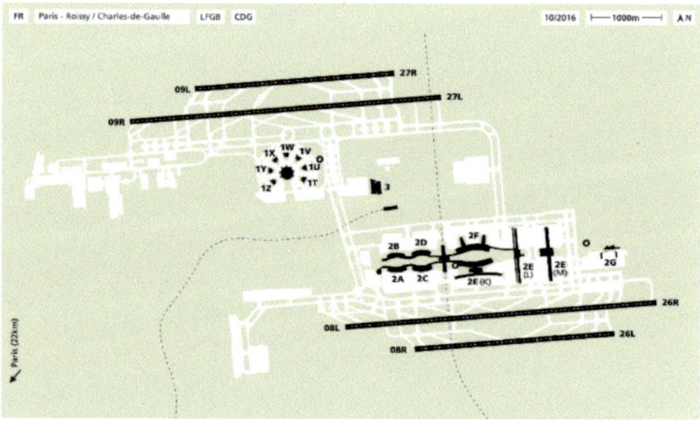

Figure 5.130 Charles de Gaulle 2015, master plan.
SOURCE: WIKIPEDIA – BY CELLARDOOR85 – 52031981

Figure 5.131 Charles de Gaulle Future Terminal 4, with T2 in the foreground and circular T1 top left.
SOURCE: ADPI VIA CDG.PARISAEROPORT.FR

In the late 1950s, Paris Orly and Le Bourget were reaching capacity, so it was realised another airport was needed and the search for a suitable location began. By 1964, that location had been found: a site positioned 15.5 miles (25 kilometres) north-east of Paris. Construction work on the then-named 'Aéroport de Paris Nord' started in 1966. Renamed 'Charles de Gaulle', the new Paris airport was officially opened on 8 March 1974. The airport's initial passenger capacity was set at 10 million per annum.

Charles de Gaulle was then the most modern airport in Europe and the only one outside the USA to adopt a master plan with two pairs of close parallel runways. The airport was an early adopter of the satellite concept in Europe when it opened in 1974. The main circular terminal, designed by architect Paul Andreu, is linked by underground walkways to seven satellites which are in turn surrounded by parked aircraft. The concept was conceived primarily to reduce walking distances, when compared to the conventional pier layouts that were common at that date, by developing the American linear unit terminal concept into a wrap-around circular configuration. This compact layout was achieved at the expense of multiple vertical passenger movements, both within the terminal building and between the terminal and satellites. Comprising a main central building and seven surrounding structures, it resembles an octopus when viewed from above. Each of Terminal 1's 10 storeys was originally given a particular purpose: for example, its second floor consisted of retail and food/drink outlets, while storeys three and four predominantly handled check-in and duty free/border control, respectively. It was given the nickname *'Camembert'* because of its round shape (Figures 5.132 and 5.133).

Passengers move from level to level on escalators located in Plexiglas tubes within a central circular light well. This created an imagery that exhibited some of the dynamic characteristics of Futurism, but the layout was unfortunately accompanied by feelings of disorientation.

Figure 5.132 Paris Charles de Gaulle, Terminal 1 section.
SOURCE: ADPI VIA THE AIRPORT 1974

Figure 5.133 Paris Charles de Gaulle, 1974.
SOURCE: ADPI VIA COPYBOOK.COM

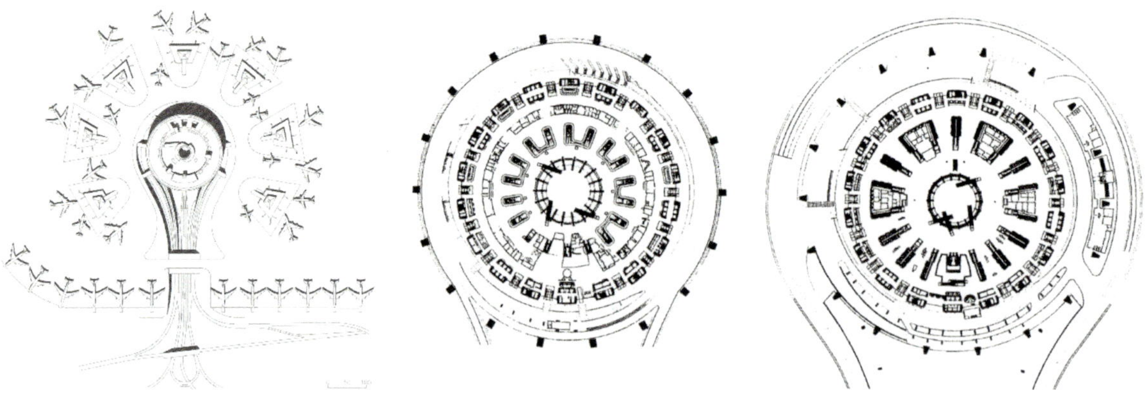

Figure 5.134 Paris Charles de Gaulle, Terminal 1 plans.
SOURCE: ADPI VIA THE AIRPORT 1974

The top three levels were dedicated to car parking; below that were an office floor, the main passenger processing floors and below that again baggage and support areas (Figures 5.134–5.136). The idea of integrating parking within the terminal building complex has since fallen out of favour as airport operators quickly recognised the inherent inflexibility of the concept. Today, it is no longer possible to consider this strategy as security requirements normally now require vehicles to be kept well away from the terminal building to prevent terrorist attacks (Figures 5.137–5.139).

There are seven satellites, accessed from the terminal by underground walkways, each with departure lounges on the 3rd level. Many travellers have commented that the layout results in very confusing wayfinding and in practice it has proved very inflexible and difficult to change (Figure 5.140).

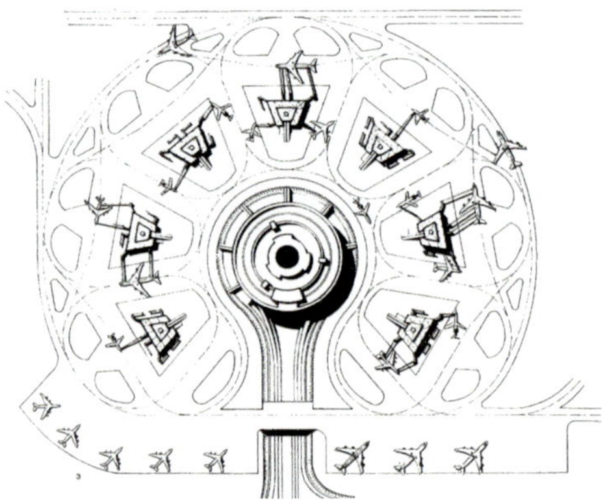

Figure 5.135 Paris Charles de Gaulle, 1974.
SOURCE: ADPI VIA ARCHITECTURAL DESIGN

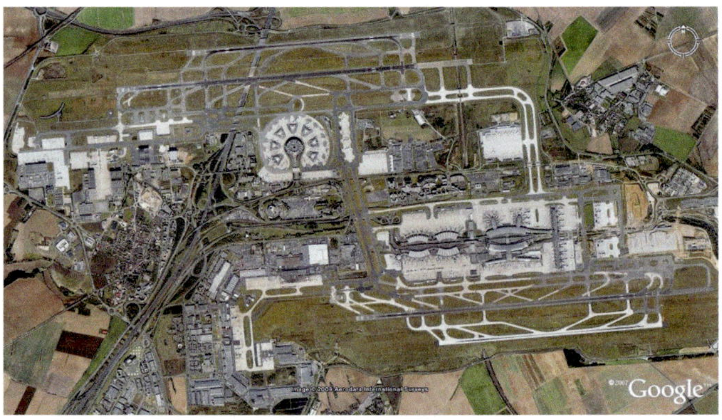

Figure 5.136 Paris Charles de Gaulle, aerial view.
SOURCE: GOOGLE EARTH

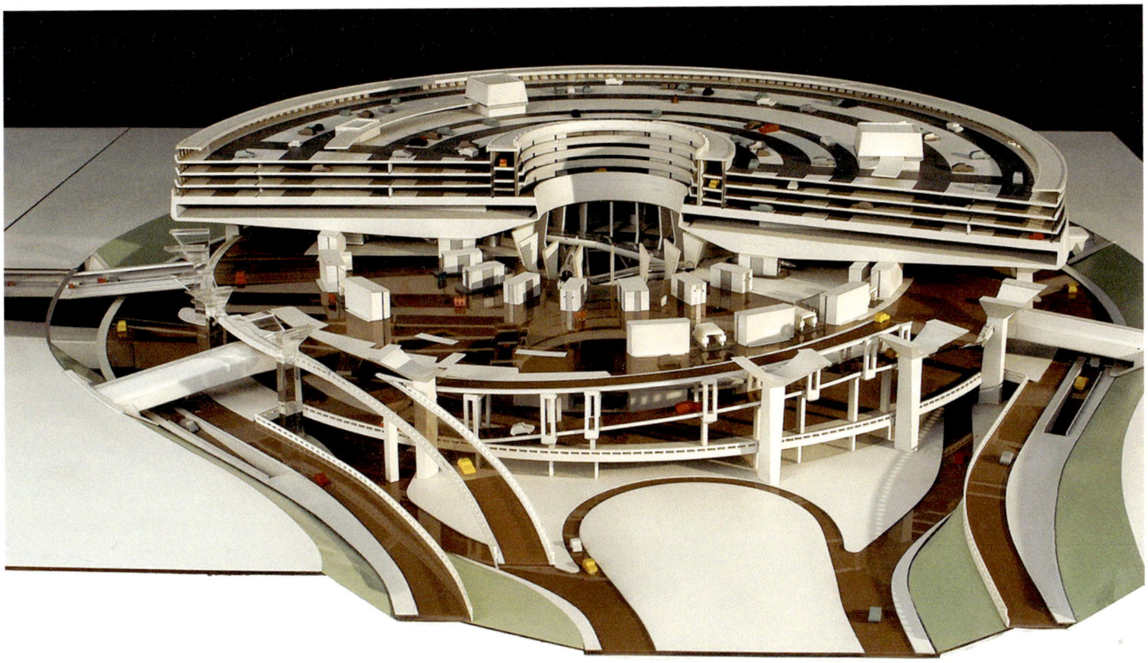

Figure 5.137 Paris Charles de Gaulle model of terminal with a car park at high levelover the over the top of the terminal.
SOURCE: ADPI VIA I.PINIMG.COM

Figure 5.138 Paris Charles de Gaulle, Terminal 1, aerial view.
SOURCE: ADPI VIA FROMABOVE.ALTERVISTA.ORG

Figure 5.139 Paris Charles de Gaulle, Terminal 1, 1974 with a car park at high level over the top of the terminal.
SOURCE: ADPI VIA WWW.FRANCE-JUSTFORYOU.COM

Figure 5.140 Charles de Gaulle, Terminal 1, central void with criss-crossing passenger conveyors in glazed tubes.
SOURCE: ADPI VIA ARCHITECTURE DAILY – PHOTOGRAPHER; PAUL MAURER

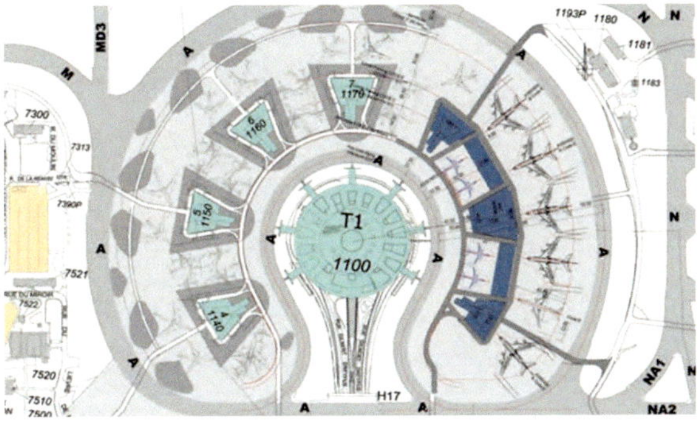

Figure 5.141 Paris Charles de Gaulle, aerial view showing proposed reconfiguration satellites.
SOURCE: ADPI VIA PARIS CHARLES DE GAULLE

Figure 5.142 Aerial view of demolition of existing satellites.
SOURCE: ADPI VIA PARIS CHARLES DE GAULLE

A recent project has started replacing three of the remote satellites, that each handled seven aircraft, with a single much larger, curved structure that accommodates 10 much larger aircraft including five A380 gates. This is an investment of doubtful value now that the A380 aircraft has ceased production, but it shows the inflexibility and limitations of the original concept (Figures 5.141–5.143).

Figure 5.143 Paris Charles de Gaulle, photomontage of proposed replacement satellite.
SOURCE: ADPI VIA PARIS CHARLES DE GAULLE

Figure 5.144 Paris Charles de Gaulle, Terminal 2A, 1981, 2B 1982, 2C 1993, 2D 1989 and circular satellite 1999.
SOURCE: ADPI

Terminal 2 was conceived as an expandable linear terminal, solely for Air France's use, but is now much more widely used. It's made up of seven sub-terminals which are lettered 'A' through to 'G'. The latter subsequently suffered a partial collapse on 23 May 2004, causing four fatalities. Construction issues were later blamed for this incident. A reconfigured Terminal 2E was opened on 30 March 2008. The original linear terminals are now supplemented by two satellites L and M and a more remote satellite G to the east, all accessed by an underground transit system (Figures 5.144–5.147).

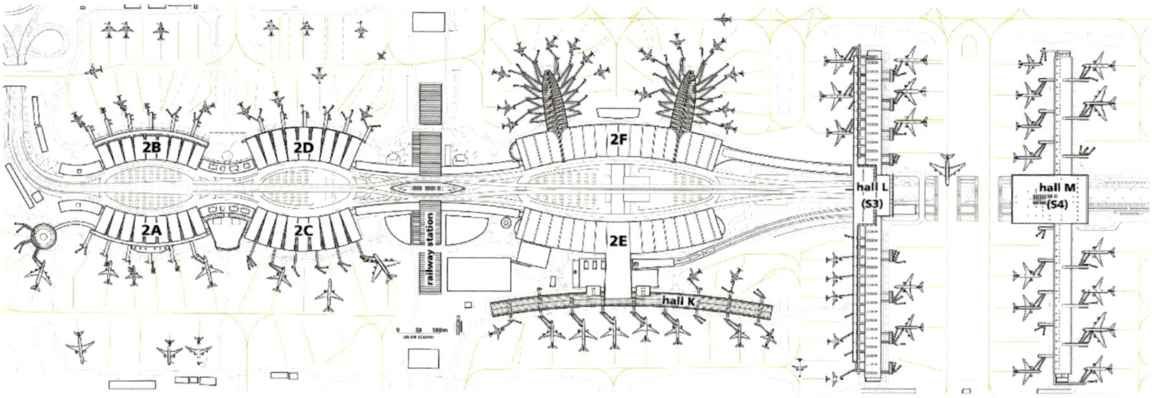

Figure 5.145 Paris Charles de Gaulle, Terminal 2.
SOURCE: WIKIPEDIA BY JULIEN.SCAVINI – OWN WORK, CC BY-SA 4.0, HTTPS://COMMONS.WIKIMEDIA.ORG/W/INDEX.PHP?CURID=76322670

Figure 5.146 Paris Charles de Gaulle, landside view of Terminal 2.
SOURCE: ADPI

Figure 5.147 Paris Charles de Gaulle, aerial view of expanded T2 with satellites in the foreground.
SOURCE: ADPI

Charles de Gaulle airport promotes itself as an international transfer hub to support Air France and incorporates a major public transport interchange. The rail station links the airport directly with the French TGV international rail network and Eurostar services. The airport is also directly linked to the Paris Metro network and the RER regional network (Figures 5.148–5.150).

Figure 5.148 Paris Charles de Gaulle, aerial views of TGV Station 1994.
SOURCE: GOOGLE EARTH AND ADPI

Figure 5.149 Paris Charles de Gaulle Rail/Air Interchange, interior.
SOURCE: WIKIPEDIA

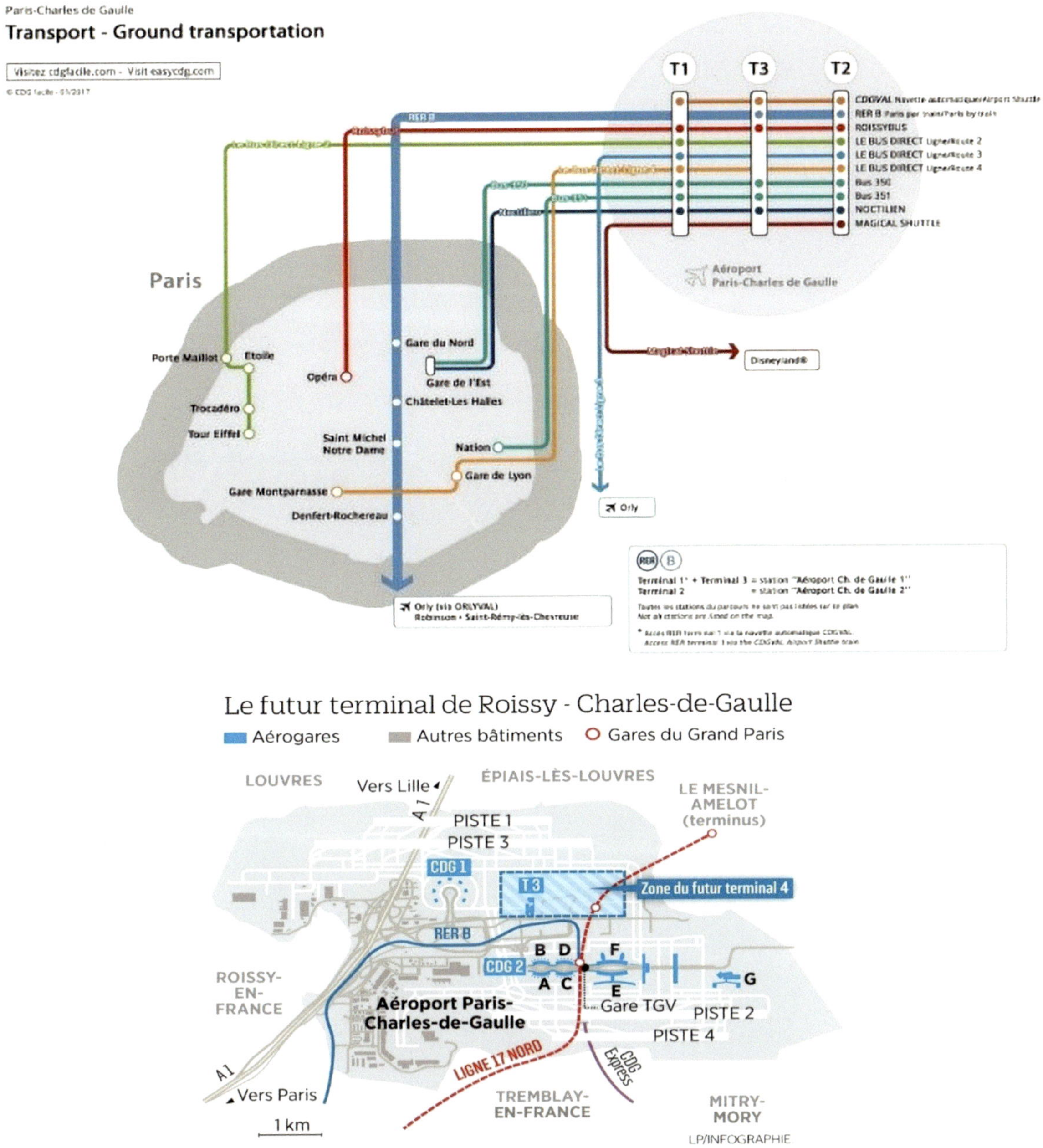

Figure 5.150 Paris Charles de Gaulle Rail/Air Interchange and rail network.
SOURCE: ADPI & LE PARISIEN

Terminal T3 handles Charles de Gaulle's budget carriers/low-cost airlines and charter fights. It opened in 1990 with five gates. The transit between Charles de Gaulle's terminals takes place via the airport's CDGVAL light-rail shuttle service. Charles de Gaulle airport's annual passenger capacity is now set at 80 million per annum.

Terminal 4 is the airport's most recent project with a central processor and conventional piers. It aims to increase the terminal processing capacity from the current declared capacity by 50%, from 80 mppa to 120 mppa. The runway capacity is however declared as only increasing from 480,000 movements per annum to 680,000 movements, or 42% per annum. No new runways are proposed, so the airport is relying on a significant increase in aircraft size and improvements to air traffic control to achieve that growth (Figures 5.151 and 5.152).

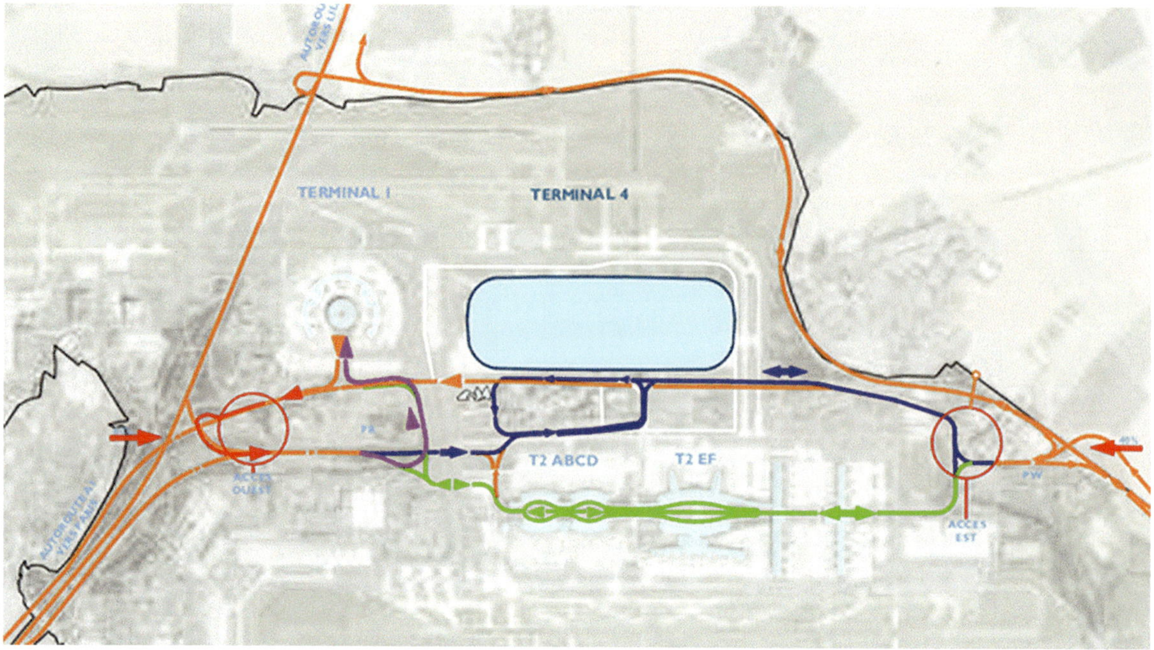

Figure 5.151 Paris Charles de Gaulle – location of future T4.
SOURCE: ADPI

Figure 5.152 Paris Charles de Gaulle, proposed T4.
SOURCE: ADPI

CHAPTER 6

TRANSFER INTERCHANGE 1980–1999

Atlanta introduced a new era in airport design when it opened in 1980, as the world's largest passenger terminal complex and initially designed to handle 55 million passengers a year. The previous airfield and terminal designs had followed the trends of the day, initially with multi-directional runways supported by a unit terminal, to be replaced with a terminal with a linear pier, followed by a terminal with radiating finger piers based on the Chicago O'Hare concept.

The 1980 Atlanta concept, with parallel runways, but no cross-wind runway and a mid-field terminal supporting a series of linear satellites linked by an underground automated people mover set a new trend. Atlanta has an exceptionally high proportion of transfer traffic at around 80%, so was specifically designed to make transferring as simple and speedy as possible. Atlanta, although designed to handle primarily domestic traffic and predominantly smaller aircraft, can be seen as the prototype for a series of later international terminals designed to facilitate transfer traffic.

The concept has a number of acknowledged benefits, including maximising efficiency of land use and optimising aircraft circulation, and has been copied by a number of subsequent designs, including Denver, and the updates to Heathrow and Washington Dulles. Set against those virtues, the layout requires an expensive people mover and stations to link the terminal and satellites. The people mover can never be 100% reliable so the designers wisely included walkways and passenger conveyors to act as a back-up in the event of failure. The other limitation is that the narrow satellite gives limited opportunities for retail and concessions. This was later addressed by widening the satellite, at the expense of aircraft stands, at the centre to allow a better commercial offer. However, the concept has been sufficiently robust to allow Atlanta to be progressively expanded to handle over 100 million passengers per annum and remain the busiest airport in the world.

Other satellite concepts have proved less durable. For example Paris Charles de Gaulle circular terminal 1 (1973) surrounded by series of small satellites, proved impossible to expand and has unusually counter-intuitive passenger movement. The original master plan envisaged a series of similar circular terminals, but the second phase of development abandoned this concept and substituted a series of linear terminals, with T2 A & B opening in 1982, but this too was later abandoned and the latest expansion is in the form of linear satellites.

Throughout this period, terminals following other concepts continued to be built; for example, Amsterdam Schiphol continued to expand their single terminal concept by adding additional finger piers, while Heathrow's Terminal 4 (1986) followed a linear concept, as did the much larger Osaka Kansai (1994). The latter incorporated transit systems to carry passengers to the end of the very long pier. Kansai also marked a radical

DOI: 10.4324/9781003323624-6

change of direction in the siting of airports that has influenced many subsequent designs. The Japanese Airport Authority, scarred from the bitter experience of facing the protest movement that had grown up in opposition to Tokyo Narita, decided to create an artificial off-shore island in Osaka Bay for Kansai Airport. They have continued with this approach with all subsequent new airport developments.

In 1986, The British Airports Authority was privatised starting a worldwide trend that required airports to improve their return on investment. This has resulted in commercial offers absorbing an increasing percentage of investment and requiring a larger share of floor space in the terminal. Heathrow now aims to support its day-to-day operating costs entirely from commercial revenues, while airport charges support investment in capital expenditure. Similarly, airport master plans have evolved to support increased levels of airport-related development to generate higher revenues from parking, hotels, offices and other property developments, giving impetus to the Airport City concept. However, during this period, Airport Cities continued to evolve in a relatively unplanned way around existing airports, which proved to be a catalyst for growth in their surrounding hinterland. Examples of the evolution of Airport Cities around existing airports include Heathrow, Gatwick, Schiphol, Frankfurt and Charles de Gaulle in Europe and Atlanta, Dallas Fort-Worth, Chicago O'Hare and New York JFK in the United States. The very success of airport-related development has, however, now become a major constraint to airport expansion, with noise, air quality and congestion affecting the surrounding communities, which has led to ever better organised and more vocal protests against airport expansion.

A reversal to this trend of ever larger and more complex airports came when American domestic carrier Southwest introduced the concept of a low-cost airline in 1979, following deregulation in the United States. They developed it throughout the following decades with the sole objective of offering cheap airfares to consumers. This created a situation where already established flag ship carriers or legacy airlines lost a significant amount of their market share to these newly formed low-cost airlines, purely because of their ability to charge a lower price than traditional full-cost airlines.

The low-cost carriers (LCCs) typically used existing airports, which had formerly been the major airport of the city but were now underused and relegated to secondary status by the construction of newer airports. Initially, Southwest was prohibited from issuing tickets that involved a transfer but soon realised that they could sell passengers two separate tickets, allowing them to self-transfer and avoid the costs incurred by legacy airlines. These secondary airports, because of the relative simplicity of their infrastructure, were able to offer lower landing charges. Examples include Chicago Midway, Houston Hobby and Dallas Love. The operational concept that was to form the model for later low-cost airlines. Their business model featured the following characteristics:

- Network – Point-to-point high-frequency routes
- Distribution – Travel agents and call centres, no tickets
- Fleet – High utilisation, same type of aircraft across the fleet
- Airport – Secondary airports with short turnaround times
- Sector length – Short (around 400nm)
- Staff – High productivity with competitive wages and profit sharing.

LCCs introduced a number of innovations aimed at reducing the cost of airfares. For example, Southwest introduced internet ticketing in 1996 and Alaska Airlines internet check-in in 1999, essential moves that have later affected terminal design, initially reducing the demand for airline ticketing desks and then check-in desks. The airports selected to illustrate the key characteristics of the era are:

- Atlanta Hartsfield (ATL) 1980
- Singapore Changi (SIN) 1981
- Jeddah, King Abdulaziz (JED) 1981
- London, Stansted (STN) 1991
- Osaka Kansai (KIX) 1994
- Denver International (DEN) 1995
- Hong Kong International (HKG) 1998
- Kuala Lumpur (KUL) 1999

22. Atlanta, Hartsfield (ATL) 1980

The evolution of Atlanta Airport can be seen as a microcosm of the evolution of airports in general. It started with passenger services handled out of a lean-to hangar (1929) and then was successively rebuilt as a simple terminal design (1931), then a linear terminal (1948) which was expanded with single-storey finger piers. It was then rebuilt as a Chicago-like multiple pier concept (1961) and then reinvented itself as a hub airport with multiple mid-field satellites fed from terminals at each end of the airport (1980). An evolution of the 'Toast Rack' master plan has served the airport well ever since and allowed it to expand progressively with additional runways and terminal capacity (Figures 6.1 and 6.2).

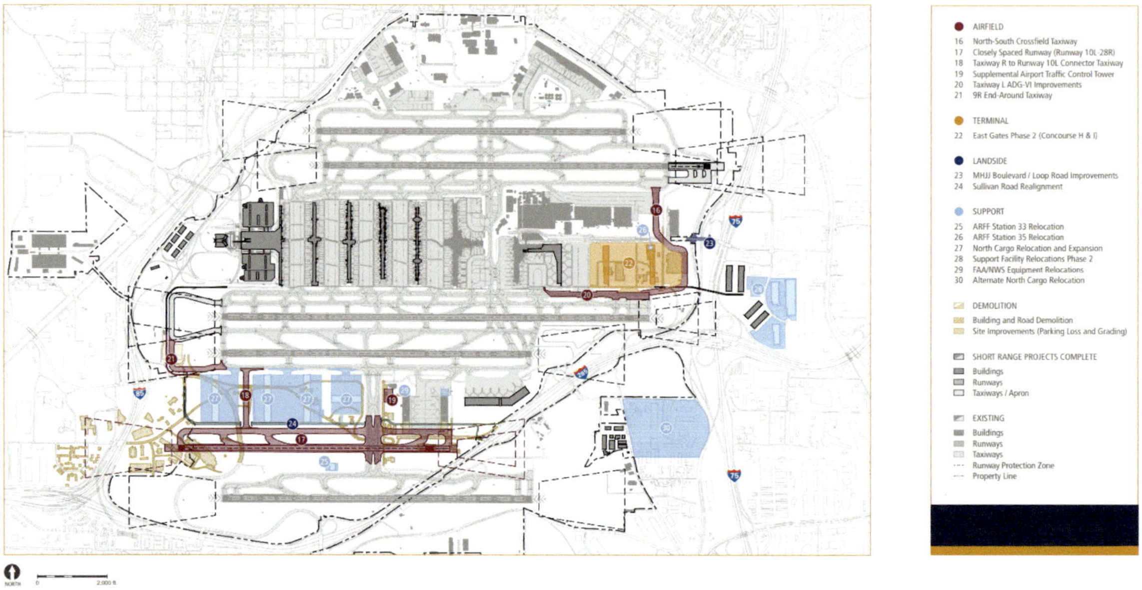

Figure 6.1 Atlanta, Hartsfield Long Range Development Plan to 20-Year Time Frame, showing 6th runway, developed mid-field terminal facilities (22) and relocated cargo facilities (27).
SOURCE: CITY OF ATLANTA, DEPARTMENT OF AVIATION

Figure 6.2 Atlanta Hartsfield aerial view (5th runway at top of photo).
SOURCE: FLICKR MFMINN 2012

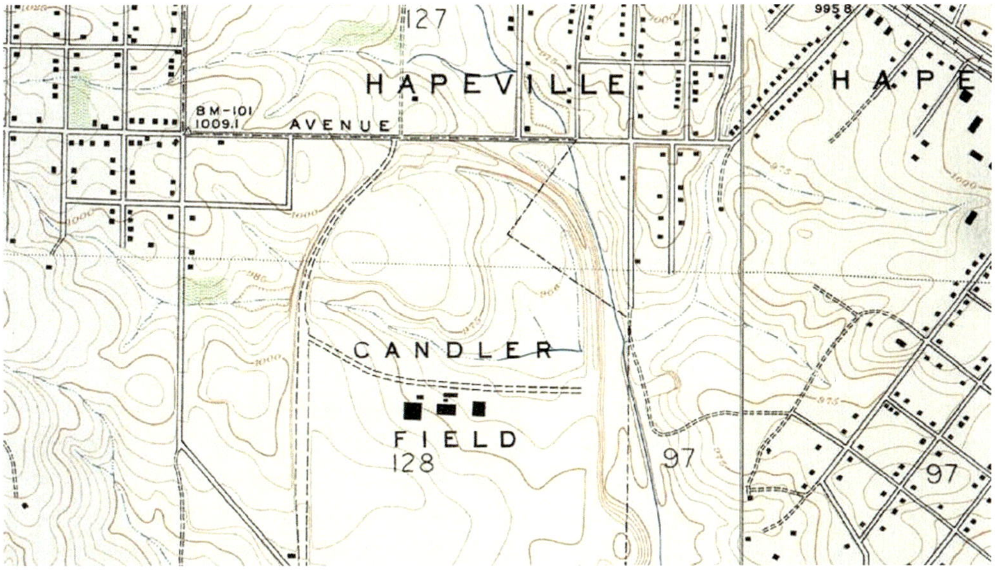

Figure 6.3 Atlanta, Candler Field, oval racetrack converted to airfield (hangars in black).
SOURCE: SUNSHINESKIES.COM

The airfield layout is recognised as being the most space-effective and operationally efficient for aircraft movement of any in the world, but it requires a complex and expensive transit system to link the terminals with the satellites. Atlanta is currently, and has been for some years, the busiest airport in the world, measured by passenger numbers, handling 110 mppa in 2019.

Atlanta, Hartsfield-Jackson was originally known as Candler Field. The airfield was created out of a former motor racing speedway circuit, whose oval shape can be recognised in the map above (Figure 6.3). The first scheduled airline service from Atlanta occurred on 15 September 1926, when a Florida Airways plane departed for Jacksonville and Miami. Charles Lindbergh visited Atlanta Airport on 11 October 1927, during a cross-country tour following his solo flight across the Atlantic. His visit to Atlanta was a major catalyst for the rapid expansion of Candler Field in the 1930s.

The original 'lean-to' terminal building and hangars can be seen below (Figure 6.4).

Figure 6.4 Atlanta, Candler Field – a 1929 view of the airport showing T.A.T. Flying Service, Pitcairn Aviation (with Atlanta Airport on the roof) and the Beeler Blevins hangar. T.A.T. was a predecessor of American Airlines.
SOURCE – SUNSHINESKIES.COM

Figure 6.5 Atlanta, Candler Field, postcard showing first terminal building 1931, for Eastern Airlines.
SOURCE: SUNSHINESKIES.COM

A mid-1930s postcard shows an Eastern Air Lines Douglas DC-2 at the first passenger terminal which opened in 1932. Eastern began flying DC-2s in late 1934 (Figure 6.5). The 2nd terminal building and control tower shown in the postcard opened in 1932 (Figure 6.6).

Atlanta airport saw the beginnings of the hub concept in 1939. In the hour between 11.45 pm and 12.40 am eleven of Atlanta's 30 daily flights were handled in an operation known as the *'Midnight Merry-Go-Round'* (Figure 6.7).

The evolution of Atlanta Hartsfield is illustrated below (Figures 6.8–6.12).

Figure 6.6 Atlanta, Candler Field – Second terminal and new control tower 1932.
SOURCE: SUNSHINESKIES.COM

Figure 6.7 Atlanta, Georgia, 'Midnight Merry-Go-Round' – 'Hubbing' at Atlanta 1939.
SOURCE: A DREAM TAKES FLIGHT (COURTESY OF JACK GRAY AND H.W. TOFFLEMIRE)

Figure 6.8 Atlanta, Candler Field, aerial view.
SOURCE: SUNSHINESKIES.COM

Figure 6.9 Atlanta, Candler Field, third terminal with linear pier 1948.
SOURCE: SUNSHINESKIES.COM

Figure 6.10 Atlanta, Candler Field third terminal with finger pier extensions.
SOURCE: SUNSHINESKIES.COM

Figure 6.11 Atlanta, Candler Field (Hartsfield) fourth Jet Age terminal 1961.
SOURCE: SUNSHINESKIES.COM

Figure 6.12 Atlanta Hartsfield evolution, a view of the first four terminals – A. Eastern Terminal, 1931, B. Terminal, 1932 to 1948, C. Temporary Terminal, 1948–1961, D. Jet Age Terminal 1961 to 1980, E. Delta hangars 1940, F. Air Host Inn, G. Virginia Avenue Entry.
SOURCE: SUNSHINESKIES.COM

Figure 6.13 Atlanta, Hartsfield, Jet Age terminal with second parallel runway.
SOURCE: SUNSHINESKIES.COM

Figure 6.14 Atlanta, Hartsfield 1976, aerial view.
SOURCE: SUNSHINE SKIES – PHOTO: JIM DOANE

After nearly a decade of planning and several years of construction, Atlanta's new Jet Age terminal opened on May 3, 1961 (Figures 6.13–6.15). The complex is claimed to be the first in the world to be built specifically for jet aircraft. At the time, it was the largest passenger terminal in the country covering 60 acres, with nearly a mile of concourses and 48 gates with parking for 52 aircraft. The enormously light-filled lobby and modern architecture were a considerable improvement on the cramped

Figure 6.15 Atlanta, Hartsfield Terminal Layout in early 1980.
SOURCE: SUNSHINESKIES.COM

and dreary temporary terminal that preceded it. The layout was fairly typical of large airports of the early '60s, similar to Miami or Chicago O'Hare. The central 2-storey terminal featured a 500-ft-long ticketing lobby upstairs and baggage claim downstairs with long concourses extending from the main building. The turquoise-panelled administration building, and control tower was one of Atlanta's most identifiable landmarks during the following two decades.

A major change had occurred when Delta Airways introduced the 'Hub and Spoke' concept in 1955, with aircraft converging simultaneously on the Atlanta Hub from a variety of regional airports, increasing the volume of traffic handled in a short period of time to new highs and requiring a new higher-capacity master plan.

The primary driver of the new layout was the efficient use of scarce runway capacity. The existing layout necessitated aircraft crossing the main East-West runway closest to the terminal to reach the further parallel runway, raising safety concerns and greatly reducing runway capacity as gaps had to be left in the schedule to accommodate runway crossings.

Atlanta Airport's 1980 rebuild is also internationally recognised as the first airport that was specifically designed as a *'Hub'* airport to facilitate the new business model developed by Delta Airlines that was based on linking many smaller cities through a central hub where passengers transferred from one flight to another. Like Washington Dulles, the Atlanta configuration involved placing the terminal between parallel runways, but its key innovation involved linking the terminal and satellites by an underground Automated People Mover (APM). The layout was adopted to supersede the former Chicago-like finger pier layout to reduce aircraft movement congestion and avoid runway crossings that limited the capacity of the former layout (Figures 6.16 and 6.17).

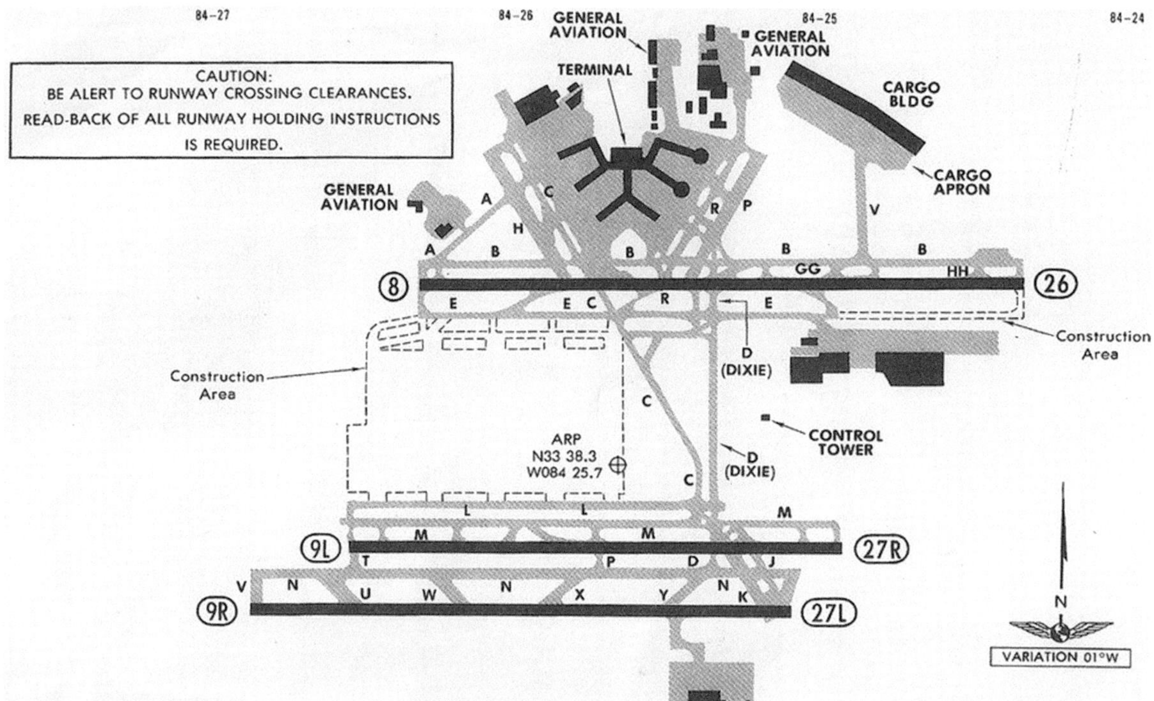

Figure 6.16 Atlanta Hartsfield Airfield Layout in early 1980 during construction of the midfield terminal, showing the third close parallel runway that was added to increase the capacity of the airfield.

SOURCE: SUNSHINE SKIES.COM

Figure 6.17 Atlanta Hartsfield, aerial view of Mid-Field Terminal and Satellites, views from the west, 1980.
SOURCE: SUNSHINESKIES.COM – PHOTO JIM DOANE

The evolution of the masterplan to support the transition from the Jet Age Terminal (1961) to the Mid-Field Terminal (1980) is illustrated (Figure 6.18). To accommodate subsequent growth the Master Plan, the Mid-Field Terminal, its transit system and retail offer has continued to develop (Figures 6.19–6.24).

The proposals included a much more ambitious terminal concept that had been developed by architects Stevens & Wilkinson, Smith Hinchman & Grylls (now called SmithGroupJJR), where several much larger and longer concourses (each over 1 km long) provided access to a large number of aircraft. The satellites were linked to the terminal building by an underground transit or automated people mover (APM), using a pinched loop APM configuration which allowed the transit to turn around at the end of the route and so increase capacity beyond that of a simple shuttle system. In case of failure, it is backed up by a tunnel walkway that incorporates passenger conveyors.

The Atlanta master plan layout, of a mid-field terminal supported by a *'Toast Rack'* arrangement of parallel satellites, directly influenced the later concept at Denver and the ongoing updating of Heathrow and Washington Dulles. The layout has been demonstrated to be both the most space-effective master plan and the most efficient aircraft circulation pattern for a large hub airport. The *'Toast Rack'* master plan of 1980 has served the airport well ever since and allowed it to expand progressively with additional runways and terminal capacity.

Finally, the introduction of multiple runways has led to another innovation pioneered at Atlanta, the addition of end-around taxiways (EATs) (Figure 6.25). Although this innovation increases taxiing time and distance it allows aircraft to reach the further runway without crossing intermediate runways, the latter reduces the capacity of the runway and is potentially more dangerous. This, in combination with the construction of a fifth widely spaced runway, together with plans for a sixth, supported by terminal expansion and additional satellites, will allow the airport to expand even further, potentially with the capacity to handle some 150 mppa.

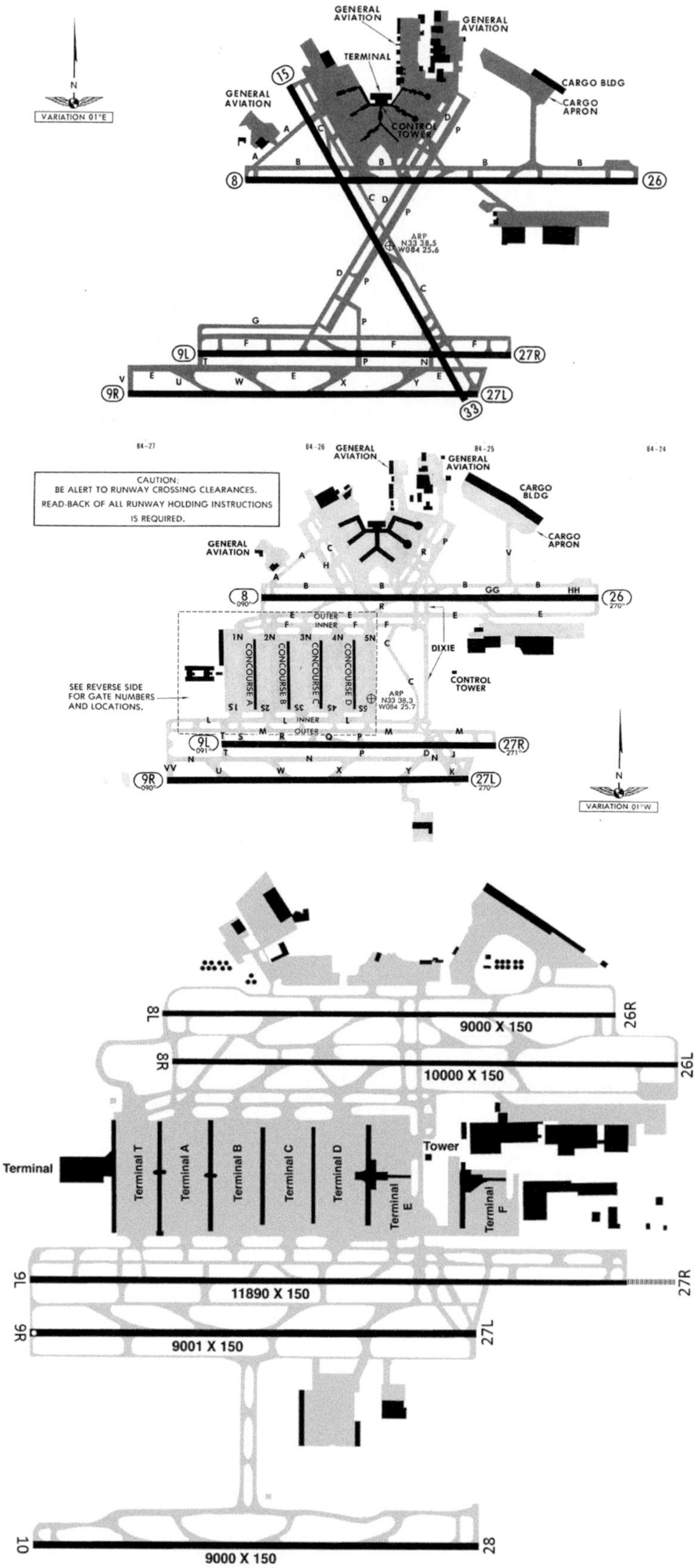

Figure 6.18 Atlanta Hartsfield Master Plan Evolution.

SOURCE: ATLANTA HARTSFIELD

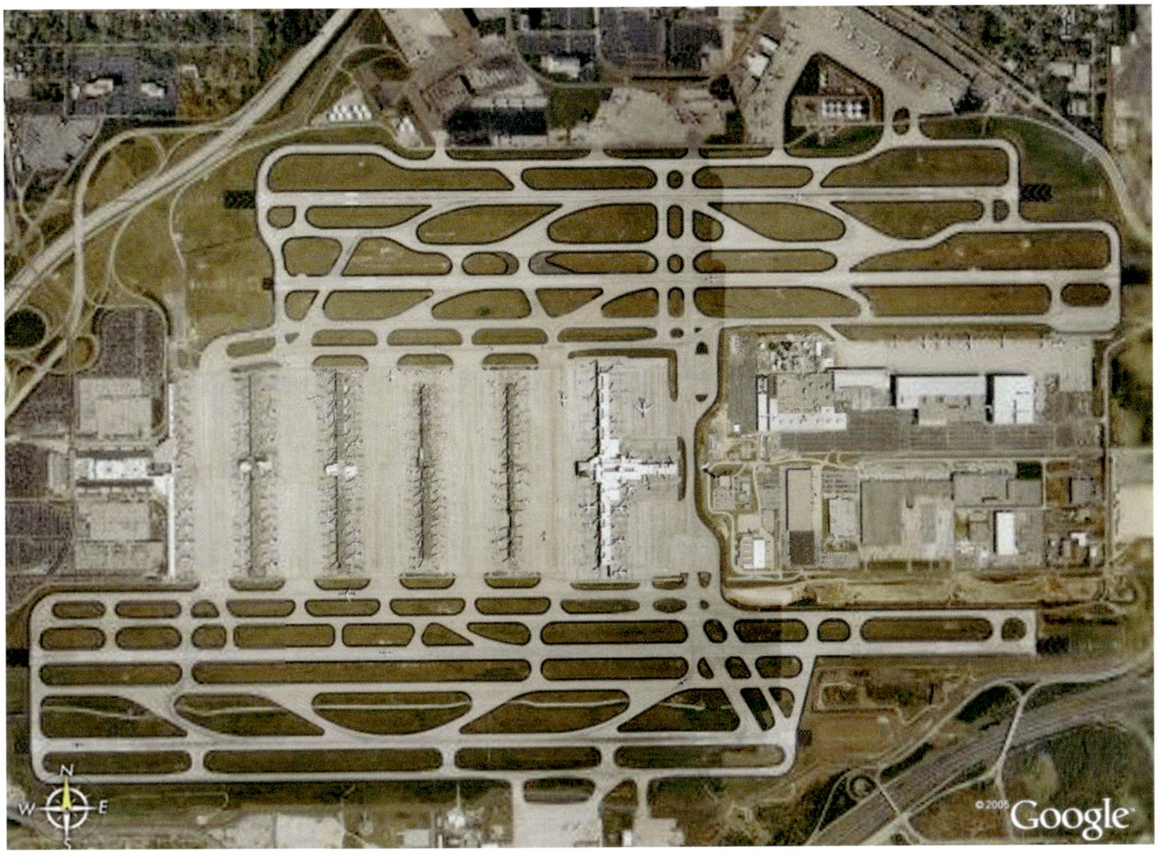

Figure 6.19 Atlanta Hartsfield aerial view after the construction of Concourse E.
SOURCE: GOOGLE EARTH C2010

Figure 6.20 Atlanta showing the later Eastern Terminal and Satellites.
SOURCE: CITY2MAP.COM

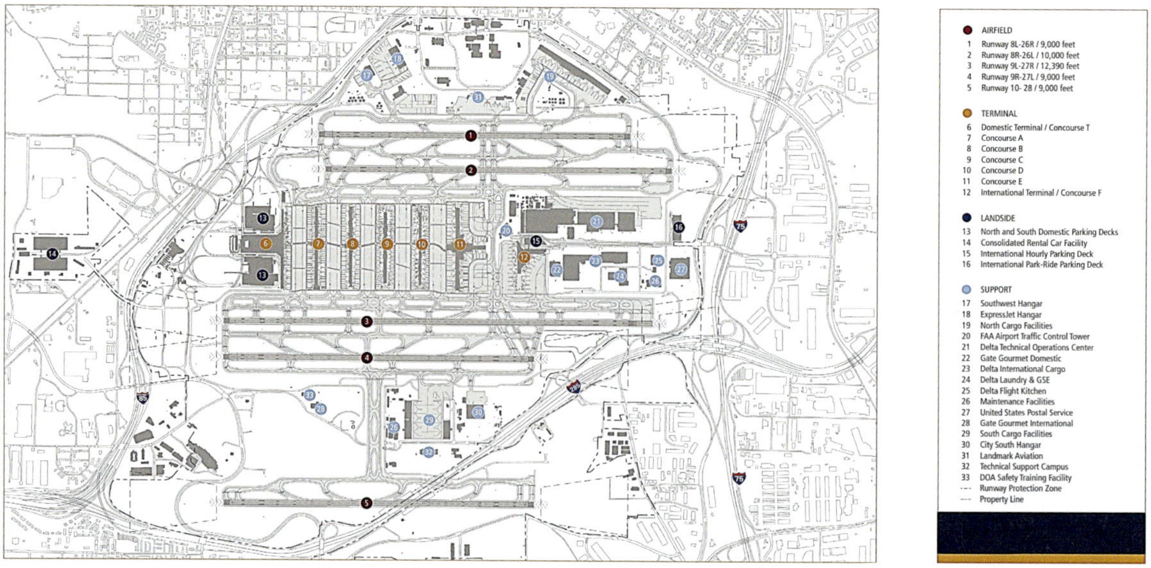

Figure 6.21 Atlanta, Hartsfield Existing Master Plan 2015.
SOURCE: CITY OF ATLANTA, DEPARTMENT OF AVIATION

Figure 6.22 Atlanta Hartsfield underground automated people mover maintenance area and station (APM).
SOURCE: ATLANTA HARTSFIELD AND WIKIPEDIA

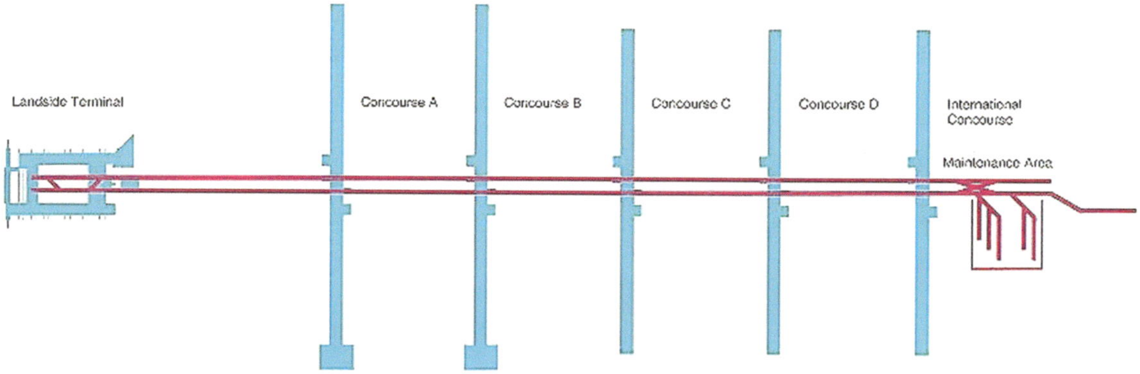

Figure 6.23 Atlanta Hartsfield underground APM system.
SOURCE: WESTINGHOUSE

Figure 6.24 Atlanta Hartsfield Retail Atrium 2016.
SOURCE: PEDROCARRION.COM

Figure 6.25 Atlanta Hartsfield, End-Around Taxiway.
SOURCE: GOMACO WORLD

23. Singapore, Changi (SIN) 1981

Singapore, Changi has successfully persevered with the pier concept, with three terminals planned in a mid-field configuration, joined at a later date by a fourth terminal. The longer routes between terminals are supported by transit systems. Changi has always built capacity in advance of demand, and this feature, together with a particular design concentration on landscaping and provision of leisure facilities for transfer passengers, has resulted in the airport being regularly recognised in passenger satisfaction surveys as first or second in the world. The Jewel, a domed fully glazed indoor landscaped garden is the latest addition. For the future, Changi plans further expansion with a fifth terminal and third runway in a manner similar to Hong Kong International (Figures 6.26–6.28). The terminal layouts follow a simple orthogonal strategy with finger piers (Figures 6.29 and 6.30).

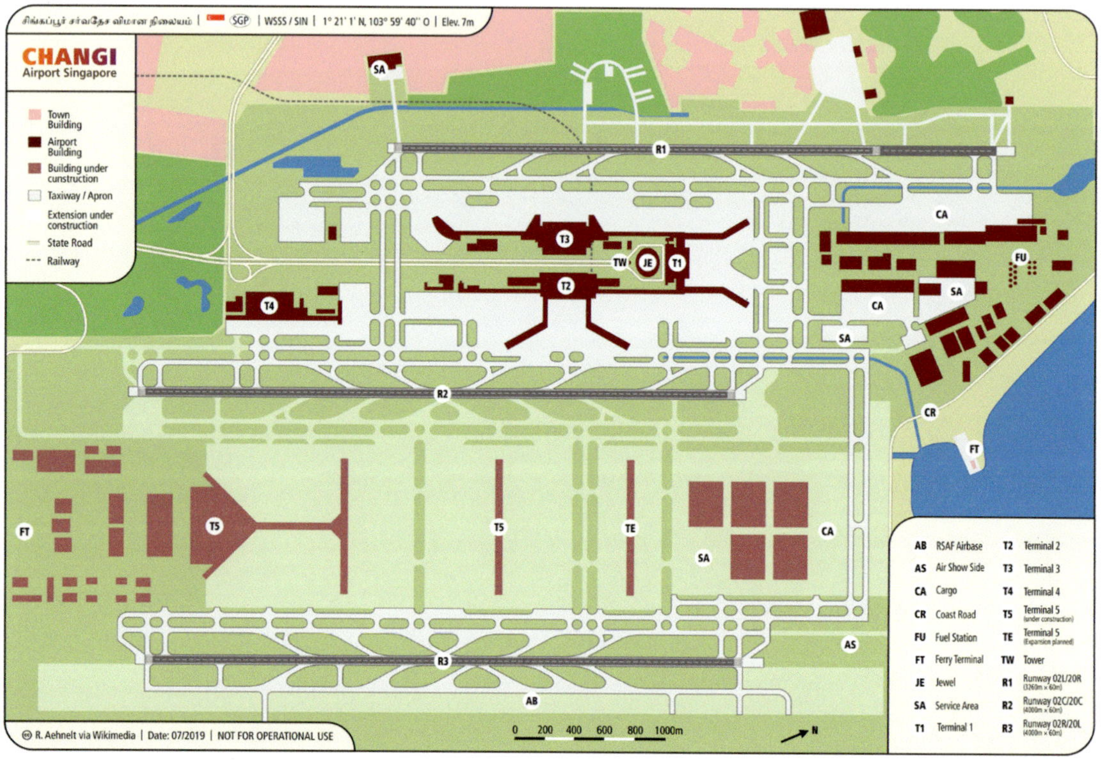

Figure 6.26 Singapore, Changi Future Master Plan.
SOURCE: WIKIPEDIA

Figure 6.27 Changi, Orchid Garden.
SOURCE: CHANGI AIRPORT VIA WORLDLANDSCAPEARCHITECT.COM

Figure 6.28 The Jewel
SOURCE: SAFDIE ARCHITECTS LLC

Figure 6.29 Model of Terminal 1.
SOURCE: CHANGI AIRPORT VIA SGBLUESKY.WORDPRESS.COM

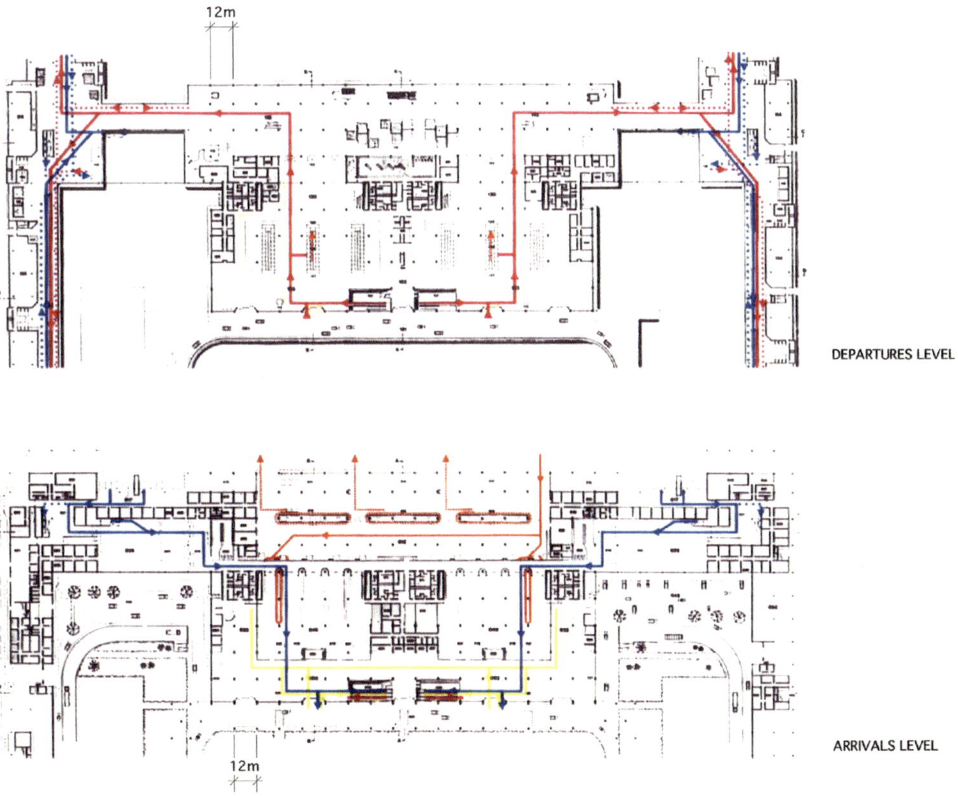

Figure 6.30 Terminal 1, Arrivals and Departures Floor Plans.
SOURCE: CHANGI AIRPORT

Figure 6.31 Butterfly Garden.
SOURCE CHANGI AIRPORT VIA THE INDEPENDENT

Figure 6.32 Atrium Cinema.
SOURCE: CHANGI AIRPORT VIA STRAITS TIMES

Changi is particularly known for its extensive interior landscaping and lush planting (Figures 6.31–6.37). It is also noted for its leisure facilities including airside hotels and a swimming pool (Figure 6.38 and 6.39).

The Jewel at Changi Airport re-imagines the centre of an airport as a major public realm attraction and an opportunity to relax. The Jewel offers a range of facilities for landside airport operations, indoor gardens and leisure attractions, retail offerings and hotel facilities, all under one roof. A distinctive dome-shaped façade made of glass and steel reinforces Changi's appeal as one of the world's leading Global Hub Airports (Figures 6.40–6.43).

Figure 6.33 Koi Carp Pond.
SOURCE: CHANGI AIRPORT VIA THE INDEPENDENT

Figure 6.34 Cactus Garden.
SOURCE: CHANGI AIRPORT VIA THE SUN

Figure 6.35 Orchid Garden.
SOURCE: CHANGI AIRPORT VIA LIVING + NOMADS

Figure 6.36 Orchid Garden and Koi Carp Pond.
SOURCE: CHANGI AIRPORT VIA LIVING + NOMADS

Figure 6.37 Check-in Concourse.
SOURCE: CHANGI AIRPORT IN VIA GQ.COM

Figure 6.38 Airport Hotel.
SOURCE: CHANGI AIRPORT VIA NEW YORK TIMES

Figure 6.39 Singapore, Changi roof top swimming pool.
SOURCE: CHANGI AIRPORT GROUP

Figure 6.40 Singapore, Changi, The Jewel 2019.
SOURCE: THE JEWEL VIA ARCHITECTURAL RECORD

Figure 6.41 Changi, The Jewel Plan.
SOURCE: SAFDIE ARCHITECTS, LLC

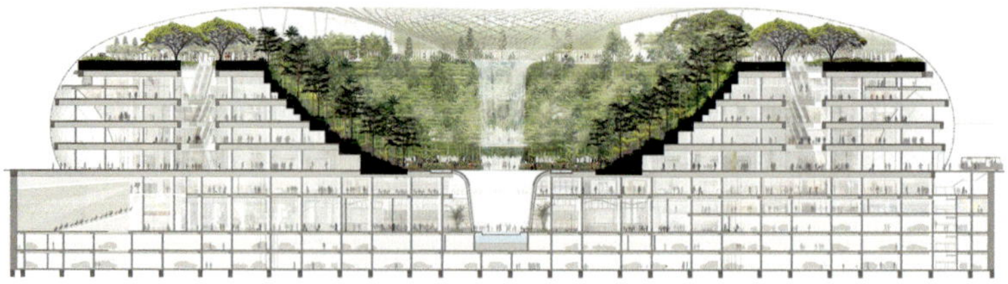

Figure 6.42 Changi, The Jewel Section 2019.
SOURCE: SAFDIE ARCHITECTS, LLC

Figure 6.43 Singapore Changi, The Jewel, Rain Vortex with transit passing through – 2019.
SOURCE: WIKIPEDIA PHOTO: MATTEO MORANDO

24. Jeddah, King Abdulaziz (JED) 1981

The Hajj terminal, designed by SOM, was built for the opening of the airport in 1981. It is only used during the Hajj season and caters to the one-way flow of pilgrims to Mecca. The terminal is one of the largest in the world, with a floor space of five million square feet (465,000 m^2) and is positioned on a site of over 100 acres (405,000 m^2) of ground area and is known for its signature roof, shaped like a Bedouin tent, to respond to the desert context (Figure 6.44). While the planning for Hajj traffic is unique, its influence can be seen in subsequent designs such as Stansted which adopts a similar structural and servicing principle and Denver whose tent-like roof echoes the imagery of the Hajj Terminal at Jeddah (Figure 6.45).

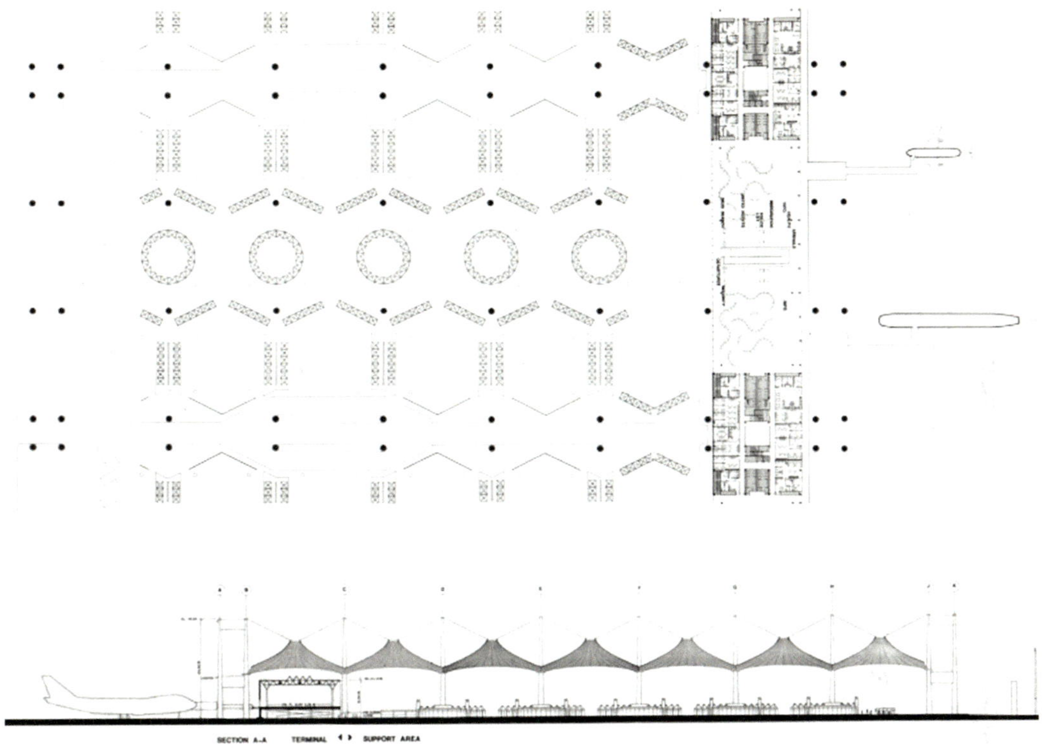

Figure 6.44 Jeddah, King Abdullaziz, Hajj Terminal, plan, and section.
SOURCE: SOM VIA AD CLASSICS

Figure 6.45 Jeddah, King Abdullaziz, Hajj Terminal, images.
SOURCE: SOM VIA AD CLASSICS

King Abdul Aziz International Airport in Jeddah is the third largest and busiest airport in Saudi Arabia. It handled 17.8 million passengers in 2010. The airport is the gateway to Saudi Arabia for a large number of pilgrims. It is close to the cities of Mecca and Medina, two of the holiest sites in Islam. The airport provides facilities required for the service of Hajj pilgrims and those who visit the country to perform Umrah during the rest of the year.

Jeddah was one of the first airports to be designed with three widely spaced runways; although the third runway was initially planned for the Royal Saudi Air Force, it has since been brought into use to support a new terminal. The Hajj terminal is dedicated to those arriving for the pilgrimage to Mecca and is highlighted in yellow, the international terminal is highlighted in red and the domestic terminal in green on the original master plan below (Figure 6.46).

SOM designed a linear terminal building and a separate, large-tented support complex where travellers can comfortably prepare for their journey to Mecca. The complex contains facilities for sleep, food preparation and various support services. The processing element itself consists of two rectangular, two-storey, jetty-served and comfort-cooled buildings (these contain airport services such as immigration, customs, baggage reclaim areas and check-in counters) which are covered by the 'tent form' roofs, which also cover the waiting area beyond making it one of the most environmentally sustainable airport terminal designs to date. The naturally ventilated building is topped by 210 semi-conical, Teflon-coated fiberglass roof units that are contained within a total of 10 modules. The modules are supported by 45-metre-high steel pylons (Figures 6.47 and 6.48). Because the fabric has low heat transmission, it allows the sun to cast a warm light over the support area. At night, it becomes a great reflective surface, as uplighters bounce light from the roof to the ground below (Figures 6.49 and 6.50).

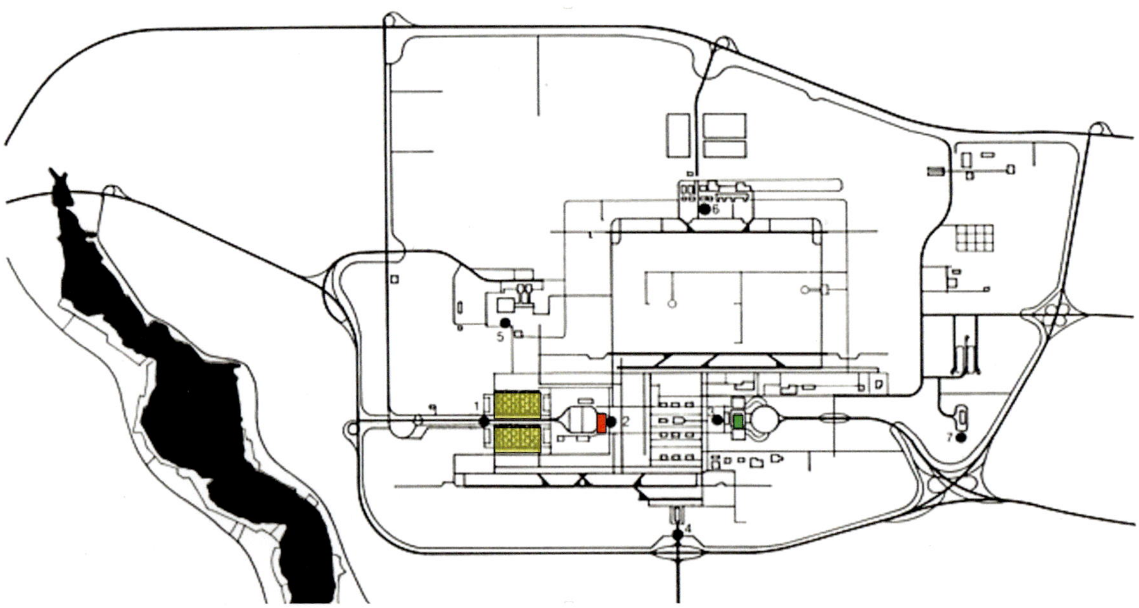

Figure 6.46 Jeddah, King Abdulaziz, initial master plan.
SOURCE: SOM VIA AD CLASSICS

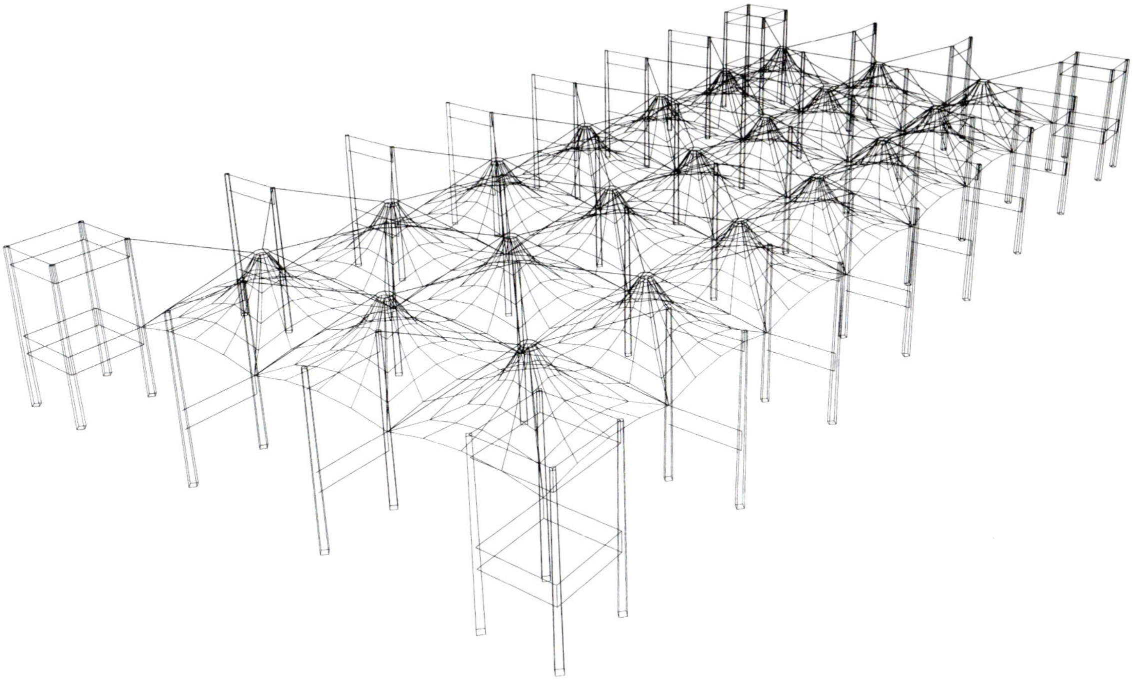

Figure 6.47 Jeddah, King Abdullaziz perspective diagram of the structure and roof layout.
SOURCE: SOM VIA AD CLASSICS

Figure 6.48 Jeddah, King Abdulaziz Hajj Terminal boarding gates.
SOURCE: SOM VIA AD CLASSICS

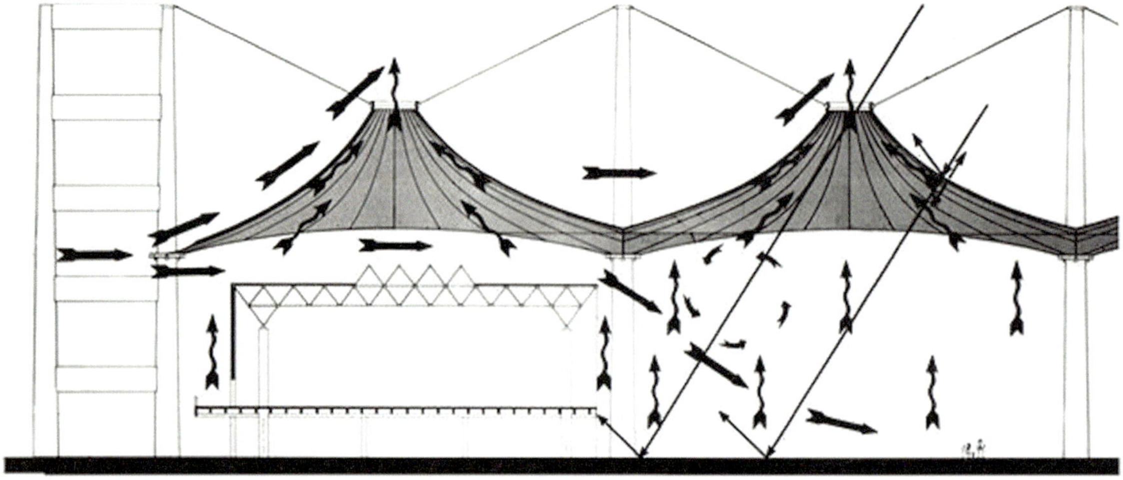

ENVIRONMENTAL DIAGRAM

Figure 6.49 Hajj Terminal Environmental diagram showing how natural convection is used to cool the terminal.
SOURCE: SOM VIA ARCHITECTURAL RECORD

Figure 6.50 Jeddah King Abdulaziz Hajj Terminal roofscape.
SOURCE: HOCHTIEF.DE

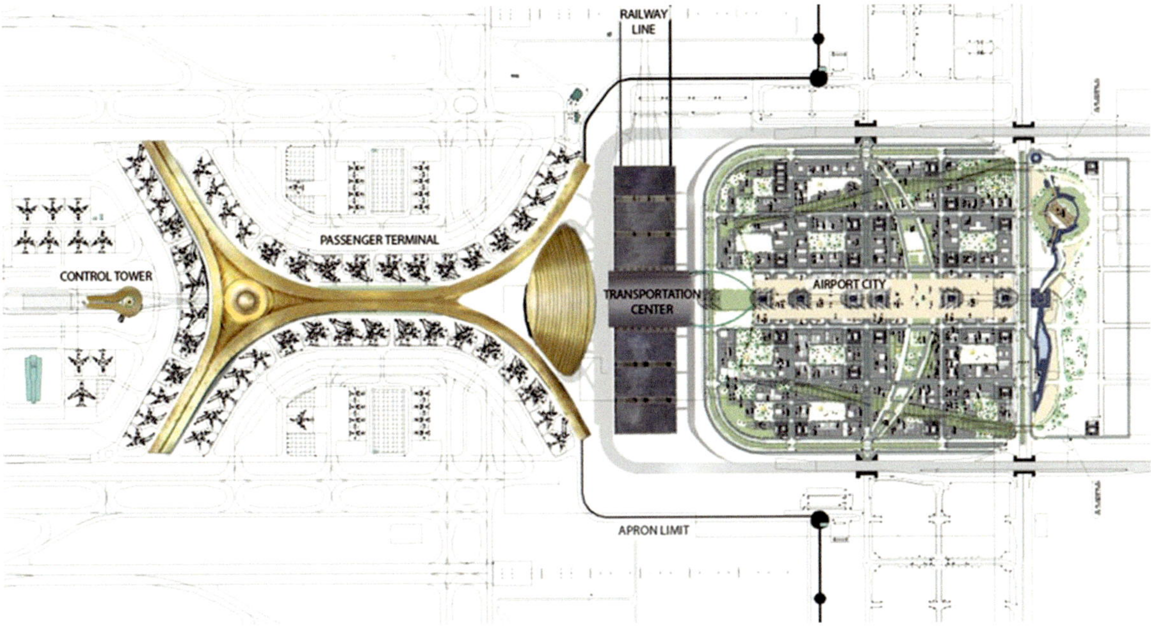

Figure 6.51 Jeddah, King Abdulaziz, site plan of the new mid-field terminal for 30 mppa 2019.
SOURCE: ATKINS

The new mid-field terminal for 30mppa is phase 1 of a major expansion programme developed by Atkins from an initial master plan by ADPI. It opened in 2019 (Figures 6.51–6.53).

Figure 6.52 Jeddah, King Abdulaziz, aerial CAD image of new mid-field terminal.
SOURCE: ATKINS

Figure 6.53 Jeddah, King Abdulaziz, aerial views of new mid-field terminal.
SOURCE: PASCALL + WATSON

25. London, Stansted (STN) 1991

The philosophy, promoted by Norman Foster, that underpins the design of Stansted Airport's terminal is immensely important in setting a new trend in terminal design (Figure 6.54). He later codified these ideas in his essay 'Re-inventing the Airport' of 1996. The ideas he popularised include the following:

- The airport should be thought of as an interiorised experience with a flexible layout under an umbrella roof, derived from his experience of working with Buckminster-Fuller on the Climatron Office.
- Services should be removed from the roof and replaced in an undercroft.
- There should be flexibility to reconfigure the layout as growth and change dictate.
- Daylight should be filtered through the roof, as well as the facades and artificial lighting designed to be bounced back from the ceiling to avoid a dark ceiling at night.
- Passengers should, as far as possible, walk in straight lines towards their destination, drawn to the light.

Many of the features that were pioneered at Stansted figure widely in subsequent airport terminal designs: the use of a floating roof as the main symbol of flight, the integration of a railway station under the terminal with direct access, the adoption of a rapid transit system to take passengers to the gates, and the use of freestanding cabins to enclose shops, offices, toilets and support areas (Figure 6.55).

Figure 6.54 London, Stansted, Stage 1 expansion of terminal.
SOURCE: NEW CIVIL ENGINEER

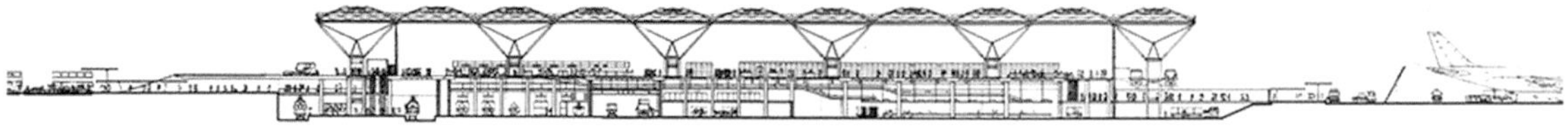

Figure 6.55 London, Stansted cross section through terminal.
SOURCE: FOSTER + PARTNERS

For the first stage of expansion, additional departures and arrivals modules have been added seamlessly, but for the most recent stage, the philosophy has been abandoned and a separate arrivals building has been shoehorned expediently into a gap between the terminal and adjacent hotel that has been built too close to allow for long-term expansion.

Stansted is London's third international airport. It is located some 42 miles (68 km) north-east of London and is currently the base for a number of low-cost carriers. Like Gatwick, Stansted only has a single runway (Figure 6.56) and the apron consists of a series of cul-de-sacs for the aircraft to access the satellites that are laid out at right-angles to the runway. In 2008, a planning application was submitted for a second runway and second terminal, but this was later withdrawn and Stansted continues to operate with a single runway and single terminal.

While some of these ideas can be seen in earlier designs such as St Louis Lambert and Washington Dulles, these were concrete buildings that couldn't offer the same level of flexibility as the tubular steel frame of Stansted. In practice, Stansted has other limitations: the service modules and roof structure at 36-m intervals limit some of the opportunities to re-configure the interior, and the pressures of increased security and retail growth mean that passengers do not walk in straight lines towards the daylight, so wayfinding has become confused. Stansted is notable for integrating the train station under the main concourse (Figure 6.57).

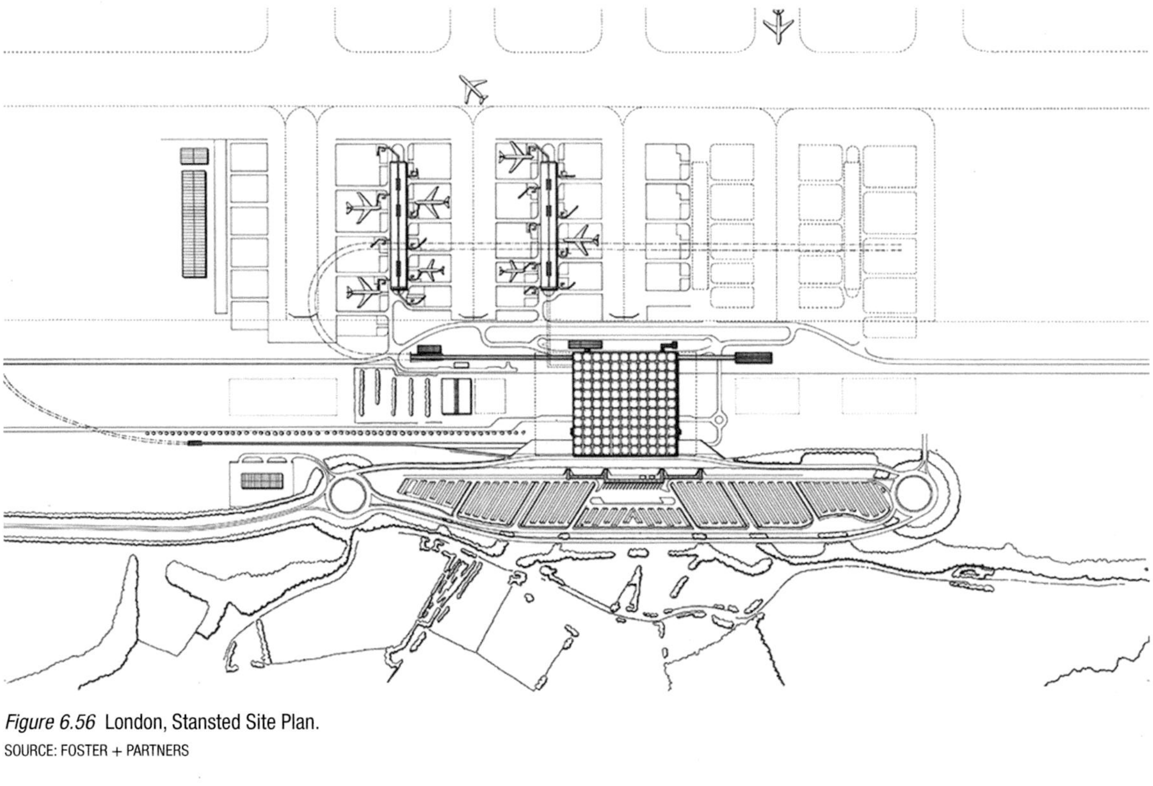

Figure 6.56 London, Stansted Site Plan.
SOURCE: FOSTER + PARTNERS

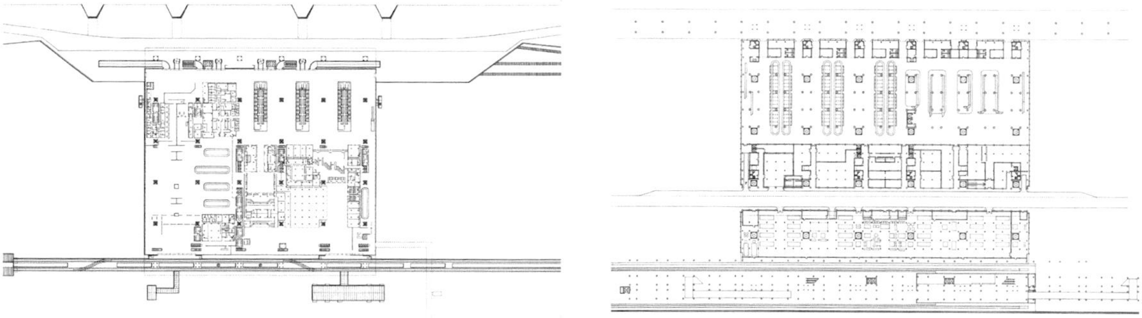

(a) (b)

Figure 6.57 London, (a) Stansted, first (passenger level) and (b) ground (baggage and support level) plans.
SOURCE: FOSTER + PARTNERS

Norman Foster at Stansted airport provided the intellectual basis for the current generation of large steel-framed terminals, which have subsequently become the norm, Norman Foster wrote about Stansted in 1996 (Figure 6.58):

Stansted represents a departure point, one that was achieved by demonstrating that the old order of the 'serviced shed' could literally be turned on its head. At Stansted, the heavy engineering of mechanical plant rooms, metal ducts and supporting structure that made up the traditional roof are all relocated in an undercroft below the concourse level……. Flexibility for change is a vital consideration in such a volatile and expanding industry. This is another reason for the services undercroft. The principle of a single lightweight roof flowing freely over a multitude of different activities makes a quantum leap in the new generation of airports….

The essence of these new large terminals, following the Stansted pattern, is a single roof, flowing freely over a fertile ground plane, on which fully-serviced instant buildings can grow within a tempered climate of unbroken space. The possibilities opened up by this evolutionary response to the realities of mass air travel can be grasped as a civic opportunity, or merely exploited for their commercial potential. In the tradition of the great nineteenth-century railway stations these new terminals are the noble halls of our age, evoking a sense of occasion and bringing a new thrill to air travel.

(Foster, 1996)

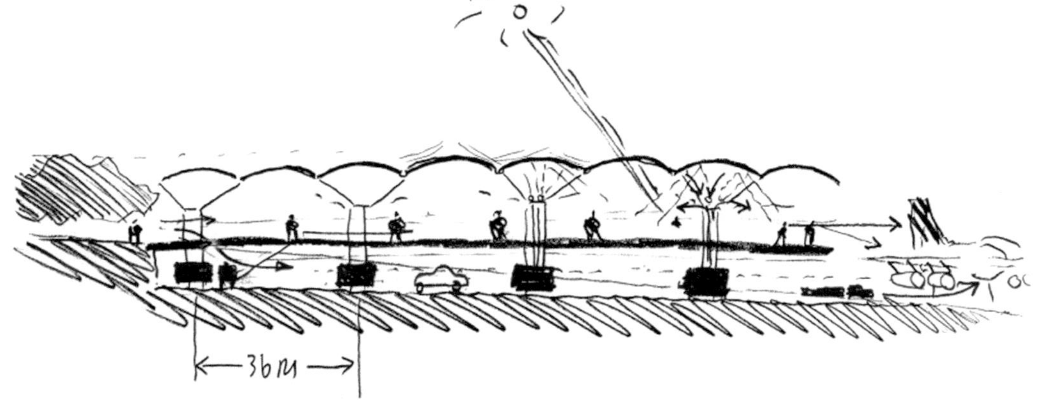

Figure 6.58 London Stansted, cross section concept.
SOURCE: FOSTER + PARTNERS

Figure 6.59 London, Stansted, terminal and forecourt.
SOURCE: PHOTOGRAPHER: KEN KIRKWOOD

The uniqueness of the contribution of Stansted to airport design has, however, sometimes been rather exaggerated, as roofs flowing over open concourses can be found at Yamasaki's St Louis Terminal and Saarinen's JFK TWA and Washington Dulles terminals, although these were in lightweight concrete shell construction. The structural concept of Stansted also owes much to the structure of SOM's tented Haj terminal at Jeddah and to Arup, who had previously designed a number of laboratories that exhibited similar integration of structure and service characteristics (Figures 6.59 and 6.60).

Figure 6.60 Stansted 1991.
SOURCE: FOSTER + PARTNERS – PHOTOGRAPHER: KEN KIRKWOOD

Figure 6.61 London, Stansted, Stage 2.
SOURCE: NEW CIVIL ENGINEER

Much longer span steel roofs at Stuttgart and Hamburg Airports, architect Meinhard von Gerkan, were also being designed in parallel with Stansted. All have in their own way sought to embrace in their designs the lightness and elasticity the Futurists sought. The latest proposed expansion envisages a new separate new arrivals building which marks a departure from Foster's original concept (Figure 6.61).

26. Osaka Kansai (KIX) 1994

Osaka Kansai was the first of a new generation of very large international airports. While Stansted had been designed for 7.5 million passengers, Kansai was designed to handle 25–30 million passengers per annum. The master plan by ADPI and terminal design by Renzo Piano broke new ground by creating the airport on an artificial island in Osaka Bay. It was the first planned for the new generation of Super Jumbo aircraft with wingspans of up to 84 metres and has now been extended to provide a second runway (Figure 6.62). The learning from this project has been used in subsequent airports reclaimed from the sea, including Chubu Centrair in Japan, Hong Kong Chek Lap Kok and Seoul Incheon.

The shape of the roof of the terminal arises from work on the dynamic lines of the air flows circulating in the building and the curves of the "wings" of the building are the result of the application of a toroidal geometry allowing vision of the apron from the control tower.

The influence of Kansai on later architectural designs can be seen most clearly at Heathrow Terminal 5 which was designed by Richard Rogers, the former partner of Renzo Piano for the Pompidou Centre, particularly in the landside 'canyon', the airside gate lounges and in the long, clear span terminal roof over the main check-in concourse (Figure 6.63).

The policy decision to build Kansai as an artificial island was influenced by the protests that surrounded the earlier construction of Narita Airport to serve Tokyo which was constructed on valuable farmland. This change of policy acknowledged the extreme shortage of land in Japan as well as the determination of protestors.

Figure 6.62 Osaka Kansai aerial view after expansion showing second runway to provide resilience.
SOURCE: WIKIPEDIA

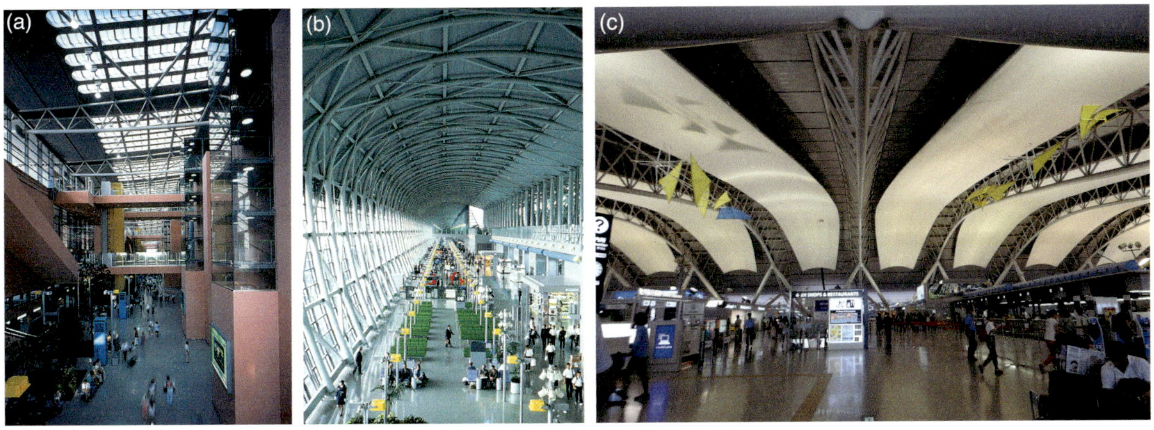

Figure 6.63 Osaka Kansai, landside canyon, (a) airside (b) gates and (c) check-in.
SOURCE: FONDAZIONE RENZO PIANO

The original master plan envisaged a single runway with a large linear terminal aligned parallel to the runway. The airport was the first to be designed for the new generation of very large aircraft and allowed for aircraft with a wingspan of up to 84 metres or more than sufficient for the Airbus A 380. Subsequently a second runway has been constructed along with a small terminal.

The site presented new engineering challenges to overcome the sinking of the infrastructure into the soft sea bed which were addressed by constructing the terminal on piles which could be extended. The island had been predicted to sink 5.7 metres (19 ft), by the most optimistic estimate, as the weight of the material used for construction compressed the seabed silts. However, by 1999, the island had sunk 8.2 metres (27 ft) – much more than predicted. The airport was also designed to withstand earthquakes and typhoons which it has already successfully resisted (Figure 6.64).

Figure 6.64 Osaka Kansai aerial view of original artificial island.
SOURCE: KANSAI VIA ADPI

Figure 6.65 Renzo Piano's initial concept diagrams 1988.
SOURCE: FONDAZIONE RENZO PIANO

The terminal is 1.7 kilometres in length, making it the longest linear terminal in the world. It has a sophisticated people mover system called the Wing Shuttle, which moves passengers from one end of the pier to the other. Renzo Piano describes his design as follows (Figures 6.65–6.72):

> *Kansai airport rests upon the island like a glider seen in plan – the main body of the airport forming its fuselage and the boarding gates positioned in its wings. A notable feature and one of primary importance in the organisation of the airport, is the unobstructed visibility of the planes themselves thanks to the uninterrupted lines of vision through the open departures level Main Terminal Building. The departure level is covered by a large, clear-span, undulating roof of asymmetrical form. It is perhaps this shape that is the project's main innovation.*
>
> *In the main terminal building the geometry of the roof's undulating cross-section is formed of a series of arcs of different radii connected at tangent points. Three-dimensional beams spanning 80m follow the cross-sectional asymmetrical form of the roof, supported at their extremities by pairs of inclined columns.*
>
> *The 42 boarding gates are housed within the "wings" of the glider. Their glazed facades address the runway, while their opaque, curved roof sweeps down to turn its back on the distant coastline. The height of the "wings" decreases to the buildings' extremities, with the roofs following an almost imperceptible curve, just sufficient to ensure the control tower's lateral line of vision*
>
> *The shape of the roof of the terminal arises from a long work on the dynamic lines of the air flows circulating in the building. It has been designed to convey*

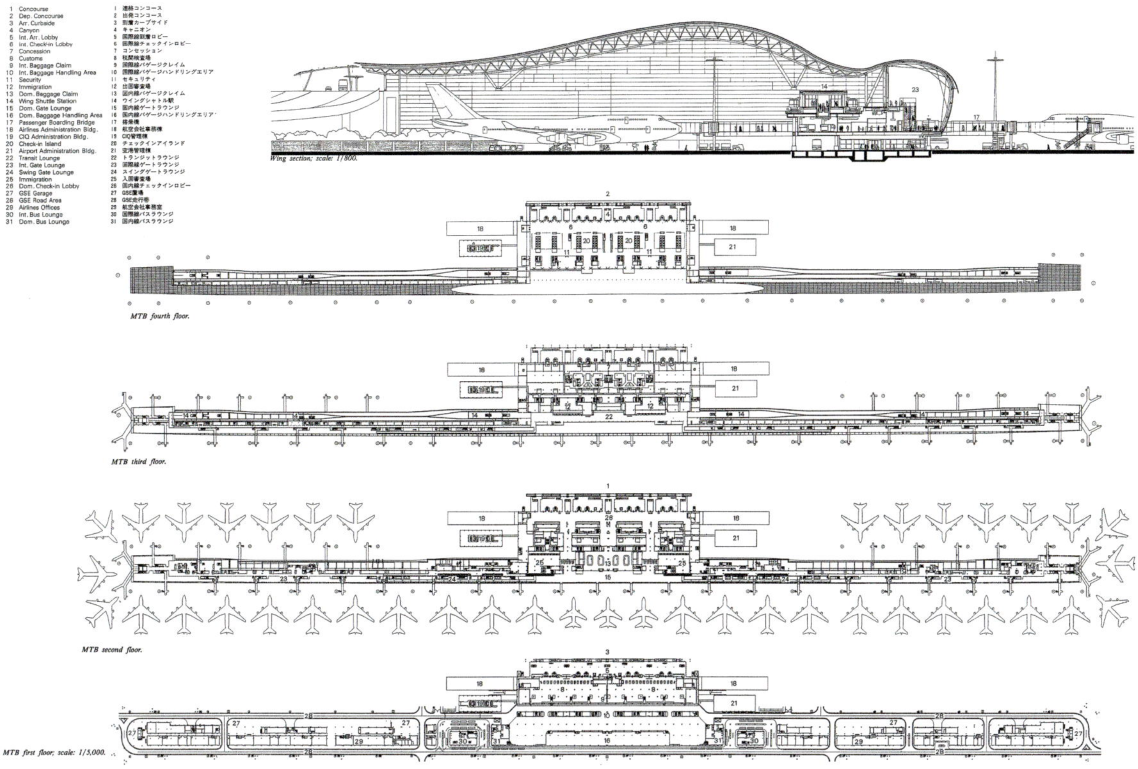

Figure 6.66 Osaka Kansai Sectional Elevation and Floor Plans.

SOURCE: FONDAZIONE RENZO PIANO

> *air from the passenger side to the track side without requiring the use of closed conduits. Blade-like baffles guide the airflow along the ceiling and reflect the light coming from above. This eliminates all the elements that would have prevented the structure from being seen. Even the curves of the "wings" of the building are the result of the application of a strict law: they have been defined on the basis of a toroidal geometry. The curvature is almost imperceptible, but it was necessary to favour the side view from the control tower.*
>
> (Renzo Piano, 2020)

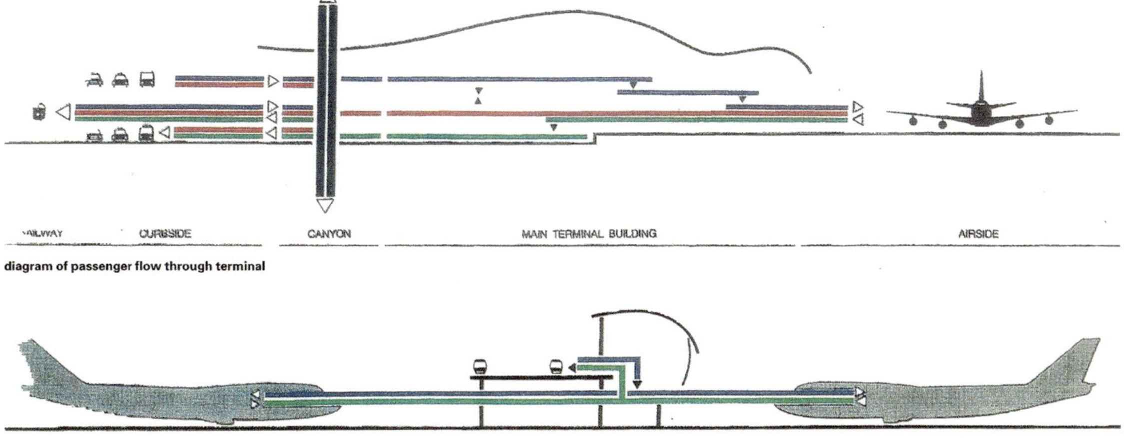

Figure 6.67 Osaka Kansai passenger flow. *Key*: Blue – International Departures, Green – International Arrivals, Red – Domestic Departures and Arrivals

SOURCE: FONDAZIONE RENZO PIANO VIA ARCHITECTURAL REVIEW

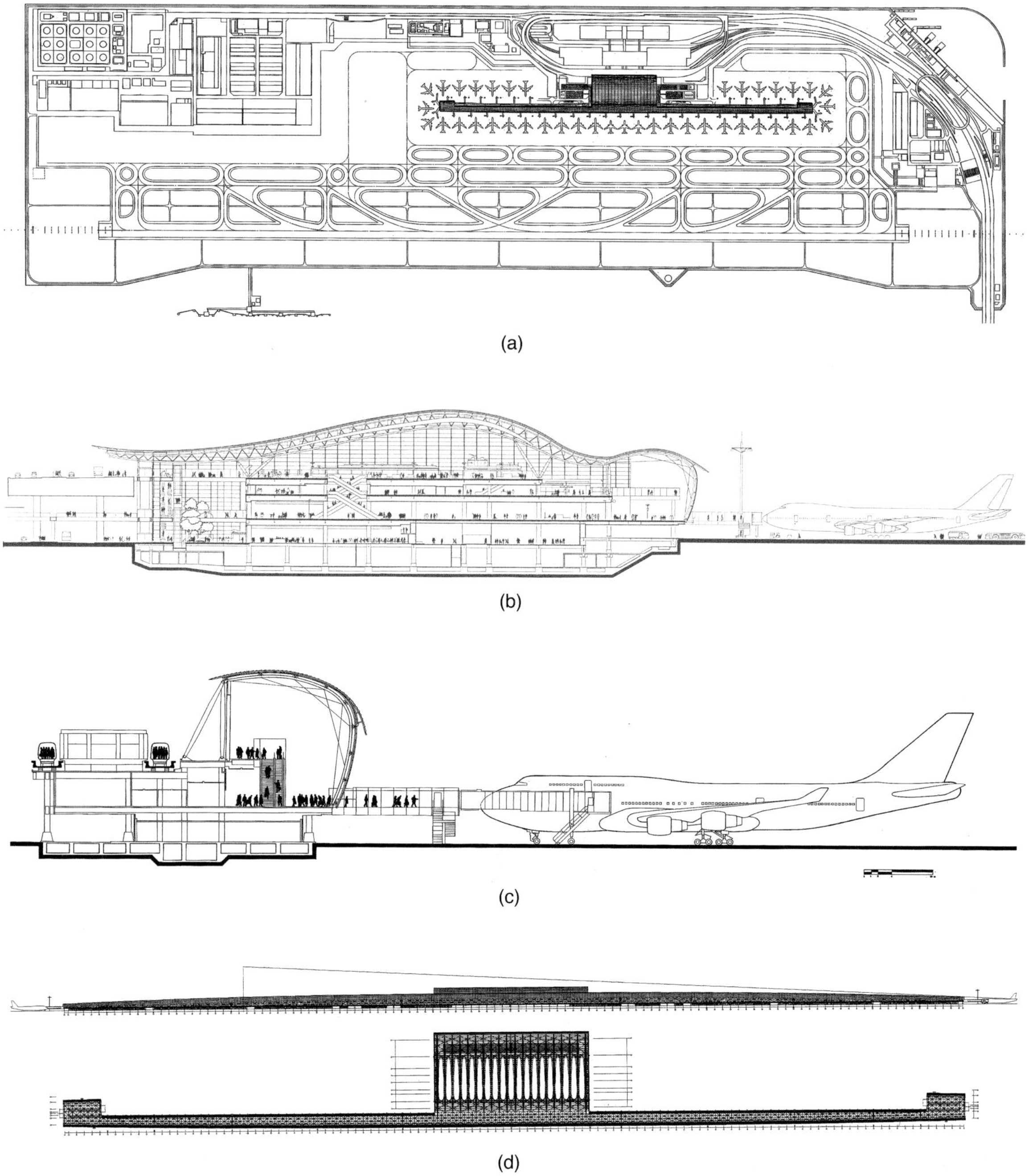

Figure 6.68 Osaka Kansai 1994, showing the linear configuration with access to the end of the pier provided by an automated people mover: Note the curvature of the roof towards the end of the pier to assist visibility of the apron the linear pier concept is most often found where the terminal is located parallel with the main runway, particularly where the airport only has a single runway, but what makes Kansai unique is the sheer scale which necessitates the provision of a transit. (a) Kansai master plan. (b) Terminal section. (c) Pier and transit section. (d) Roof plan and elevation.

SOURCE: FONDAZIONE RENZO PIANO

Figure 6.69 Osaka Kansai, 1994 aerial view of terminal.
SOURCE: FONDAZIONE RENZO PIANO

Figure 6.70 Osaka Kansai 1994.
SOURCE: KANSAI

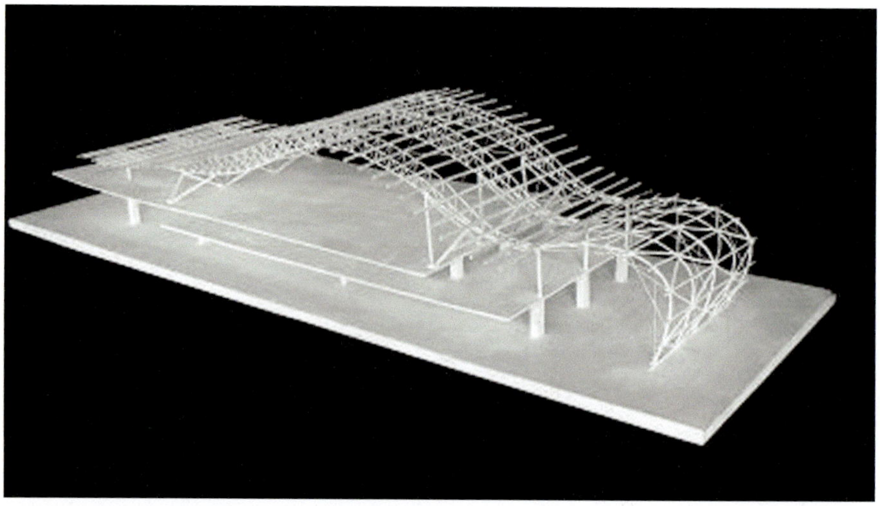

Figure 6.71 Osaka Kansai 1994 model of roof structure.
SOURCE: FONDAZIONE RENZO PIANO

Figure 6.72 Osaka Kansai 1994 showing curved and tapering pier and transit system.
SOURCE: KANSAI

Figure 6.73 Osaka Kansai 1994 to Kobe Sea Transfer.
SOURCE: WIKIMEDIA HTTP://EARTHOBSERVATORY.NASA.GOV/NEWSROOM/NEWIMAGES/IMAGES.PHP3?IMG_ID=16451

The Ferry Terminal includes an unusual air-to-sea-to-air connection with a high-speed ferry that links it to Kobe Airport and provides a more direct route to the city of Kobe than travelling by road. The route is shown in red on the map (Figure 6.73).

27. Denver International (DEN) 1995

Fentress Architects, embraced contextual design and sought to create an identity for the airport in response to the local landscape, climatic conditions and vernacular architecture, taking inspiration from the imagery of the Rocky Mountains that form the backdrop to the airport (Figure 6.74). The airport was planned as a major transfer hub based on learning from Atlanta (Figure 6.75).

The airport is located on the western edge of the Great Plains and within sight of the Front Range of the Rocky Mountains (Figures 6.76–6.78). It occupies a site area greater

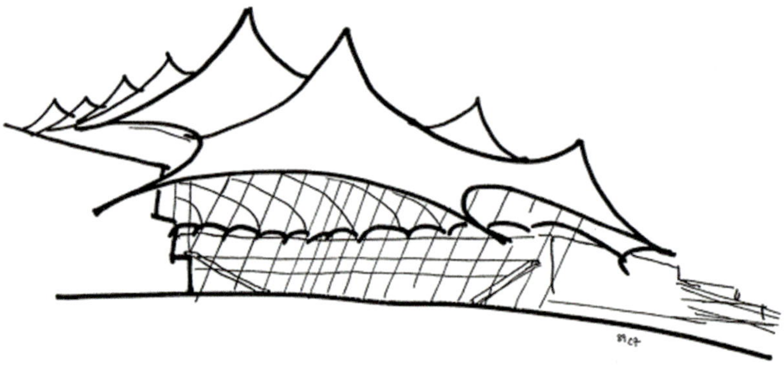

Figure 6.74 Denver, Architect's sketch of tent roof concept.
SOURCE: FENTRESS ARCHITECTS

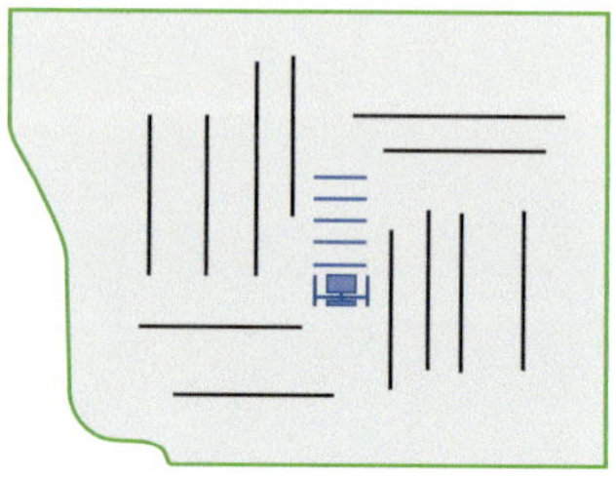

Denver International (DIA) 2030	
Site area	132.4 km^2
Pax	94 mppa
Area/Pax	1.4
ATMs	1,187,900

Figure 6.75 Denver master plan 2030.
SOURCE: ATKINS

Figure 6.76 Denver Roofscape with Rocky Mountains in the background 1995.
SOURCE: ARCHITONIC.COM

Figure 6.77 Night-time view of the Denver Jeppesen Terminal.
SOURCE: PARKDIA, MELISSA LUTZ

than that of Manhattan and is the largest airport in the USA and the third largest in the world. The terminal is organised in a 'toast rack' configuration like Atlanta (Figure 6.79). Unlike Atlanta, there is no back-up passenger tunnel to support the transit system which has resulted in operational disruption mitigated to a limited extent by the pedestrian bridge link from the terminal to the first satellite (Figure 6.80). The original pioneering automated DCV baggage system was a notorious failure and a conventional conveyor system had to be retrofitted before the delayed opening. The airport is now a hub for both United Airlines and Frontier Airlines and a base for Southwest Airlines.

Figure 6.78 Denver Roofscape of Jeppesen Terminal that was clearly influenced by the SOM Hajj Terminal in Saudi Arabia.
SOURCE: PARKDIA, MELISSA LUTZ

Figure 6.79 Denver airside underground transit linking satellites.
SOURCE: DENVERPOST.COM

Figure 6.80 Denver central landside atrium in the terminal.
SOURCE: INTERNATIONAL AIRPORT REVIEW

Figure 6.81 Boeing 737 taxiing under Denver pedestrian bridge between terminal and satellite 1.
SOURCE: FLICKR

Denver was the first airport to introduce an elevated pedestrian bridge over a taxiway. It was designed to allow only the small Boeing 737 aircraft to pass underneath. The damage caused by a larger B757 aircraft which tried to pass underneath is clearly visible. This approach influenced later bridges at airports such as London Gatwick, Hong Kong and Brussels, though these bridges were much higher and designed to allow Jumbo aircraft to pass underneath (Figure 6.81).

28. Hong Kong, Chek Lap Kok (HKG) 1998

Hong Kong International Airport was built on a large artificial island formed by flattening and levelling the Chek Lap Kok and Lam Chau islands and reclaiming the adjacent seabed, effectively demolishing a mountain and pushing it into the sea to create the airport (Figures 6.82 and 6.83).

The Hong Kong International design was the result of collaboration between Foster+ Partners, architects, Mott MacDonald, engineers and BAA airport planning. The plans were based closely on the initial master plan by the Geiner-Maunsell consortium which envisaged a large terminal linked to a Y-shaped pier and X shaped satellite with an underground people mover. Foster and team, however, proposed a number of improvements, including the integration of the railway station with the terminal by ramps and the adoption of a single continuous vaulted roof, which integrated the terminal into a single form.

Hong Kong is the first airport planned from the outset as an 'Aerotropolis', with an integrated commercial, residential and leisure district. As Norman Foster described in his essay 'Reinventing the Airport' the design for Hong Kong builds on his experience of designing Stansted. Key differences in approach include the fact that the terminal is planned on two levels with Departures above Arrivals instead of the single-level solution at Stansted and the structure is largely concrete to suit local availability of materials, with only the roof in steel. Unlike Stansted, the building services have been disassociated from the structure.

Most recently, the new master plan for 2030 shows that the airport is capable of further expansion to include a third runway and second terminal. A major innovation subsequent to the opening is the construction of the SkyPier to cater for cross-border ferries from China creating a unique interchange between land, sea and air travel (Figure 6.84).

Figure 6.82 Hong Kong International, Master Plan 2030 showing the ultimate development of the three-runway airport with a satellite to Terminal 2 on the left and the existing Foster designed terminal to the right.
SOURCE: HKIA

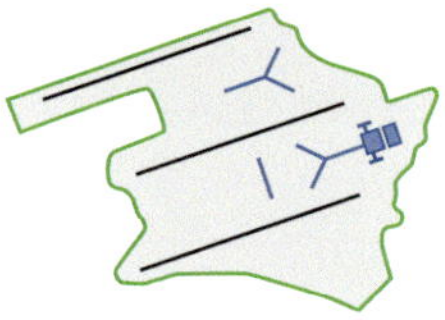

Hong Kong (HKG) 3R 2035	
Site area	20.6 km^2
Pax	120 mppa
Area/Pax	0.17
ATMs	620,000

Figure 6.83 Hong Kong 3R master plan 2030.
SOURCE: ATKINS

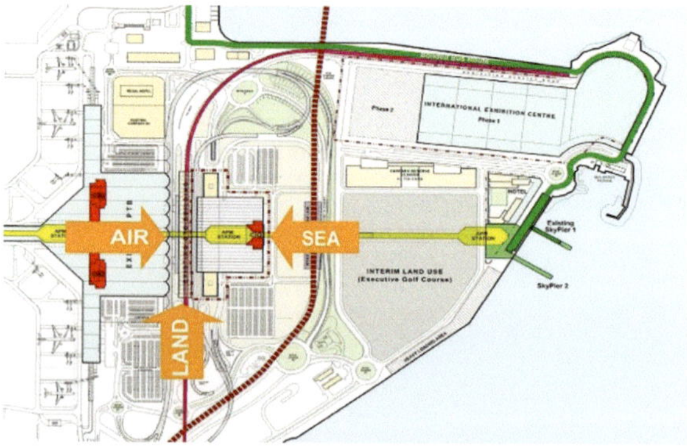

Figure 6.84 Hong Kong, SkyPier and Sky City land, sea–air interchange.
SOURCE: SOM.COM

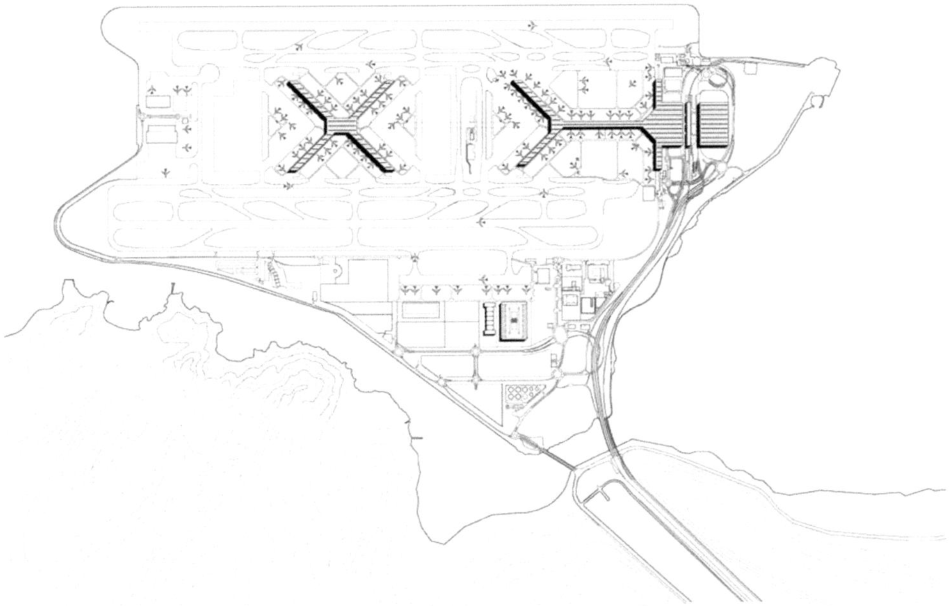

Figure 6.85 Hong Kong International, winning competition entry, site plan.
SOURCE: FOSTER + PARTNERS

Hong Kong International at Chek Lap Kok was designed as a replacement for the former Hong Kong International Airport (commonly known as Kai Tak Airport) built in 1925. The original master plan envisaged two widely spaced parallel runways, with a large Y-shaped mid-field terminal and an X-shaped cruciform satellite (Figure 6.85). Foster & Partners won the competition to design the terminal with the following design (Figures 6.86–6.90). The design makes provision for airline lounges at a mezzanine level about the departures lounge (Figures 6.91 and 6.92).

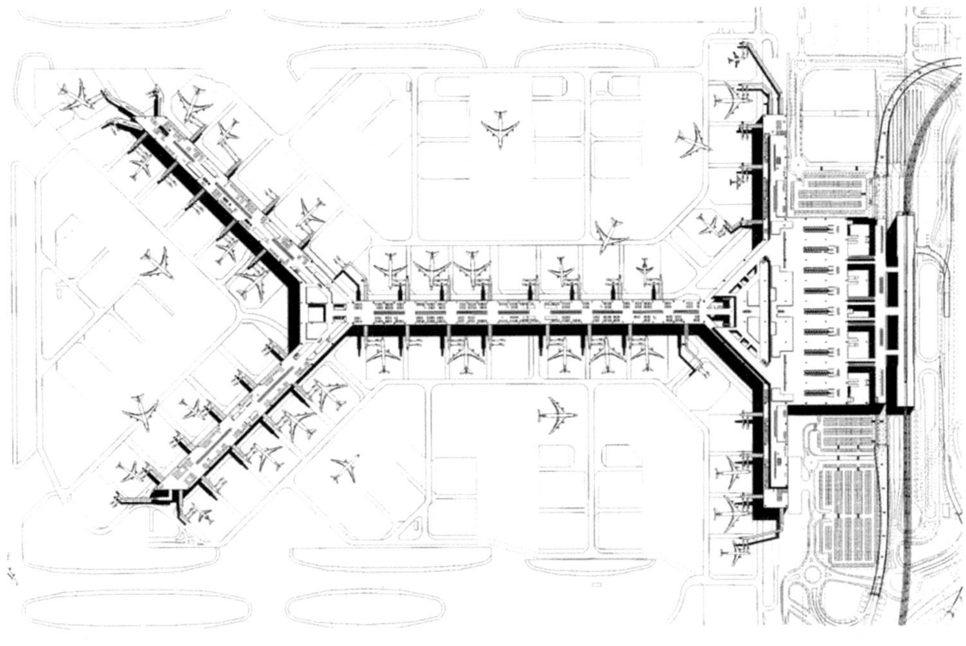

Figure 6.86 Hong Kong International, winning competition entry, departures level.
SOURCE: FOSTER + PARTNERS

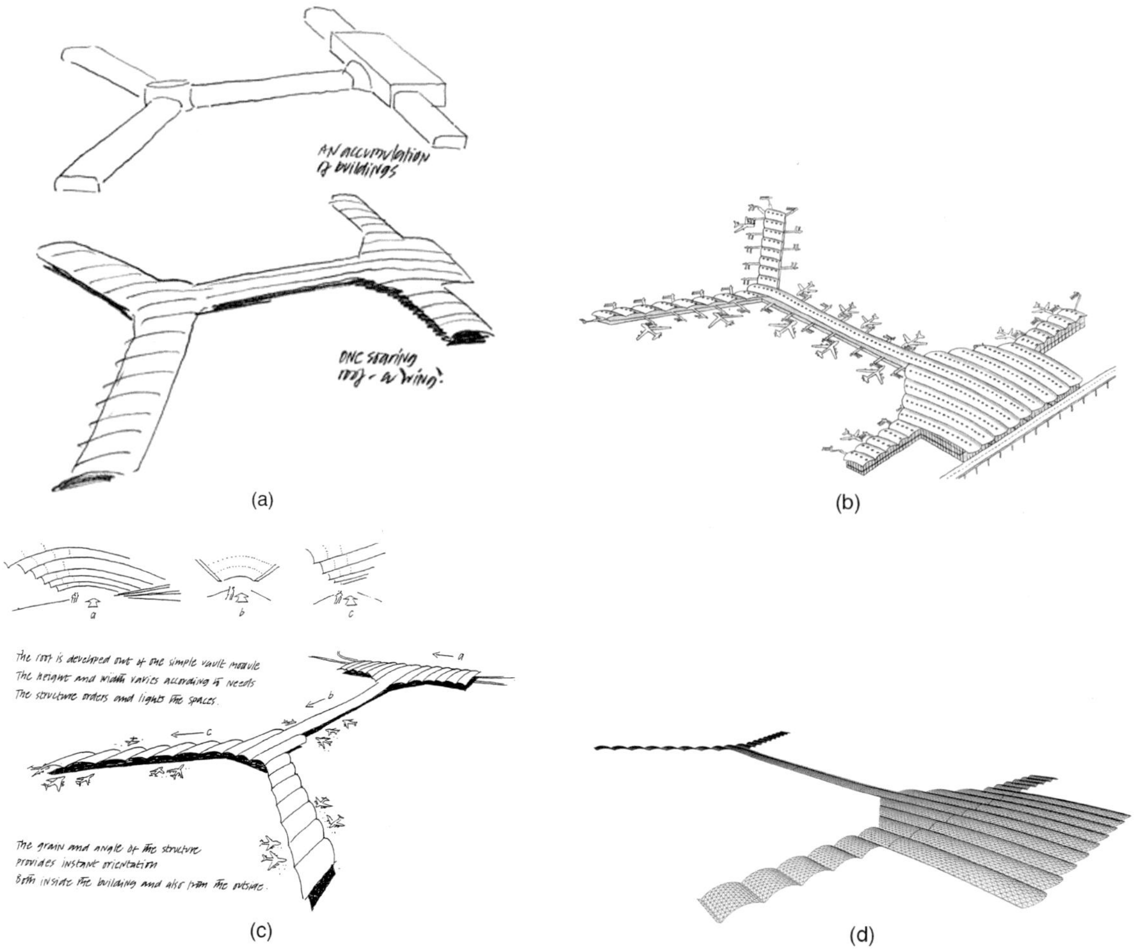

Figure 6.87 Hong Kong International, evolution of roof design by Norman Foster.
SOURCE: FOSTER + PARTNERS

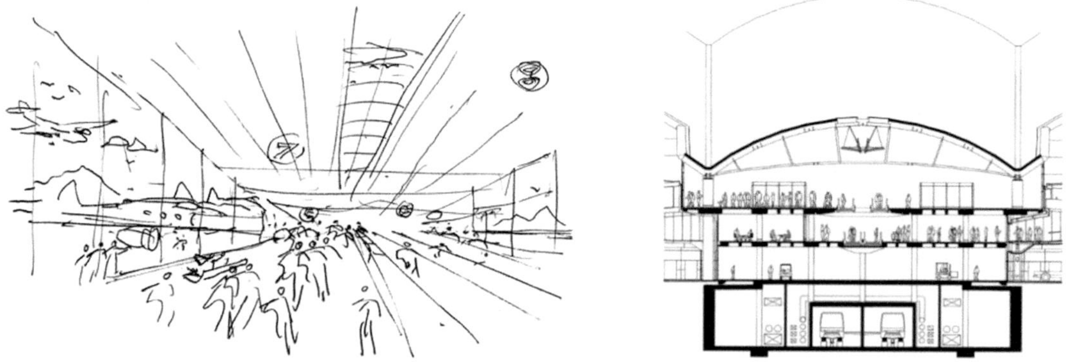

Figure 6.88 Hong Kong International, gate-lounge concept by Norman Foster and section through pier.
SOURCE: FOSTER + PARTNERS

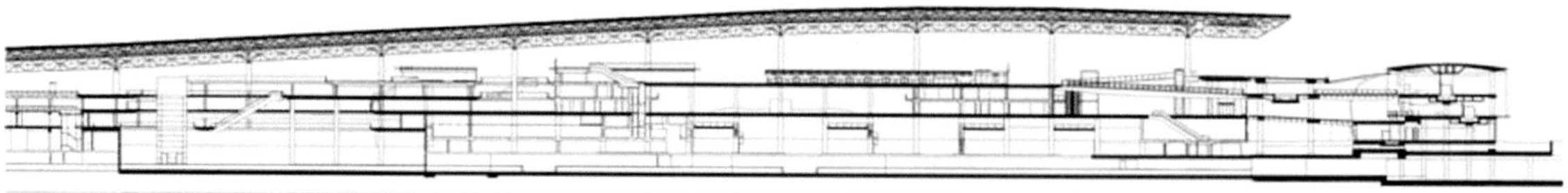

Figure 6.89 Hong Kong International, Long Section through terminal and interchange.
SOURCE: FOSTER + PARTNERS

Figure 6.90 Hong Kong International 1998, Departures and Arrivals Concourses.
SOURCE: FOSTER + PARTNERS — PHOTOGRAPHER: MICHEL PORRO

Arguably, Hong Kong International Airport is the most developed transfer Mega Hub is as it has the most ways of integrating journeys of any airport, including car, taxi, bus, rail and air and unusually major sea links to mainline China. It also serves as an effective hub for Cathay Pacific and the One World Alliance, although the latter did not come into existence until a year after the airport first opened. Hong Kong Airport includes a unique air–sea transfer facility known as SkyPier (2003) which supports a very successful sea route to nine ports in the Pearl River Delta (PRD) of mainland China and Macau. Passengers arriving by ferry from China are routed through a dedicated transfer processing facility that includes check-in and security checks to allow passengers to access the terminal airside via a dedicated transit system. The SkyPier has eight berths and is designed to handle up to 8 million passengers a year. It currently has over 110 scheduled sailings a day (Figure 6.93).

Hong Kong Airport Express is the fastest way to travel from Hong Kong Airport to downtown; it only takes about 24 minutes from Hong Kong airport to Hong Kong Central. If you take Airport Express to Hong Kong airport, you could also enjoy the free In-Town Check-In Service at Hong Kong or Kowloon stations as late as 90 minutes before the

Figure 6.91 The Deck, Cathay Pacific Lounge 2018 at Hong Kong International.
SOURCE: CATHAY PACIFIC

Figure 6.92 Enclosed interior, Cathay Pacific Lounge at Hong Kong International.
SOURCE: CATHAY PACIFIC

Figure 6.93 Hong Kong International with the SkyPlaza and SkyPier in the foreground and Aerotropolis to the top left.
SOURCE: HKIA VIA YOUTUBE

departing flight or as early as the previous day (Figure 6.94). Later phases of development include a short haul satellite, the SkyPier sea terminal serving routes from the Pearl River Delta (PRD) in mainland China, expanded Commercial Districts such as the Sky City and the Tung Chung Aerotropolis (Figures 6.95–6.100).

For the next stage of development, Hong Kong International Airport has constructed a third runway on reclaimed land and is re-build terminal 2 to serve a new large remote satellite (Figure 6.101). The airport is the focus of major investment in new transport infrastructure, with a bridge link under construction that will link Hong Kong with Macau. What is just as important is that all this development just occupies a thin northern strip of Lantau Island, preserving over 90% of the island as a tropical rain forest. Although it was designed before John Kasarda popularised the term, Hong Kong Chek Lap Kok exhibits

Figure 6.94 Hong Kong International Airport 1998 – Air, Road, Rail and Sea Interchange.
SOURCE: HKIA

Figure 6.95 Hong Kong International Airport with later phase short haul satellite and bridge link over the taxiway in the foreground.
SOURCE: HKIA

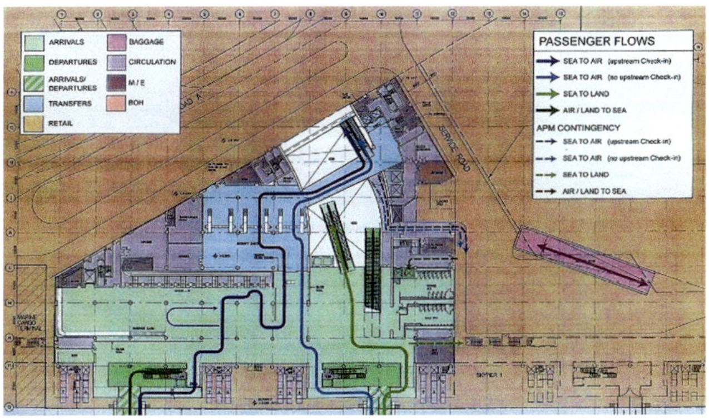

Figure 6.96 HKIA SkyPier 2003 Passenger Flow Layout.
SOURCE: AIRPORT AUTHORITY HONG KONG

Figure 6.97 Ferry routes from the PRD.
SOURCE: HKIA MASTERPLAN 2030

Figure 6.98 Hong Kong Chek Lap Kok and Tung Chung and Lantau Island.
SOURCE: WIKIPEDIA

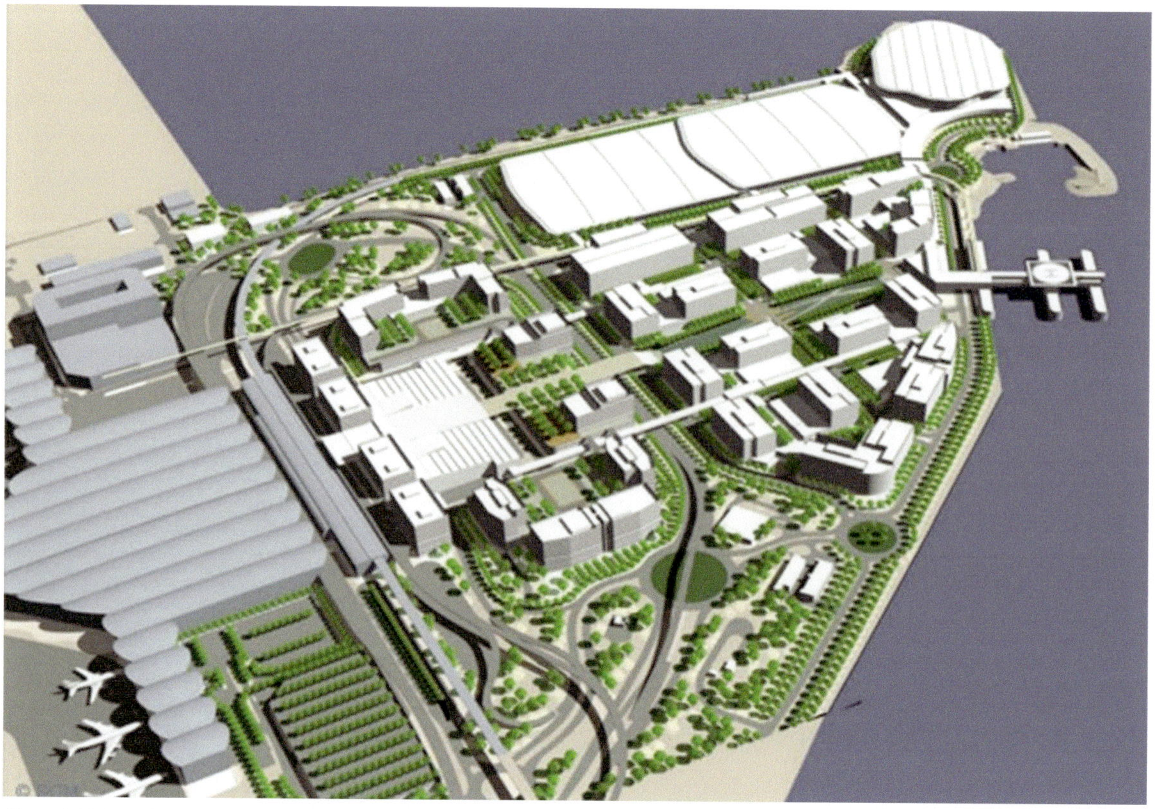

Figure 6.99 Hong Kong International Airport aerial view of SkyCity and SkyPier Ferry Terminal.
SOURCE: HKIA VIA 720 COLLABORATIVE STRATEGIC MASTER PLANNING CONSULTANCY

Figure 6.100 Hong Kong Tung Chung Aerotropolis.
SOURCE: WIKIPEDIA

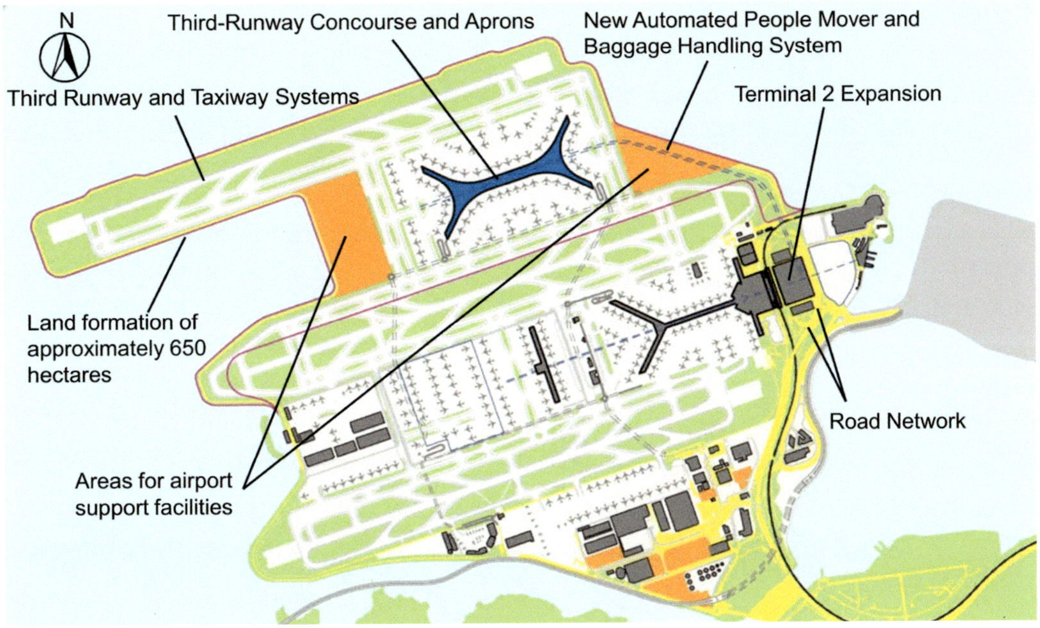

Figure 6.101 Hong Kong International Airport, Three Runway Master Plan 2030.
SOURCE: HKIA INFORMATION PACK 20.06.2014

many of the characteristics of an Aerotropolis, but with an added element of rationality, practicality and contextual and environmental sensitivity. The airport campus itself contains a 30-hectare South Commercial District which focuses on logistics, including Tradeport Hong Kong Limited, HACTL, the world's largest stand-alone air freight and express cargo facility, Asia Air Freight Terminal, Mixed Use freight forwarding centre and DHL air express hub, a 10-hectare East Commercial District with an office park development of 300,000 metres^2, including Cathay Pacific's head office and training facility, 57-hectare North Commercial District known as SkyCity adjacent to the airport terminal and served directly by an extension to the airport express train containing:

- SkyPlaza, a multi-purpose commercial complex
- IMAX 3D theatre
- International exhibition and trade centre
- Hotels and entertainment facilities
- Sky Pier Ferry terminal providing a cross border interchange between the airport and the Chinese mainland.

Hong Kong already has the expansion of the airport under way, with the addition of a third runway being reclaimed from the sea. Many of these features can be found in major international airport cities, but what makes Hong Kong different are the adjacent facilities on Lantau Island that have been developed as a part of an integrated regional master plan to create an Aerotropolis:

- The new town of Tung Chung currently houses 80,000 people but is planned to expand to 250,000 people, many of whom work at the airport. It has been carefully sited to the side of the airport so that the noise from take-off and landing aircraft will not intrude.
- A separate cross-border high-speed ferry terminal links Hong Kong with the Chinese Cities of the Pearl River Delta adjacent to but by-passing the airport.
- A Disney Theme Park.

29. Kuala Lumpur (KUL) 1999

Phase 1 of Kuala Lumpur International Airport, by architect Kisho Kurokawa, featured in the original master plan a linear terminal for domestic and short-haul traffic linked to two cruciform satellites by a transit system, although to date only one satellite has been constructed (Figure 6.102). The master plan was updated in 2017 to incorporate a second terminal and third runway (Figure 6.103). The cruciform configuration offers improved passenger connectivity and shorter walking distances of just 250 metres, combined with the opportunity for a better consolidated retail offer, but is less efficient in taxiway layout and land use. The transit is unusual in that it rises out of the ground at the satellite end

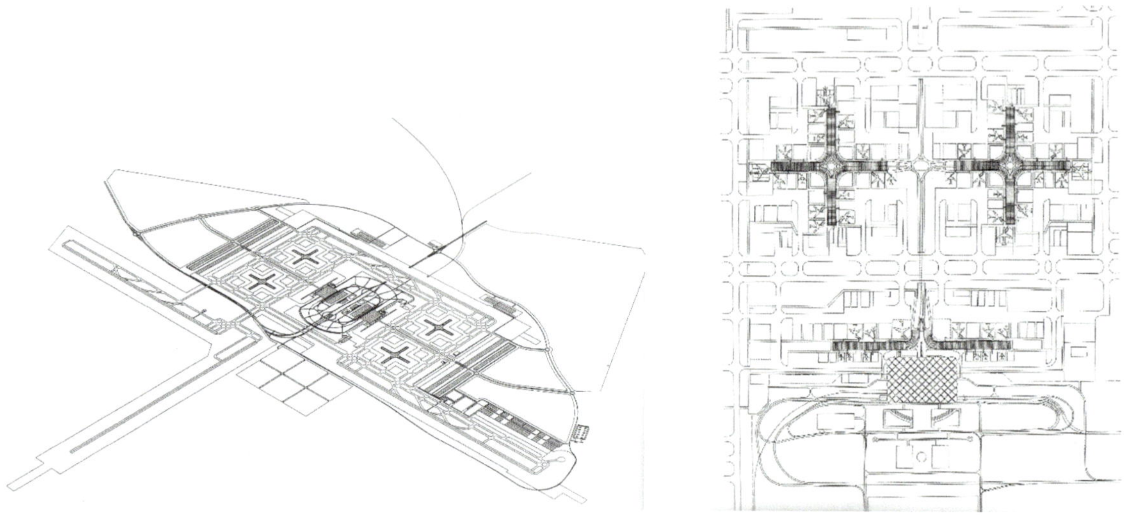

Figure 6.102 Original KLIA Master Plan.
SOURCE: KLIA VIA SKYSCRAPERCITY.COM

Figure 6.103 Kuala Lumpur International Master Plan 2017.
SOURCE: MALAYSIA AIRPORTS @ FIATA CONFERENCE

Figure 6.104 Kuala Lumpur Satellite 1999.
SOURCE: KLIA VIA PARADISEINTHEWORLD.COM

to reach a passenger level at first floor level. At the centre of the cruciform satellite is a tropical garden that brings the jungle into the centre of the airport (Figures 6.104–6.107).

Terminal 1 is designed for full-service airlines while Terminal 2, part of a more recent master plan and expansion programme, has been constructed within a lower budget and without a transit system to suit the low-cost carrier operation (Figures 6.108 and 6.109). The differences between the concepts of Terminal 1, with its cruciform satellite and Terminal 2, with its attached piers, show that there can be more than one appropriate solution at an individual airport to suit different markets.

Figure 6.105 Kuala Lumpur Satellite Transit Station.
SOURCE: WIKIPEDIA

Figure 6.106 Kuala Lumpur Satellite Jungle Boardwalk.
SOURCE: KLIA

Figure 6.107 KLIA Jungle Boardwalk.
SOURCE: KLIA VIA KUABY.COM

To address the expanding market for LCCs, KLIA, like Singapore Changi, initially built a simple terminal for the market, but later replaced it with a much more ambitious terminal. Unusually, for a new terminal design aimed at LCCs, it incorporated both airbridges and a transfer baggage facility.

Figure 6.108 Model of KLIA T2.
SOURCE: KLIA VIA SAYS.COM

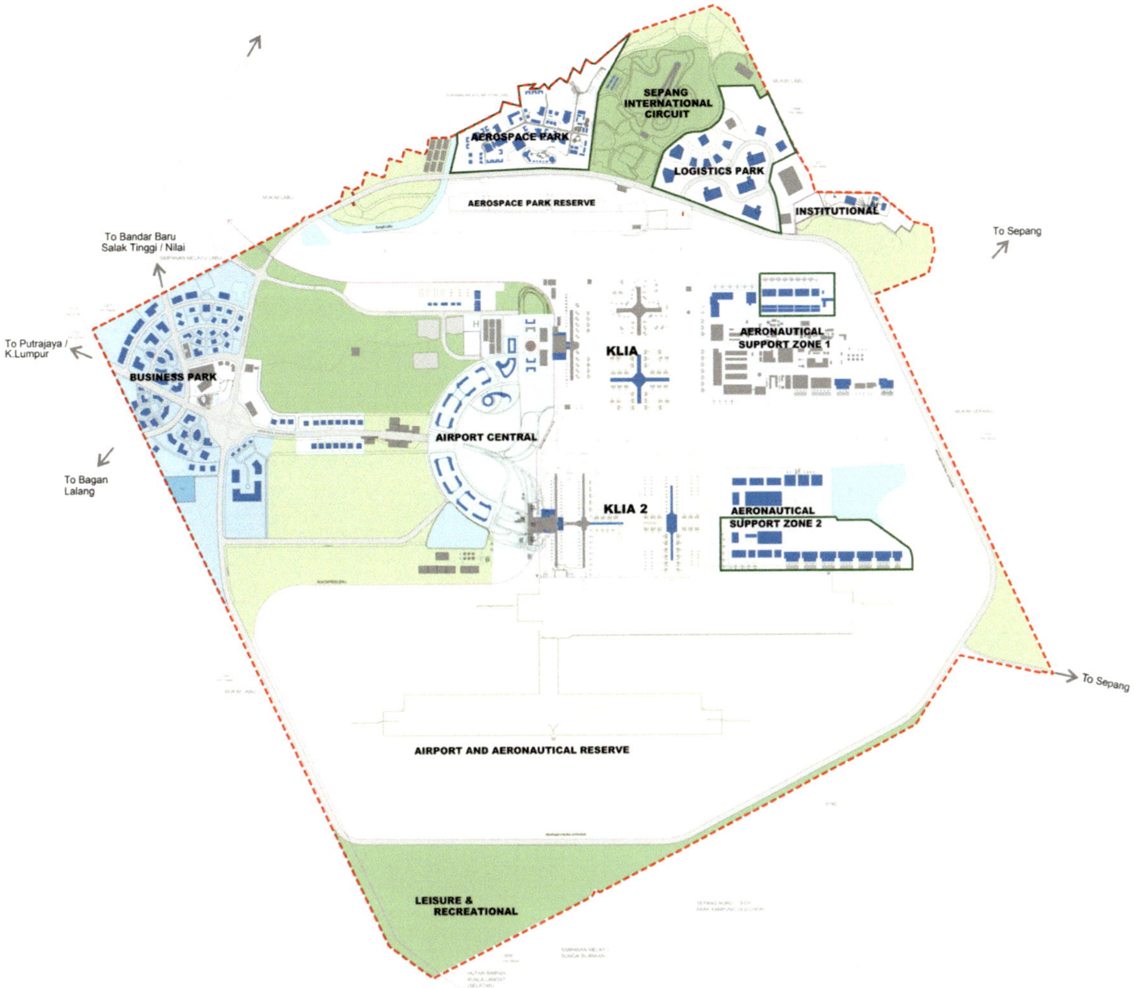

Figure 6.109 Kuala Lumpur International 'Aeropolis' 2050 Master Plan.
SOURCE: KLIA.COM.MY

CHAPTER 7

MEGA HUB 2000–2019

The current generation of designs is responding to several new influences:

- the rise of Global Alliances requiring larger and more complex interchanges
- the development of LCCs needing to cut costs
- the introduction of new aircraft types
- the expectation that an airport-focused Aerotropolis can induce growth in the surrounding area
- ever more stringent security requirements with greater space requirements for both passengers and baggage checks
- more rigorous demands for environmental mitigation
- higher expectations of commercial income

These influences are discussed in turn in the subsequent sections:

Global Alliances are an aviation industry arrangement where two or more airlines agree to collaborate in providing marketing branding and in helping travellers make inter-airline 'codeshare' connections with other airlines. They were formed either side of the Millennium; The Star Alliance was founded in 1997, One World in 1999 and Sky Team in 2000. Their foundation helped give impetus to a new generation of international hub airports specifically designed to make transfers from one flight to another as easy and as reliable as possible. To achieve this new infrastructure was required and new operating practices were needed to facilitate the transfer of both passengers and baggage. At the same time, security requirements became more onerous, requiring security screening of both transfer passengers and baggage, so the size and complexity of major terminals increased.

A new generation of international transfer hubs was designed for Asia and the Middle East including Hong Kong, Kuala Lumpur, Beijing Capital, Seoul Incheon, Dubai International and Abu Dhabi among others to achieve competitive minimum connection times (MCTs) for transfer passengers. In the meantime, major projects were initiated at established airports such as Singapore Changi, London Heathrow, Amsterdam Schiphol and Frankfurt to maintain their competitiveness in the transfer market in response to this new airline business model.

While the rise of the Global alliances led to ever more complex and sophisticated terminals, the growth of low-cost carriers (LCCs) pulled in the opposite direction towards simplicity. They focused on point-to-point operations and avoided the cost and complexity of transfer traffic. Simplification also led to the use of standardised aircraft types B737 & A320, to keep operating and maintenance costs as low as possible. This in turn has made it possible to standardise the layout of aircraft parking on the apron as can be seen

DOI: 10.4324/9781003323624-7

at Brussels Charleroi one of the few airports to be designed specifically for Ryanair. The subsidy that was required to build that new airport has now been outlawed by the EU so further new airports designed specifically for LCCs are unlikely in Europe. Some LCCs like easyJet are increasingly choosing to fly from mainstream airports even though they object to the major subsidy given to transfer traffic, often around 50%, which in their view unfairly favours legacy airlines.

LCCs, such as AirAsia X, Tiger Air and Jetstar are increasingly moving into the long-haul market where transfer traffic becomes more important. In response Gatwick, for example, is providing self-transfer facilities to allow passengers to check-in transfer bags in the reclaim hall to avoid passengers having to drag their bags through the normal check-in process in the landside concourse. Kuala Lumpur has taken this a step further by designing Terminal 2 with full transfer facilities, even though it is specifically targeted at the AirAsia LCC, while Terminal 1 continues to serve Malaysia Airlines and the One World Alliance.

The introduction of the Super Jumbo, as the Airbus A 380 is commonly described, required new airfield and terminal infrastructure. The aircraft was first introduced into regular service in 2007, but had been planned for long before. Initially, airports had to speculate on how large the aircraft might be, with airports such as Kansai, Hong Kong and Heathrow and Paris Charles de Gaulle designing for wing spans in the order of 83 to 85 metres, but eventually a new Code F standard was devised assuming the aircraft would fit within and 80 metre x 80 metre box was agreed on. This in turn generated changes to the airfield, with increased runway width, wider spaced taxiways, larger stands and new apron equipment and airbridge types to serve the upper deck of the aircraft. This investment may yet prove to have been unwise, as with less than 300 ordered, the programme was terminated in May 2021. The future of long-haul aircraft seems to be with very large twin jets that are more economical. The Boeing 777–800 and 900 series, now being tested by Boeing as a successor to the B747, have folding wingtips that allow the aircraft to manoeuvre on the ground within the space allowed for the previous largest Code E aircraft the Boeing 747-400 which had a wingspan of 65 metres.

There is also a view that other large twin jets such as the Boeing 787 'Dreamliner' and the Airbus A 350 will lead to more point-to-point flights bypassing the new Super Hubs. However, with more than 80% of all orders for these new aircraft coming from established legacy carriers, this has to date led to an increased range of long-haul destinations being served from the existing major hub airports driving their expansion even further.

There has been evidence for some time that an airport development can induce growth in the surrounding hinterland. Hong Kong International was one of the first to plan for that in a holistic way, incorporating new communities as well as businesses into their master plan. Since then, many new airports have sought to follow their example, particularly in Asia and the Middle East, including Kuala Lumpur, Seoul Incheon and, the most ambitious of all, Dubai World Central.

The current era has been described as the Digital Age, a period when the internet has opened up new opportunities. Alaska Airlines was the first to offer online check-in. The system was initially offered on a limited basis starting in the second quarter of 1999 and was available to the general public on selected flights the following quarter. Since then, a growing number of airlines have introduced the system. British Airways researched the Alaska Airlines approach for their operation at Heathrow Terminal 5. This led directly to their new design for check-in, where three check-in options are available depending on how much of the process has been completed at home online. British Airways followed this up shortly after opening with the introduction of self-service baggage drop.

Proposals for the use of autonomous vehicles are being developed, and trials at a number of airports are under way. Only Dubai World Central is planning for separate segregated infrastructure to support the operation of autonomous vehicles; others are relying on integrating autonomous and conventional vehicles into the existing airside road network.

The environment has emerged as a key business management topic for airports around the world. In response to societal expectation, environmental factors are increasing in importance in order to meet society's expectations for sustainable development.

Environmental and sustainability issues have, to date, not always been regarded as constraints to growth. However, in some cases, the scale of the impact that would accompany proposed airport developments has resulted in the plans being changed or – as in the case of the Thames Estuary airport proposal in London – contributed to the proposal being dropped altogether. Airports are only just beginning to recognise that society increasingly expects them to limit their contribution to climate change through global warming in both their construction and operating procedures.

There has been increasing objection to airport development in some parts of the world, often resulting in protests ranging from peaceful demonstrations to direct action. However, noise and air quality and traffic impact have remained the most serious concerns of local communities. This has led directly to more remote locations for new airports being chosen often on land reclaimed from the sea. All new airports in Japan have been built on artificial islands since Kansai in 1994.

Commercial revenue has continued to play an ever more important part of an airport's finances, so an ever greater proportion of space is being devoted to commercial operations. This has changed the look and feel of airside lounges, forcing them to be inward rather than outward-looking. Airports have yet to come to terms with internet shopping which with the opportunities offered by home delivery could turn retail outlets at airports into showrooms. Food and Beverage offers are becoming more ambitious, sometimes reflecting local or ethical considerations. This is in part because LCCs no longer serve meals as part of their offer and at least on short-haul routes legacy carriers are beginning to follow their example. Also, as security requirements become more onerous, airlines are requiring passengers to check-in earlier, potentially leaving passengers with more leisure time in the departure lounge.

As some terminals have become so large that they cannot be understood as a whole, there is an increasing emphasis on the bird's-eye view of the roof, a concern that originated with Walter Gropius and Le Corbusier in the 1930s. It is a view that can never be seen in practice except by a photographer hovering in a helicopter. This is complemented by a focus on the interior view, which, however, can only give a partial understanding of the terminal and becomes an experience similar to walking around a city centre, simple if the route is well known but confusing to the first-time traveller. No longer is the main entrance to the terminal the key view from which the building can be understood. While this trend has been developing for some time, it can be seen most clearly in some of the most recent designs.

The airports selected to illustrate the key characteristics of the era are:

- Seoul Incheon (ICN) 2001
- Bangkok Suvarnabhumi (BKK) 2006
- Madrid Barajas (MAD) T4 2006
- London Heathrow (LHR) T5 2008 & T2 2014
- Dubai International (DXB) 2010
- Beijing Capital (PEK) 2008
- Istanbul (IST) 2019.
- Beijing Daxing (PKX) 2019

30. Seoul Incheon (ICN) 2001

Seoul Incheon was built on land reclaimed from the sea between Yeongjong Island and Youngyu Island. The airport supports one of the most ambitious Aerotropolis concepts to date. Unusually, only international air traffic is handled through Incheon, while complementary domestic services continue to be served through the existing Gimpo airport. The rail link, accessed via a dramatic interchange, is unique in handling transfer passengers to Gimpo, while also being designed to continue to the capital Seoul.

Seoul Incheon airport has, for the last ten years, topped worldwide passenger satisfaction surveys, because provision has always kept ahead of demand and because of its wide range of leisure facilities and excellent connectivity. The airport demands careful attention as a benchmark because it achieves such high satisfaction levels. The airport supports one of the most ambitious Aerotropolis concepts to date, although the development has met a lot of criticism for being a soulless place to live, a typical reaction to most new towns and cities. However, it has helped the hinterland to build on airport-driven growth.

Seoul Incheon was designed when the existing international Gimpo airport was reaching capacity. The original master plan envisaged a mid-field terminal, located between parallel runways, with a terminal with two piers and a remote satellite and was designed to accommodate A380 aircraft from the outset (Figure 7.1). The second terminal is a large mid-field design with finger piers, phase 1 of which has been completed as stage 3 of the master plan (Figure 7.2). The engineering design of the reclamation built on the experience gained at Kansai and Hong Kong and involved building two sea walls to link the two islands and draining the land behind to form the platform of the airport which, like Schiphol is located below sea level.

The airport was originally planned to be built in three phases, incrementally increasing airport capacity as demand grew. This was changed, however, to four phases after the airport was opened. Stage 1 included the main terminal building and two runways while stage 2 extended the airport to include the remote satellite accessed by an underground transit system and added a third runway. Stage 3 includes a fourth runway and the first stage of the second terminal which will be completed in stage 4 along with a fifth, widely spaced runway.

Figure 7.1 Seoul Incheon, aerial view of T1.
SOURCE: IIAC VIA FROMABOVE.ALTERVISTA.COM

Figure 7.2 Seoul Incheon, aerial view of T2 phase 1 with T1 in the background.
SOURCE: IIAC

Seoul Incheon International Airport includes a dramatic rail transfer interchange designed by Terry Farrell and supports an Aerotropolis (Figure 7.3). The airport is unique in having a direct high-speed rail link to Seoul Gimpo Domestic Airport to facilitate transfer traffic before continuing to the City Centre (Figures 7.4 and 7.5). The airport unusually works in concert with the existing Gimpo airport which is now used for domestic flights only. It offers a fully integrated rail transfer facility between the two airports with through booking of flights permitted, including the rail transfer. Uniquely, Seoul quotes inter-airport minimum connection times for transfer passengers. The terminal plans and exterior and interior imagery exhibit strong spatial concepts, very generous space standards and high quality of fit out (Figures 7.6–7.8). These have helped Seoul Incheon receive high passenger satisfaction ratings regularly.

Figure 7.3 Seoul Incheon Interchange.
SOURCE: KEN ECKERT/WIKICOMMONS

Figure 7.4 Seoul Incheon to Seoul Gimpo AREX rail link.
SOURCE: THE GLOBAL AIRRAIL ALLIANCE

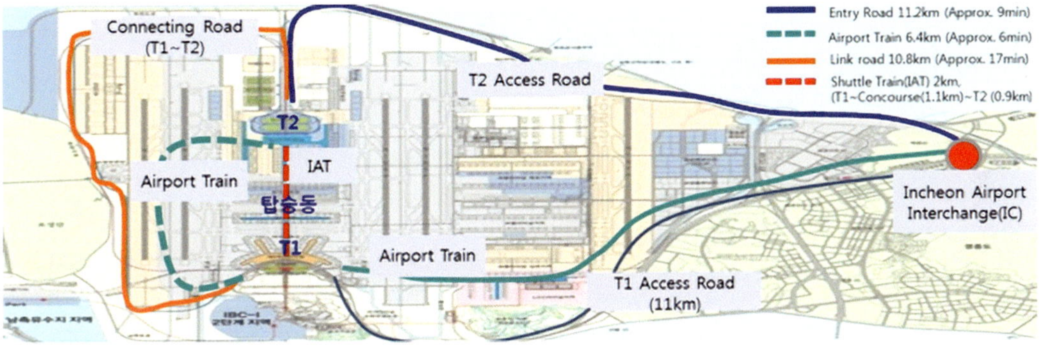

Figure 7.5 Seoul Incheon rail network.
SOURCE: IIAC VIA THE NATION

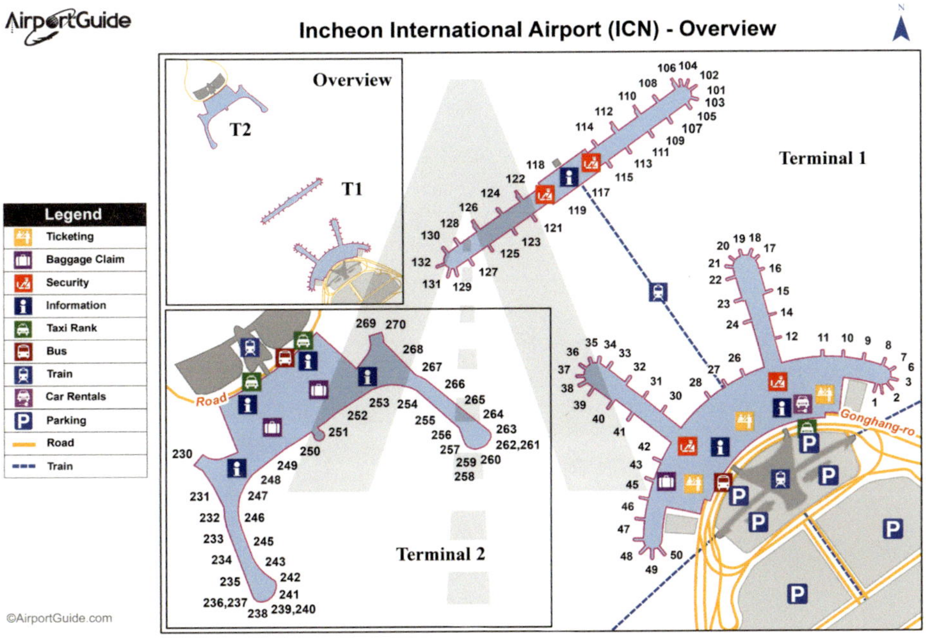

Figure 7.6 Seoul Incheon, Terminal 1 and 2 layouts showing the gates.
SOURCE: IIAC VIA AIRPORT GUIDE.COM

Figure 7.7 Seoul Incheon, CGI of second terminal.
SOURCE: GENSLER

Figure 7.8 Seoul Incheon Terminal 2 interior.
SOURCE: IIAC VIA GETBYBUS.COM

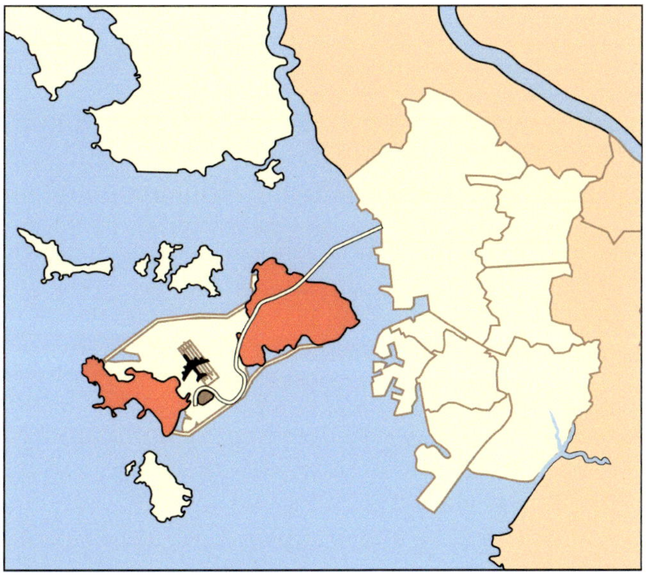

Figure 7.9 Seoul Incheon location plan.
SOURCE: WIKIPEDIA

Geographically, Incheon Airport is within reach of 61 foreign cities, each with over 1 million people and can be reached within 3.5 hours flying time. It is claimed it is possible to make a day trip from the airport and reach a population of 2.5 billion people in Asia.

The airport was created by linking two islands, Yeongjong to the east and Yongyu to the west (Figure 7.9). Dams were formed to keep the sea at bay and allow the construction of the airport on the former seabed. It was planned from the outset as an Aerotropolis and is supported by three development zones within an overall Free Economic Zone (IFEZ). An express railway links Incheon with the domestic airport Gimpo and capital city Seoul some 61 km and 43 minutes away.

The three zones of the Incheon Free Economic Zone (IFEZ) comprise (Figure 7.10):

- Yeongjong, which forms a continuum to the east of the airport, focuses on airport support facilities and logistics, tourism and leisure.

Figure 7.10 Seoul Incheon and Aerotropolis, Yeongjong, Cheongna and Songdo.
SOURCE: SONGDO

- Cheongna, further to the east on the mainland, concentrates on international finance, high-tech industry and distribution. It will have a future planned population of 90,000 people.
- Songdo international city is located to the southeast of the airport on 600 hectares of reclaimed land and is connected to the airport by a 12 kilometre bridge. It is characterised as a 'smart city'. Songdo is home to international business relating to Information Technology, Biotechnology and Research and Development. Songdo is still under development and has a current population of 100,000 people and a planned future population of 300,000 people.
- Also, Incheon Harbour which consists of cutting-edge harbour infrastructure is located a 20-minute car distance from the IFEZ. On top of this impressive connectivity, the area lies only an hour away from Seoul, Korea's metropolitan area.
- Aerial views of Songdo City show a very high level of ambition for the development of the Aerotropolis. They also show the purpose-designed road and rail networks that link the city directly to the airport via a giant causeway/bridge.

The airport master plan shows how the rail and road network linking to the capital and the Aerotropolis cities is incorporated into the airport (Figure 7.11).

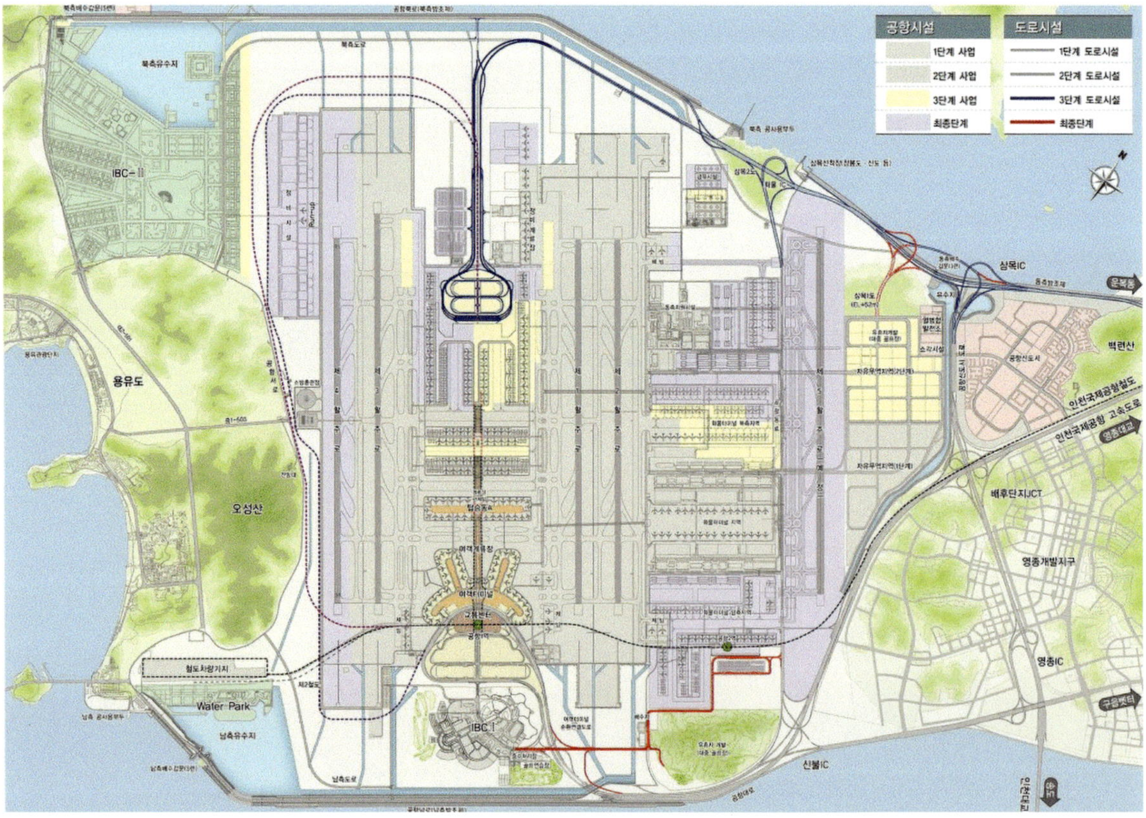

Figure 7.11 Seoul Incheon master plan, with five runways and two terminals.

SOURCE: SEOUL INCHEON

31. Bangkok Suvarnabhumi (BKK) 2006

Suvarnabhumi Airport was originally designed to operate as many as 76 flights an hour on two simultaneously operating mixed-mode runways. At opening, it had 51 aircraft stands and 69 remote parking bays for wide-bodied aircraft and was designed to handle over three million tonnes of cargo annually (Figure 7.12–7.15). With a second terminal it is now planned to handle 100 million passengers per annum (mppa) The terminal was, at the time, one of the largest to be designed without a satellite, with the attached piers forming a double cruciform.

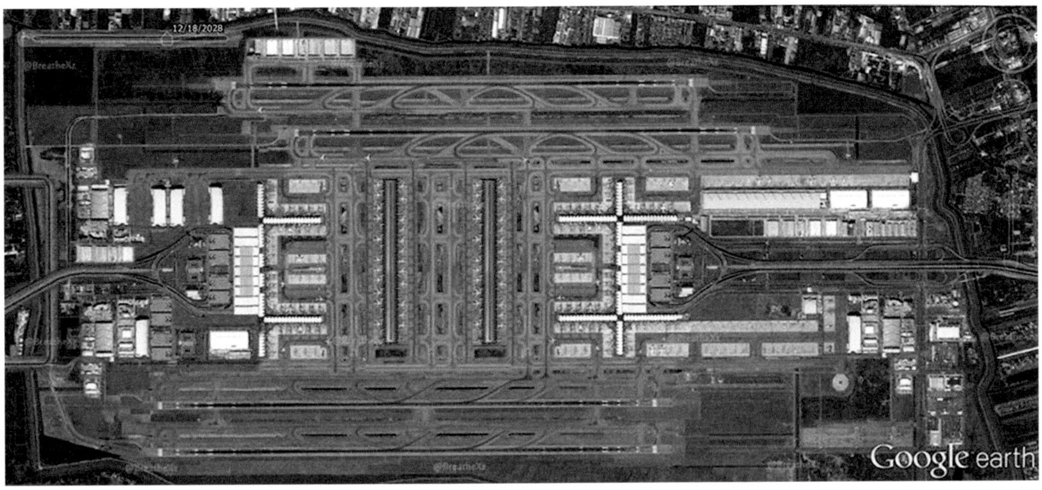

Figure 7.12 Bangkok Suvarnabhumi, 2030/2040 Master Plan.
SOURCE: BKK VIA SKYSCRAPER CITY

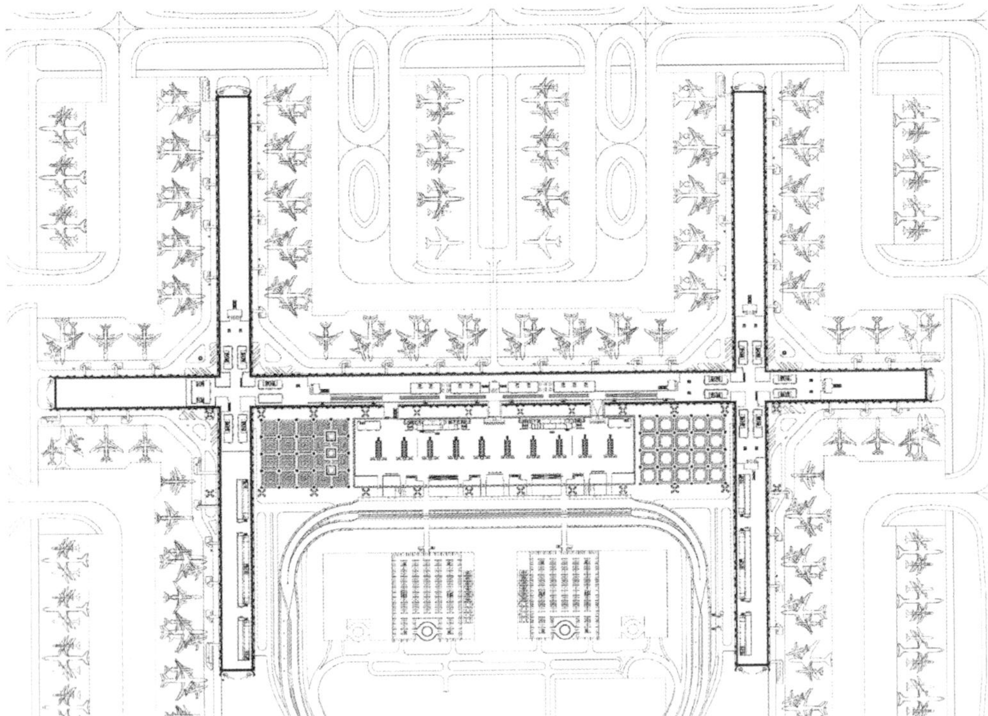

Figure 7.13 Bangkok Suvarnabhumi, terminal gate layout.
SOURCE: BKK VIA AIRPORT BUILDERS

Figure 7.14 Bangkok Suvarnabhumi aerial view.
SOURCE: BKK

With a total capacity of more than 30 million passengers per annum in the first phase and an ultimate capacity of 100 million passengers per annum, the new airport is one of the most important airports in Asia and has become a major transfer hub. However, for the transfer passenger, while the circulation pattern is very simple and clear, the pier layout can result in some very long walking distances.

Facilities that were completed in the first phase included:

- a 563,000 m^2 passenger terminal
- two parallel runways, each 60 m wide, one 3,700 m long and the other 4,000 m long with a runway separation distance of 2,200 m
- two parallel taxiways to facilitate simultaneous departures and arrivals
- 120 parking bays (51 with contact gates and 69 remote gates) and five of these capable of accommodating the Airbus A380
- 132 m air traffic control tower (the tallest in Asia)
- two five-storey parking garages with a capacity for 5,000 cars
- a 190,000 m^2 cargo terminal
- *aircraft maintenance facilities*: four fully equipped aircraft hangars to service up to 12 aircraft.

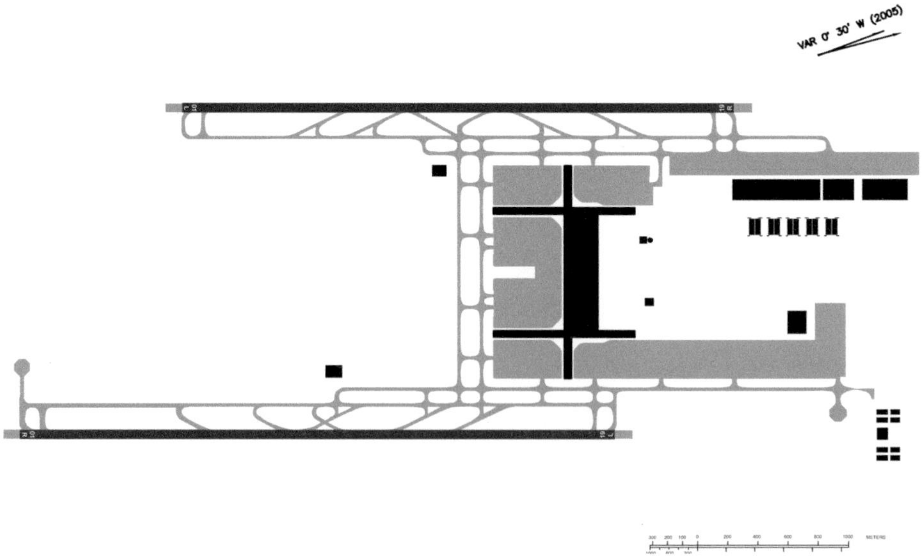

Figure 7.15 Bangkok Suvarnabhumi, phase 1 airport layout in 2006.
SOURCE: WIKIPEDIA

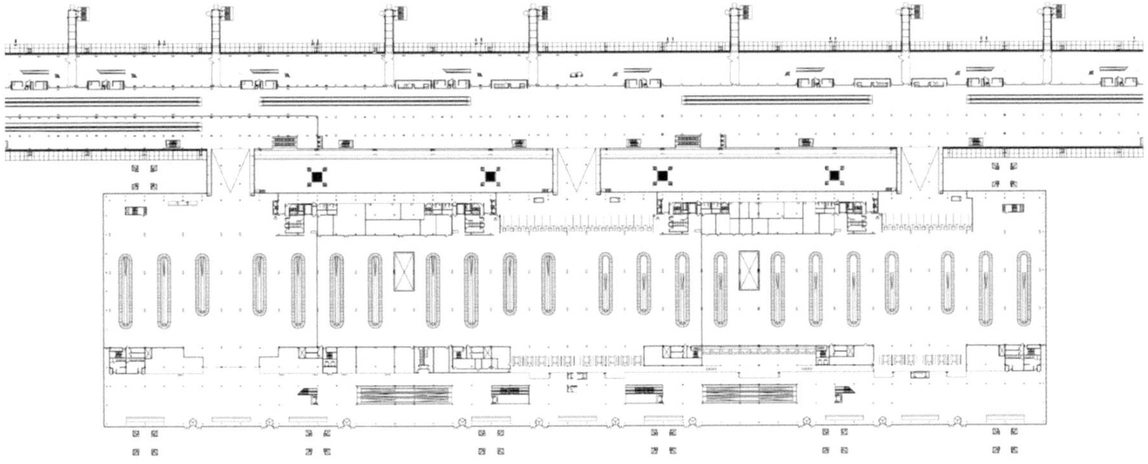

Figure 7.16 Bangkok Suvarnabhumi, terminal lower boarding and arrivals level 2.
SOURCE: BKK VIA AIRPORT BUILDERS

Designed by the MJTA group of consultants, comprising Murphy Jahn Architecture and TAMS consultants (USA) and ACT Engineering consultants (Thailand), the passenger terminal complex (comprising the terminal itself as well as the concourse) covers an area of 182,000 m^2. Seven floors and a basement gives the terminal a total floor area of over 563,000 m^2 (six million ft^2) making it the largest in the world at the time.

There are 360 check-in counters on ten check-in islands, all with connected baggage belts and another 100 check-in counters not connected to the baggage belt system. The domestic and international halls are clearly separated with the second floor as a dedicated arrival hall and the fourth floor as a departure hall (Figures 7.16 and 7.17). The structure's main materials are steel and glass. An innovative roof trellis (one of the largest in the world) designed to shade the building against intense tropical sun and reduce the cost of air conditioning, was also the largest of its kind (Figures 7.18–7.20).

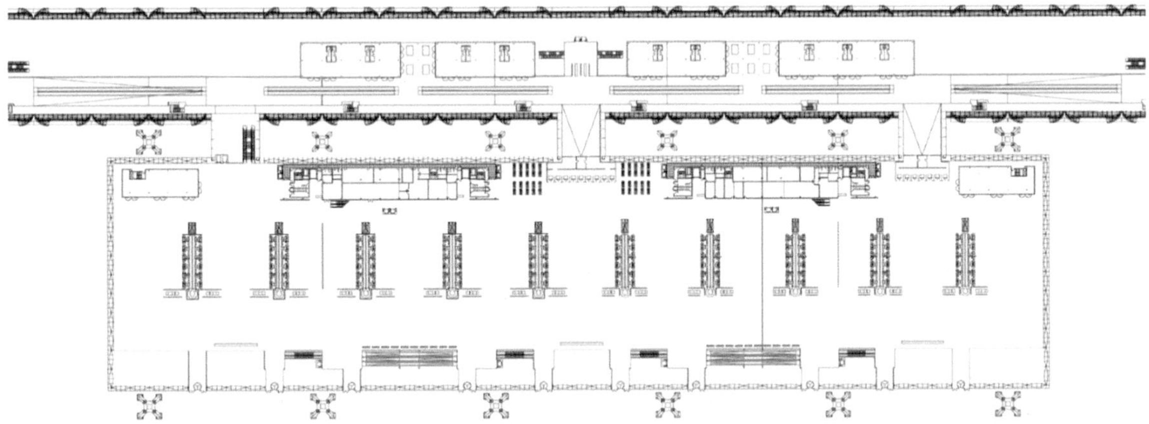

Figure 7.17 Bangkok Suvarnabhumi, terminal check-in and departures level 4.
SOURCE: BKK VIA AIRPORT BUILDERS

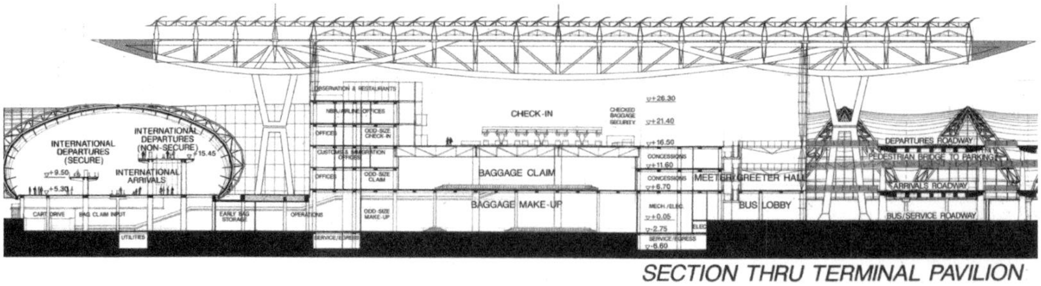

Figure 7.18 Bangkok Suvarnabhumi, cross section through terminal.
SOURCE: BKK VIA AIRPORT BUILDERS

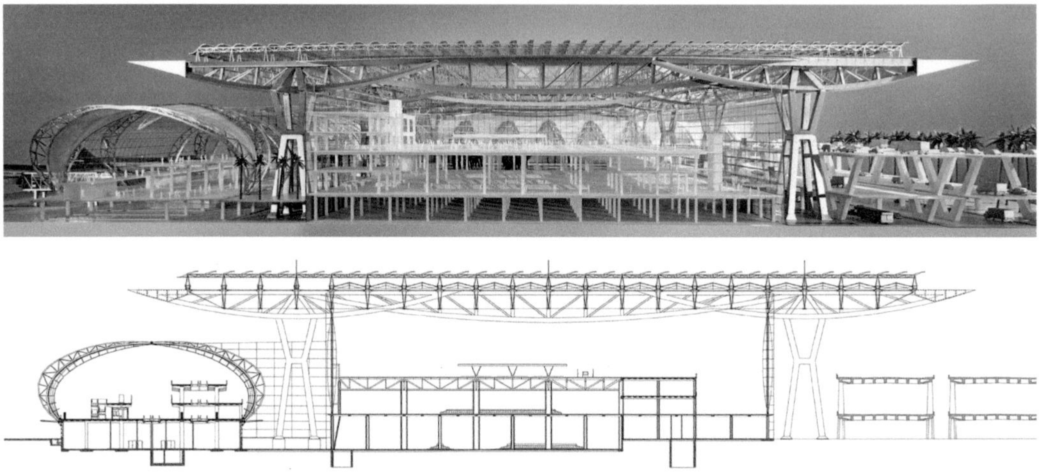

Figure 7.19 Bangkok Suvarnabhumi, sectional model of terminal.
SOURCE: BKK VIA AIRPORT BUILDERS

Figure 7.20 Bangkok Suvarnabhumi, photograph of upper level of the pier.
SOURCE: BKK VIA ARCHITECTURE DAILY – PHOTOGRAPHER: RAINER VIERTLBOECK

Figure 7.21 Bangkok 2006 airside retail in magenta.
SOURCE BANGKOK AIRPORT

The retail offer is placed along the normal passenger route, but without distorting the route and adding to the overall passenger walking distance, as can be seen in the map of the airport (Figure 7.21). However, although the retail concept was retrofitted into the concept during construction some retail designers argue that the layout is sub-optimal because the symmetrical bifurcated plan means that departing passengers only have to pass half of the retail offer on their way to the piers and boarding gates (Figure 7.22).

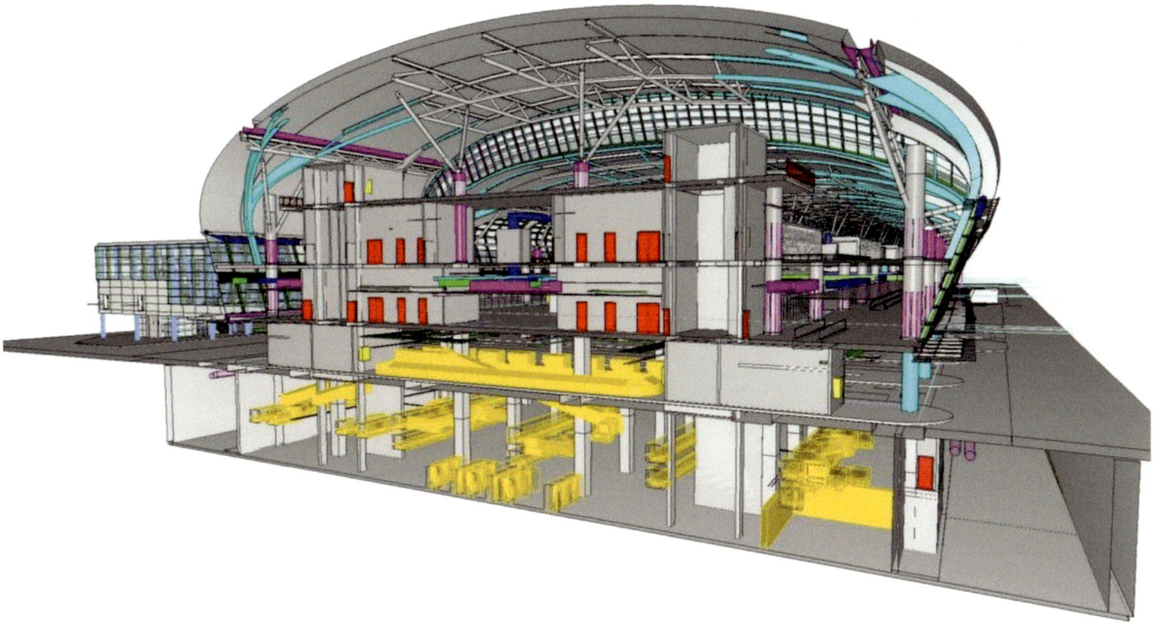

Figure 7.22 Bangkok Suvarnabhumi, sectional perspective through pier.
SOURCE: MAA GROUP CONSULTING ENGINEERING

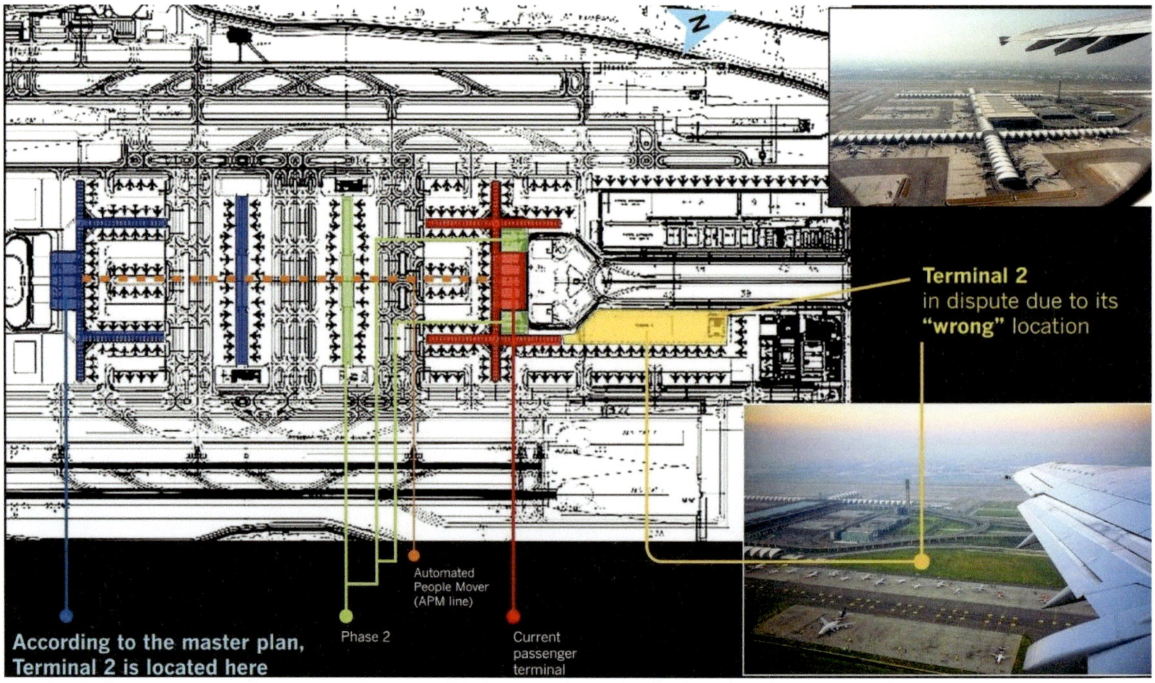

Figure 7.23 Bangkok Suvarnabhumi, proposed second terminal in blue and an alternative but disputed location for a second terminal in yellow.
SOURCE: THE NATION

The future master plan includes two satellites and a second terminal, the location of which is still being debated. At this stage, a transit will become essential to link the terminals and satellites together (Figure 7.23).

32. Madrid Barajas (MAD) 2006

Aena, the airport operator, describes Madrid Barajas as 'The Gateway to Latin America'. It was designed as a hub airport and is mainly used by the One World Alliance, including Iberia and British Airways as complementary to Heathrow. It is located 15 km northeast of Madrid and linked to the city by a regional rail line and an underground metro. It has a unique master plan with four widely spaced runways, which are slightly misaligned, with the terminals on the perimeter and a new satellite located mid-field (Figure 7.24). Terminal 4, designed by Richard Rogers was opened in 2006 and is the only airport terminal to have won the RIBA Stirling Prize (Figures 7.25 and 7.26).

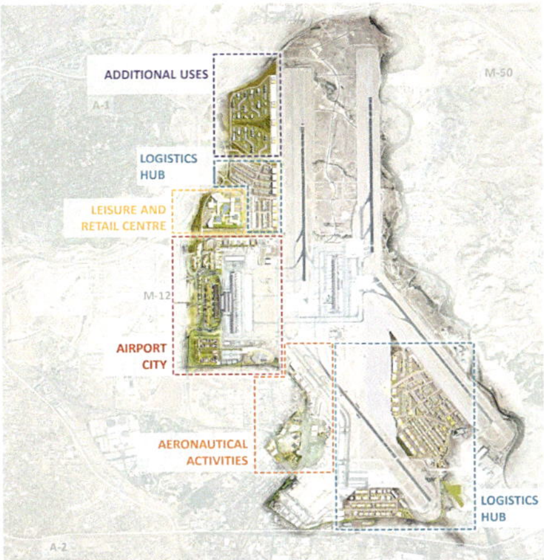

Figure 7.24 Madrid Barajas, master plan.
SOURCE: AENA

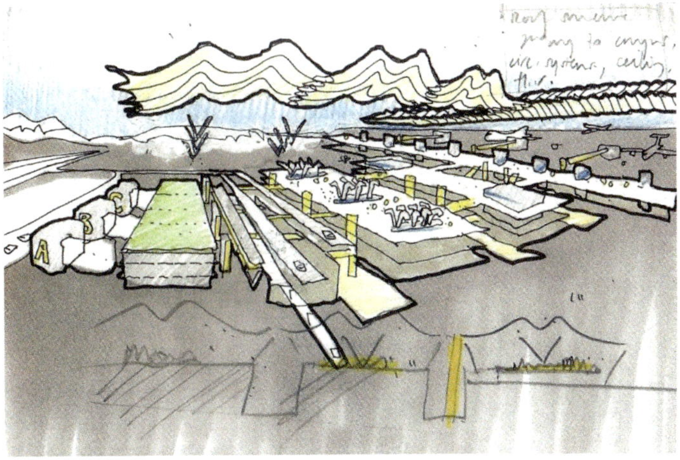

Figure 7.25 'Magic Carpet' Roof Concept.
SOURCE: © RSHP

Figure 7.26 Madrid Barajas, Magic Carpet Roof.
SOURCE: © RSHP

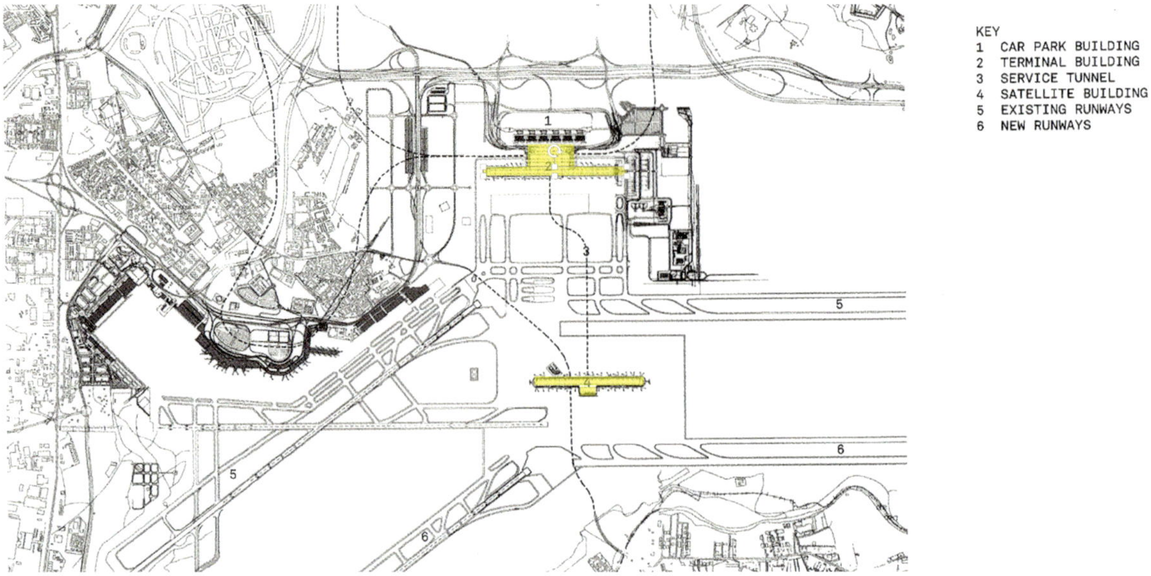

Figure 7.27 Madrid Barajas, master plan 2006 with Terminal 4 and satellite highlighted in yellow.
SOURCE: © RSHP

It is interesting to compare Madrid Barajas, with Heathrow Terminal 5 where the RSH team developed similar conceptual ideas in parallel. The main differences are that the Madrid site is far more generous, allowing the terminal to spread out and allow daylight to penetrate to the lower levels and the overall floor area of 1,158,000 m^2 is more than twice the size of Heathrow T5. There are, however, similarities; for example, both are hub airports for One World Alliance airlines, British Airways and Iberia, who engaged YRM to look after their interests, and both airports have terminals linked to remote linear satellites by an underground transit system (Figure 7.27).

Rogers, Stirk, Harbour, architects for the new terminal, describe the planning and architecture of Madrid Barajas with a particular emphasis on its environmental credentials (Figures 7.28–7.31):

> *The terminal, which is the biggest in Spain, was commissioned to enable Barajas International Airport to compete with major hub airports within Europe. The core*

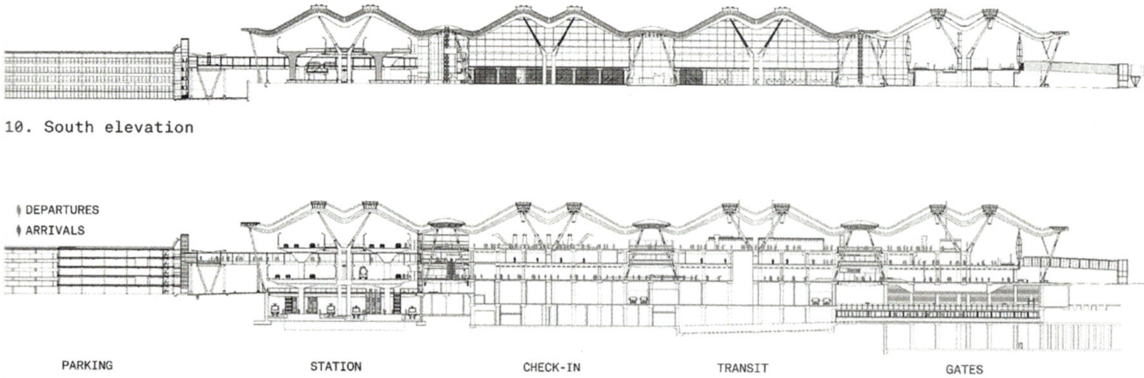

Figure 7.28 Madrid Barajas, South elevation and section through terminal.
SOURCE: © RSHP VIA AJ

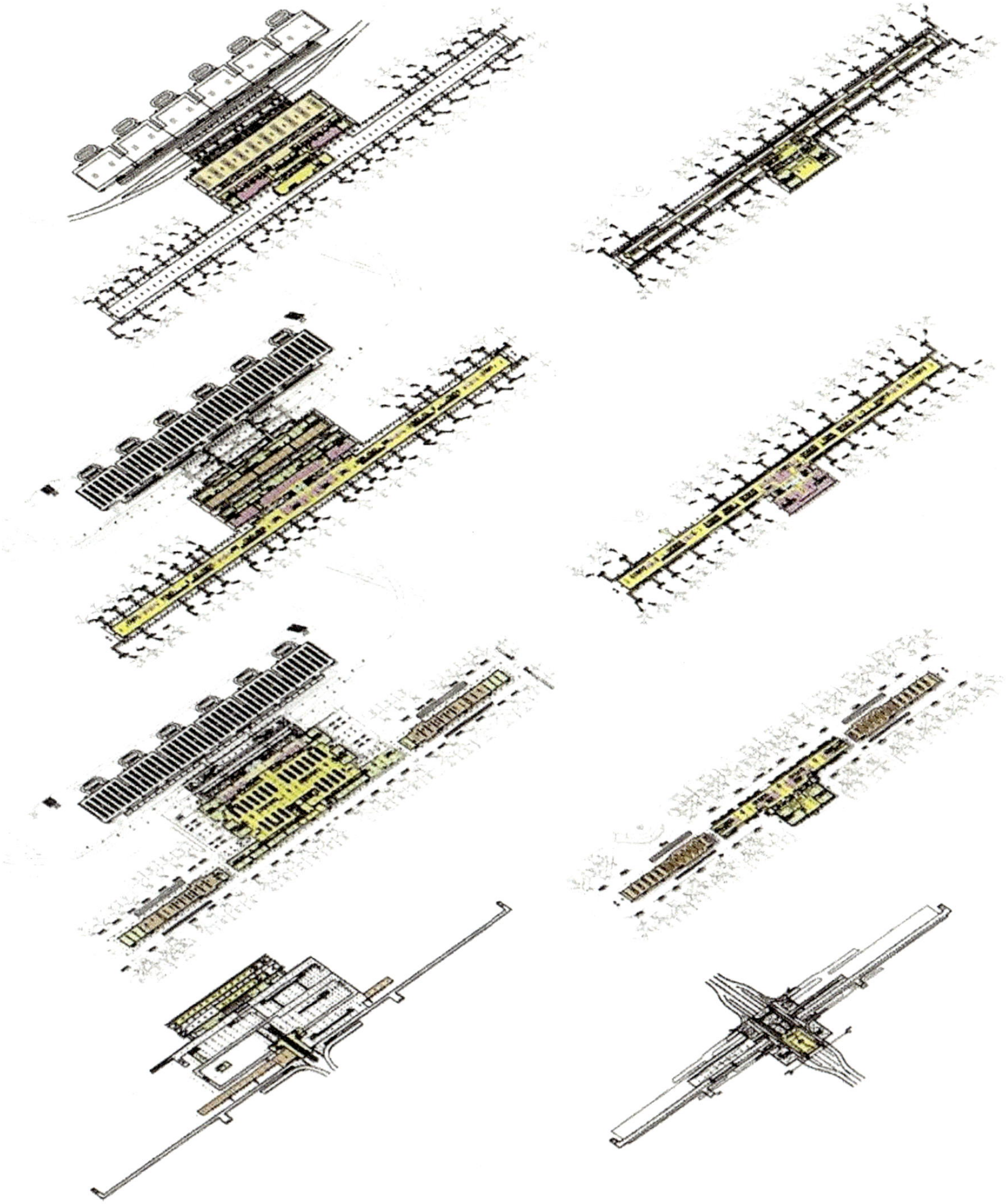

Figure 7.29 Madrid Barajas, isometric plans of terminal and satellite.
SOURCE: © RSHP VIA AJ

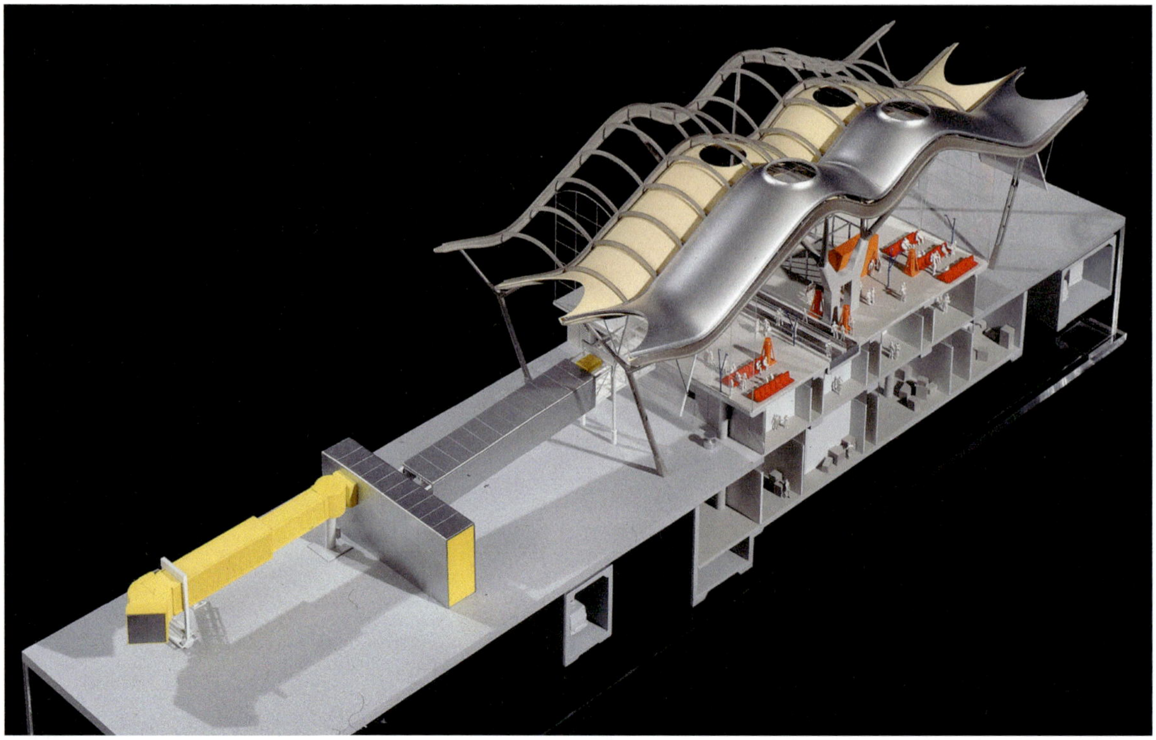

Figure 7.30 Madrid Barajas, model of satellite.
SOURCE: © RSHP

Figure 7.31 Madrid Barajas, views of terminal and satellite.
SOURCE: AENA

building comprises a sequence of parallel spaces separated by a linear block allowing daylight to penetrate deep into the interior. The same form is applied to the satellite, which is comprised of two linear blocks, one for passport control and the other containing the gates. The bamboo linear roof structure is connected above by a chain of roof lights, permitting maximum flexibility in the arrangement of accommodation on each of the floors. This enables the building to be expanded in phases.

The accommodation is distributed over six floors; three above ground for check-in, security, boarding and baggage reclaim and three below ground levels for maintenance, baggage processing and transferring of passengers between buildings.

Environmental measures, aimed at significantly reducing energy consumption, include a stratified cooling system, displacement ventilation supply to the piers, low level air supply to all other passenger areas, extensive shading to the facades and roof lights, zoned lighting and the collection of rainwater to irrigate the landscape.

The building is covered by a wave-shaped roof, supported on central 'trees' and punctuated by roof lights that provide carefully controlled natural light throughout the upper (departures) level of the terminal. The roof then oversails the edge of the building to shade the façades. Given the multi-level section, a strategy was required to bring natural light down into the lower levels. The solution is a series of light-filled 'canyons' that separate the parallel slices of space that demarcate the various stages of transit, from the arrival point to check-in, security and passport control, then on to departure lounges and finally to the aircraft.

The canyons are spectacular full-height spaces, spanned by bridges from which arriving and departing passengers, though segregated, can share the drama of the impressive space. The grand scale generates the feeling of a truly significant public space. The canyons also act as locators, underlining the clear sense of direction and legibility that is fundamental to the scheme.

Despite the extreme heat of summer in Madrid, the design team were committed to the use of passive environmental systems wherever possible, while aximizing transparency and views towards the aircraft and the mountains beyond.

The building benefits from a north-south orientation with the primary façades facing east and west, the optimum layout for protecting the building against solar gain. The design team set out to maximise natural light to all passenger areas and reduce dependence on artificial light, while providing views out but reducing solar gain with a combination of deep roof overhangs and external shading.

A low-energy displacement ventilation system is used in the pier and elsewhere in the terminal a more conventional high-velocity system is used.

(Harbour, 2020)

33. Heathrow (LHR) T5 2008, T2 2014 and the Heathrow Expansion Programme

This case study focuses on the evolution of the airport as it moved into the 21st century, first with Terminal 5, subsequently Terminal 2 and lastly the Heathrow Expansion Programme that looks forward towards 2050 with the addition of a third runway. As the site is very restricted, all these projects have been influenced by the Atlanta Hartsfield linear satellite concept which is a model of effective land use and efficient aircraft movement.

The original Heathrow war-time master plan has been discussed in an earlier case study together with its unique underground surface access concept. The underground surface access has over time proven to be one of the most intractable constraints to the evolution of the airport, around which subsequent upgrades have had to be planned. Terminal 5 and the future Heathrow Expansion Programme are illustrated below (Figures 7.32–7.35).

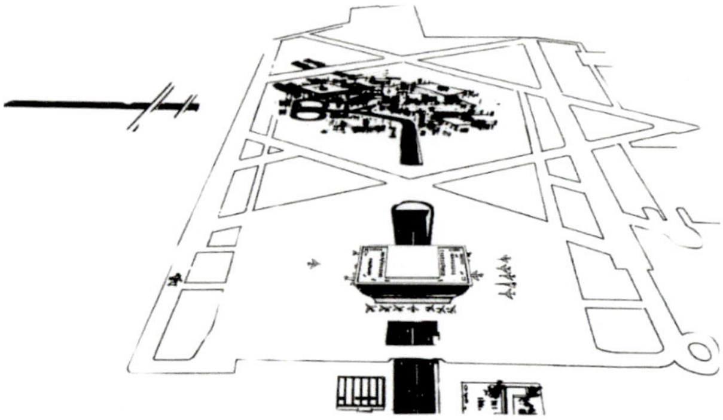

Figure 7.32 Early concept for Heathrow T5 with a motorway under the terminal (1976).
SOURCE: BAA

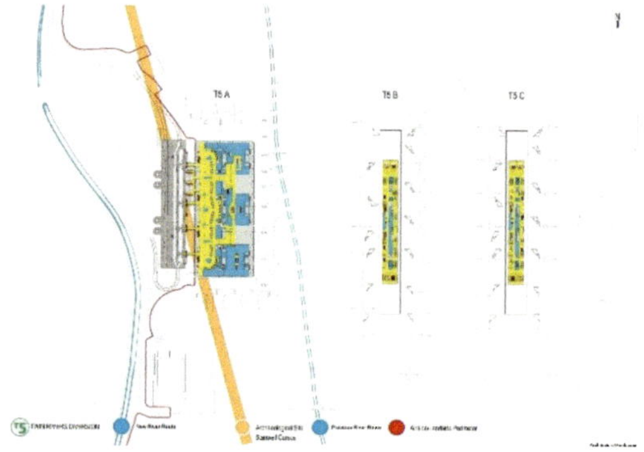

Figure 7.33 Heathrow T5 'Toast Rack' Master Plan (2008).
SOURCE: © RSHP

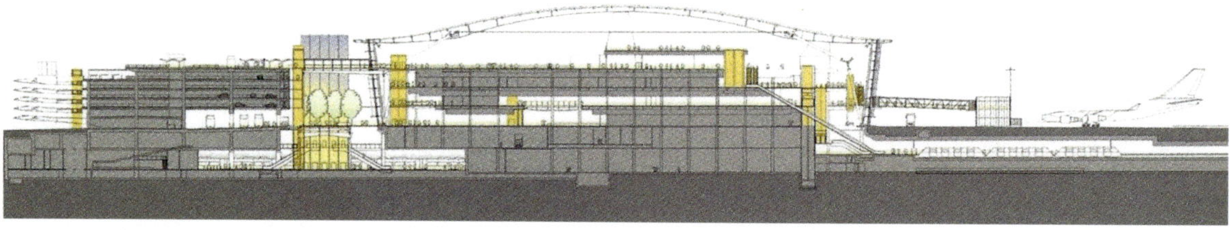

Figure 7.34 Heathrow Terminal 5 sections.
SOURCE: © RSHP VIA ARCHITECTS JOURNAL

Figure 7.35 Heathrow Expansion Project.
SOURCE: HEATHROW AIRPORT GRAPHICS BY GRIMSHAW ARCHITECTS

All these projects have followed the efficient Atlanta 'toast rack' model with linear satellites as the airport is uniquely restricted in area. It has the smallest area per passenger handled of any of the major airports. However, this has resulted in complex connectivity. The Heathrow Expansion Project is currently on hold.

Serious design began after the design competition in 1989 which saw the Richard Rogers Partnership selected, narrowly winning over YRM with jury votes split 7 to 6 in their favour. The multidisciplinary design at Heathrow T5, with co-located teams of architects and planners, was a long and complicated process aggravated by the longest public inquiry process in UK history, but illustrates well many of the contemporary issues associated with multidisciplinary airport design. The master plan layout was largely based on the 'toast rack' concept, with a mid-field terminal supporting a series of satellites, which had been developed for Atlanta airport, as this was shown to the most efficient land use on a uniquely tight site and optimised aircraft movement across the airport.

The architectural team included:

- Rogers, Stirk, Harbour + Partners (formerly the Richard Rogers Partnership) – Concept Architect and Lead Designer.
- YRM – British Airways Architects and responsible to BAA for the campus design guidelines and design of the ancillary buildings.
- HOK – Architect for railway infrastructure.
- Chapman Taylor – Retail architect.
- Pascall + Watson – Executive architect for the fit out.

Several engineering disciplines were also involved, including Arup, Mott MacDonald, DSSR, TPS, WSP and Atkins. At its peak, over 2000 designers were working on the project. Instead of having many different firms spread out across the country, sending drawings and specifications to each other, a core multidisciplinary design team was assembled that worked from a single building at Heathrow, integrating teams from the airport and the airline with the construction management team. So while design team management to produce a single integrated design concept was demanding, it was greatly assisted by co-locating key members of the team together.

The work on design and construction continued for nearly 20 years, the extreme length being partly due to the extended public inquiry process. It was inevitable that over such a long time period there would be significant changes to the brief and design approach. These included major changes to the security regime, retail concept and adjustments to the planning to suit the operation of the main tenant British Airways. The opportunity was also taken by both BAA (now Heathrow Airport Ltd.) and British Airways to improve operational processes by housing their operational staff in adjacent spaces and embracing advances in IT and other technology. The terminal concept developed over a long design period as the context and brief evolved.

Concept 1 – 'Magic Carpet'

The first concept was a single-level scheme, developed from the winning competition entry between 1989 and 1992 (Figures 7.36–7.38).

Mike Davies, the RSH+P project director explained:

> *It was built on the concepts which had previously been explored at Stansted including the notion of a building where the arriving and departing passengers were all on one level and would experience the same quality of space.*

Figure 7.36 Heathrow T5 'Magic Carpet' Concept.
SOURCE: © RSHP

> *The idea was to create a great room, covered by a flowing wavy roof, which gradually crescendoed in height over the largest hall space. We created multiple waves and the biggest wave sat over the most important public spaces. The roof was deliberately conceived as flowing and tipped up on the outside edge on the airside as a celebration of 'up, up and away'...... instead of squeezing passengers through 5–6 metre high spaces, we offered great lofty heights, more like cathedrals, in which people felt they were moving through open and airy spaces with good views – generous architectural spaces.*

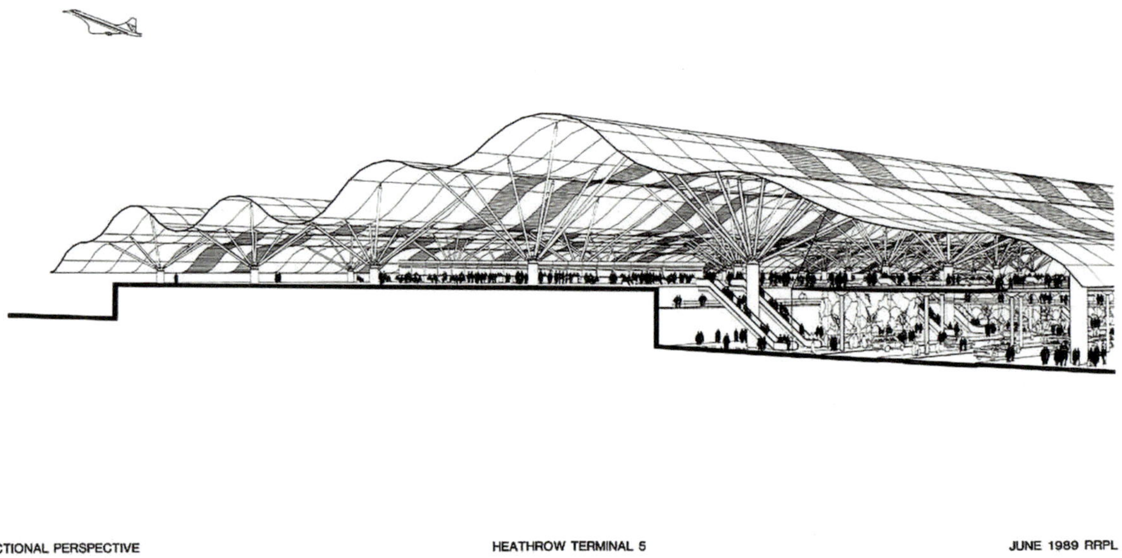

Figure 7.37 Heathrow T5 Competition Scheme (June 1989).
SOURCE: © RSHP

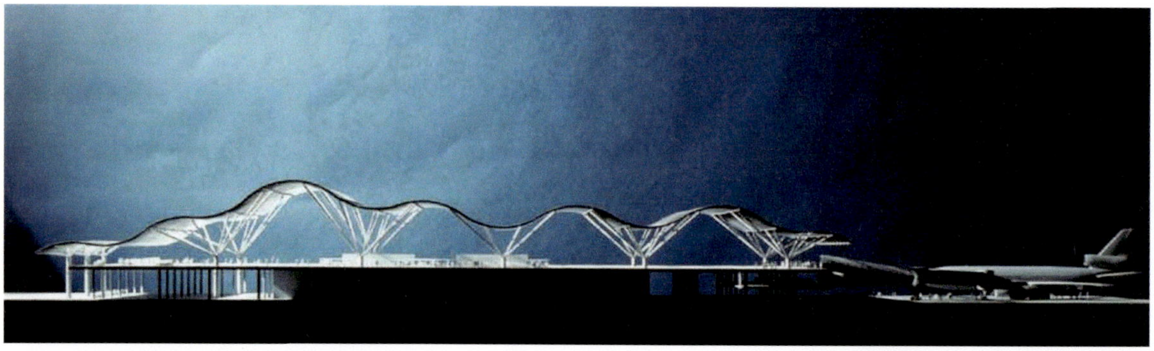

Figure 7.38 Heathrow T5 'Magic Carpet' Model.
SOURCE: © RSHP

Figure 7.39 Heathrow T5 'Canyon' Concept.
SOURCE: © RSHP

This scheme was abandoned during the public inquiry when legal advice was that building in the green belt was too contentious, requiring the terminal to be redesigned. This led directly to the development of a second, more vertically stacked concept, with 5 passenger levels above ground and 3 levels of basement supporting accommodation.

Concept 2 – 'Canyon'

This was a more vertically developed concept, resulting in a more compact footprint on a restricted site. It was developed in the period between 1992 and 2000 to address the legal advice that restricted the site. This taller scheme splits the floor plate into three, with light wells separating them (Figures 7.39 and 7.40).

Mike Davies described the design as follows:

> *The scheme was characterised by the organisation of the overall plan into a landside plate, a process plate and an airside plate, with all three bands of floor linked by bridges across two canyons, which allowed the light to penetrate down to the lower levels of the building. Essentially the landside processes fitted one plate, the security, customs and immigration processes sat in the middle plate and the airside plate accommodated the airside dwelling lounges and aircraft gates.*

Figure 7.40 Heathrow T5 'Canyon' model.
SOURCE: © RSHP

Figure 7.41 Heathrow T5 'Tied-Arch' concept.
SOURCE: © RSHP

It was eventually not pursued, in part because of the inherent inflexibility of the floor plates, which did not naturally fit the functional requirements and in part because during the long delay caused by the public inquiry RSH+P had developed another similar, but more generous and spread-out canyon scheme for Barajas Airport in Madrid, on a much less restricted site which also used a 'canyons of light' concept. No doubt they did not want to repeat a similar, but compromised, approach on a more difficult and restrictive site.

Concept 3 – 'Tied-Arch'

A long-span structure of 144 m had been proposed by YRM in their T5 completion entry in 1988, in which they came a close second to Richard Rogers. They were by now working for British Airways as a part of the integrated multidisciplinary design team and co-located with RSH+P at Heathrow. The joint team now saw the flexibility a long-span structure could offer in the light of an increasing number of changes to requirements such as more stringent security standards, following Lockerbie and 9/11, imposed by the UK government.

Other new influences, such as the emergence of Global Alliances and, in particular, the One World Alliance saw an expansion of transfer traffic, differentiated products for premium passengers, such as fast track and ever more glamorous airline lounges. Retail concepts had also evolved, with a higher proportion of the provision being airside, and when allied to new faster check-in processes, it resulted in passengers spending a higher proportion of their dwell time in the shopping mall than the airside lounge had become. The response was the 'Tied Arch' concept (Figures 7.41 and 7.42).

Mike Davies continued;

> *What became clear, in visiting many airports, Stansted included, was that whatever you do the internal column layout becomes a constraint on internal planning. So as a team, BAA and we looked at longer spans, which meant less columns and greater flexibility. We studied several longer spans solutions and created what was called the 'tied arch' scheme. The 'tied arch' scheme had quite an ambitious roof span, around*

Figure 7.42 Heathrow T5 'Tied-Arch' model.
SOURCE: © RSHP

100 metres, which dramatically reduced the number of internal columns in the building. The tied arch scheme had also incorporated the various rail and underground stations underneath the terminal, directly integrated into the terminal space.

After a series of detailed studies that incorporated learning from the Kings Cross Underground Station fire (1987) about separating the terminal from the rail station below to prevent fire from spreading from the station to the terminal, the design team decided that the legislative regime of a station was so different from that of a terminal that the two couldn't be merged into a single space. This was combined with an understanding that the tied arch concept produced a series of new constraints for interior planning as the ties couldn't be cut or eliminated and so the fourth and final concept was developed.

Concept 4 – 'Great Hall'

The final concept that evolved out of collaborative multidisciplinary work was the interchange concept. YRM had proposed a landscaped zone between the multi-storey car parks and the terminal to create a breathing space on the passenger journey. During the design, a new directive was issued from the UK DfT that vehicular traffic had to be kept 30 metres away from the terminal. The concept of a green space in front of the terminal that had lain dormant was then revived (Figures 7.43 and 7.44).

It involved moving the forecourt away from the terminal and relocating it above the car park to create a vehicle-free zone in front of the terminal. This allowed the creation of an external light-filled piazza planted with trees just outside the terminal, into which the vertical circulation to the rail stations was moved, creating an interchange. RSH+P then designed an even longer span structure of 156.6 metres for the terminal, removing many of the internal constraints to the layout of the departure concourse.

Mike Davies explained the new concept this way:

We ended up with a very clear concept which in some ways returned to the spirit of our earliest competition idea, yielding a very large departures floor plate with no constraints on it at all. Alongside this long span 'Great Hall' concept came the decision to make the interior fabric of the building below the departures levels, an independent, steel frame structure. The external structure is freestanding and does

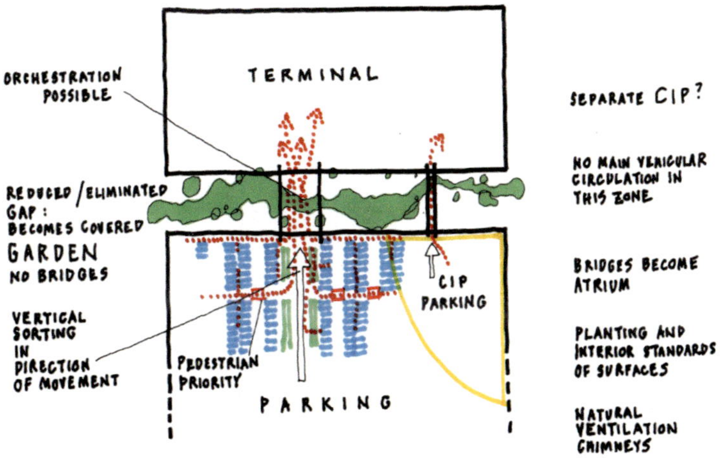

Figure 7.43 T5 landscaping concept.
SOURCE YRM (REPRODUCED WITH PERMISSION FROM RMJM)

Figure 7.44 T5 landscape photos.
SOURCE: AUTHOR'S COLLECTION

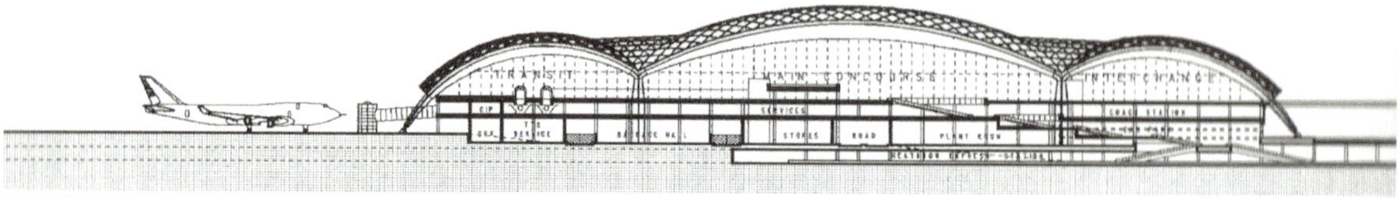

SECTION

Figure 7.45 YRM Heathrow T5 Competition Design Source.
SOURCE: YRM ARCHITECTS AND PLANNERS (REPRODUCED WITH PERMISSION FROM RMJM)

> *not join with the edges of the external building envelope at any point. You have an external wrapper with a void space all around the edges with an easily adaptable island inside it. As a result, natural light pours into the building at the facades, then beyond this void, you have an internal structure which can be changed at will, without any impact whatsoever on the main external envelope.*

This concept owed something to the collaboration with YRM, who had previously explored the idea of a very large clear span over the main terminal of 144 m in their competition proposal of 1989 (Figure 7.45).

The 'Great Hall' concept, which is also reminiscent Piano's Kansai design, on the landside and airside edges, is illustrated overleaf (Figure 7.46).

British Airways embraced the self-service flow-through concept at Heathrow Terminal 5. With the help of their architects YRM, they introduced a new way of organising the check-in concourse, with three waves of processing: self-service check-in, followed by flow-through bag drop and backed up by full service desks for more complex transactions, such as visa checks. This design responded to the *'state of readiness'* of the passengers, rather than according to class of travel or destination (Figure 7.47).

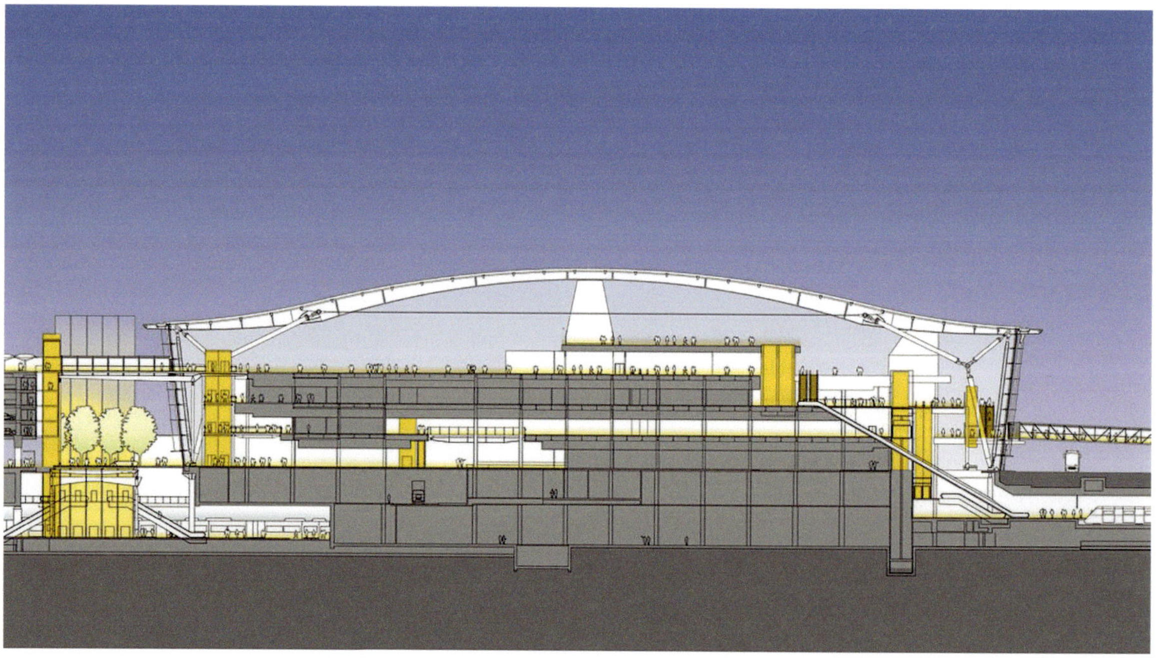

Figure 7.46 Heathrow T5 – 'Great Hall' Concept.
SOURCE: © RSHP

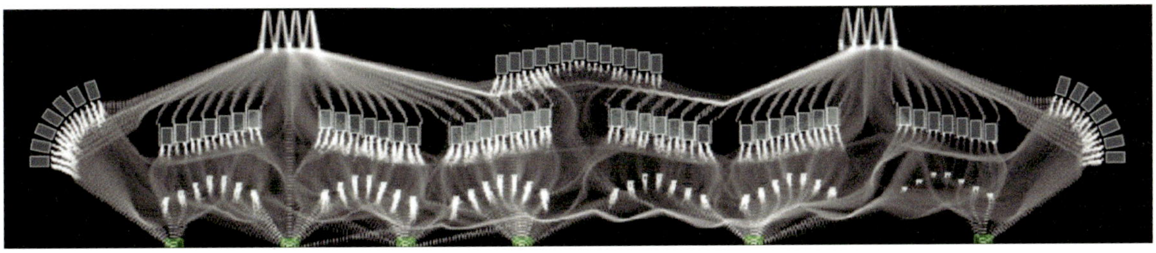

Figure 7.47 Heathrow Terminal 5, 'Skywalker' check-in simulation as exhibited at the Royal Academy Summer Exhibition 2004.
SOURCE: YRM ARCHITECTS AND PLANNERS (REPRODUCED WITH PERMISSION FROM RMJM)

Predicting how people would behave given this level of choice led YRM to develop its own interactive passenger flow modelling simulation – '*Skywalker*' which allowed the user to change inputs while the model was running. Desks could be opened and closed during the working day, transaction times varied and individual passenger characteristics altered during the simulation. This gave the opportunity to design a freeform passenger reception layout that responded directly to the new approach to passenger segmentation and to the natural desire lines for passenger movement while incorporating clear wayfinding.

The built scheme for the main terminal offers an unencumbered, long-span 'envelope' – developed with Arup – with a flexible internal space conceptually similar to that of Madrid Barajas or the practice's much earlier design for the Pompidou Centre in Paris. Departure and arrivals areas, check-in desks, commercial space, retail, offices, passenger lounges, back-up and other facilities are all contained within freestanding steel-framed structures inside the building and can be dismantled and reconfigured as future needs change.

The built, multi-level scheme is contained beneath an elegant, curved floating roof, supported by slim columns at the perimeter edges to provide the required highly flexible and visually dramatic internal space (Figure 7.48 and 7.49). In this scheme, passengers depart and arrive in a terminal building that offers generous spaces and fine views across the airport. As well as the design of the main terminal building, RSHP was also responsible for the design of two satellites and Heathrow's new control tower, which became operational in early 2007.

The Terminal 5 floor plans are illustrated overleaf (Figure 7.50). Proposals are being developed to increase the capacity of Terminal 5 and modify the layouts of facilities.

Figure 7.48 Heathrow T5 – 'Great Hall' structure.
SOURCE: © RSHP

Figure 7.49 Heathrow T5 – 'Great Hall' roof.
SOURCE: © RSHP

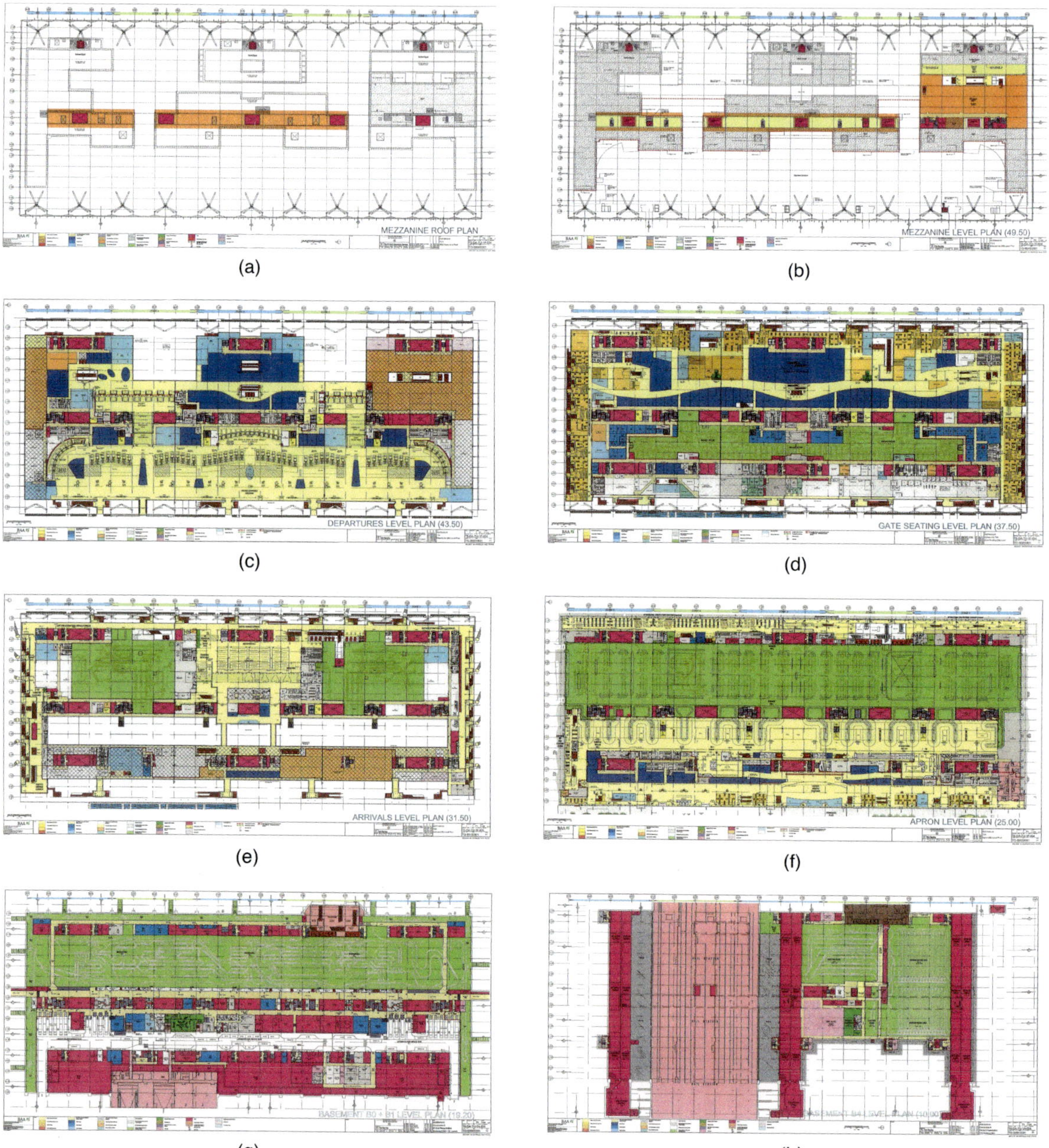

Figure 7.50 Heathrow T5 Floor Plans. (a) Heathrow Terminal 5 Floor Plans. (b) Mezzanine Roof. (c) Mezzanine Level. (d) Departures Level. (e) Gate Level. (f) Arrivals Level. (g) Apron Level. (h) Upper and Lower Basement Level.
SOURCE: AUTHOR'S OWN COLLECTION

Constraints to increasing throughput include the government-imposed requirement for more rigorous security checks which requires additional space and staffing levels. The exploded view of the multiple levels of Terminal 5, show the complexity involved in housing the various passenger, baggage and transportation functions under the single 'Great Hall' roof 2005 (Figure 7.51).

Contemporary food and beverage offers are no longer provided in separate dining rooms but have become integrated with the shopping outlets, with much less formal seating layouts, to suit the casual browser and those in a hurry. International brand

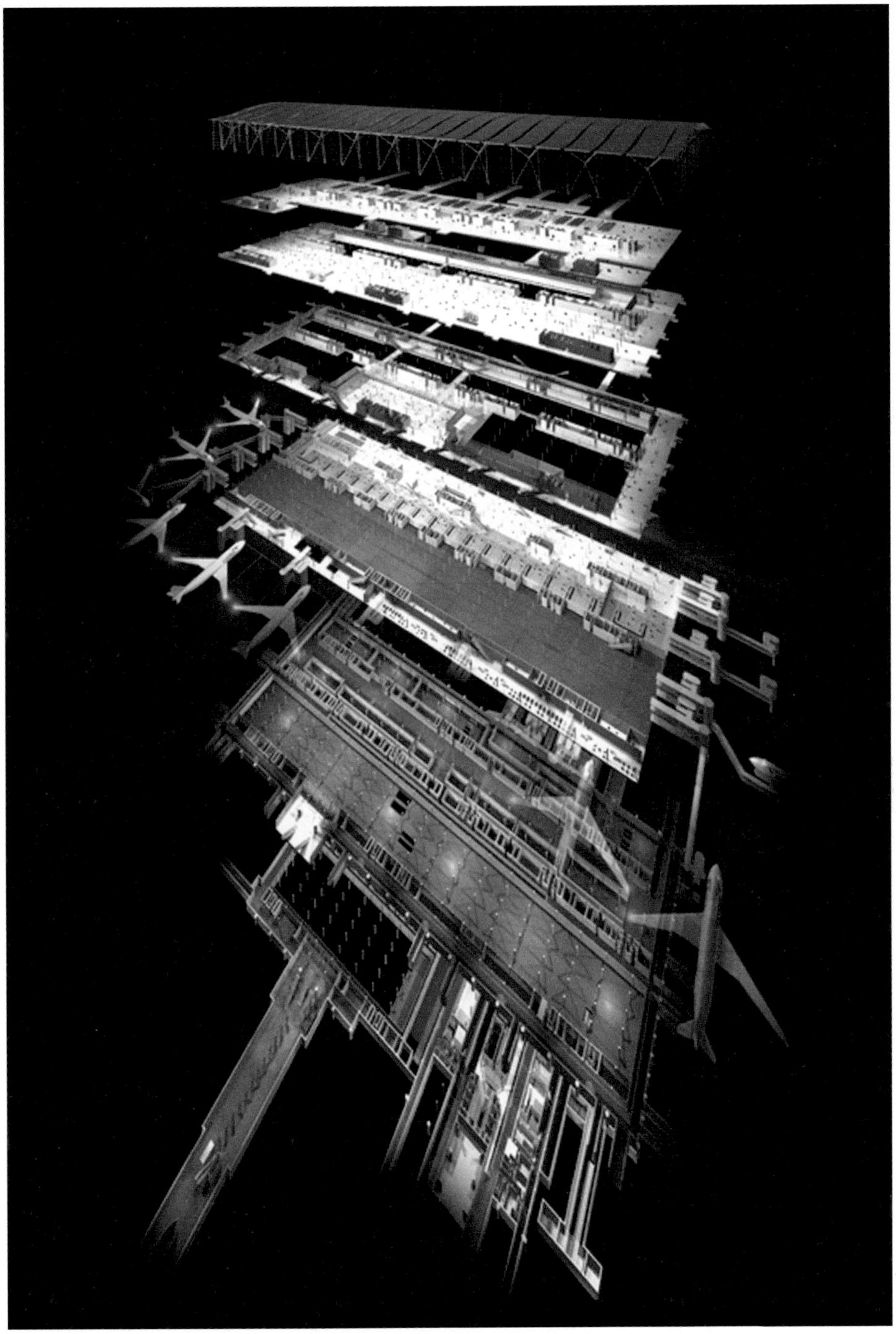

Figure 7.51 Heathrow Terminal 5 – Exploded view 2005.
SOURCE: YRM – (REPRODUCED WITH PERMISSION FROM RMJM)

Figure 7.52 Heathrow T5 2008 Airside Lounge surrounded by the retail and catering offer.
SOURCE: BAA

names have been introduced, so people know what to expect from the outlet before they sample it, as can be seen in the examples from Heathrow Terminal 5. The airline lounges are integrated with the catering outlets at the upper level of the Airside Lounge (Figures 7.52–7.55).

Figure 7.53 Heathrow Terminal 5 2008, Food and Beverage.
SOURCE: FORTNUM & MASON

Figure 7.54 Heathrow Terminal 5 2008, Food and Beverage.
SOURCE: GORDON RAMSAY

Figure 7.55 Heathrow T5, British Airways Concorde, First and Galleries Lounges 2008.
SOURCE: BRITISH AIRWAYS

Heathrow Terminal 5 was developed following the Atlanta model using a pinched loop APM configuration, supported by a back-up tunnel incorporating passenger conveyors, which helps to make it one of the most competitive transfer hubs in Europe, delivering minimum connection times of 45–60 minutes (Figure 7.56).

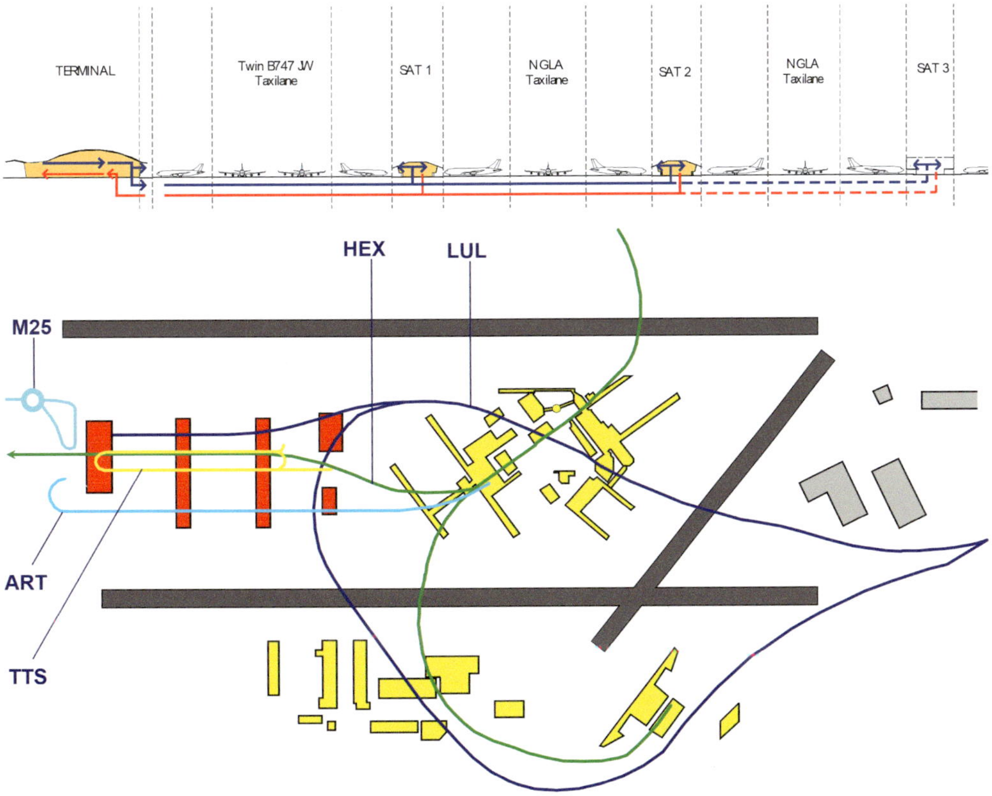

Figure 7.56 Section and site plan showing the tracked transit system (TTS) or APM 2008.
SOURCE: YRM (REPRODUCED WITH PERMISSION FROM RMJM)

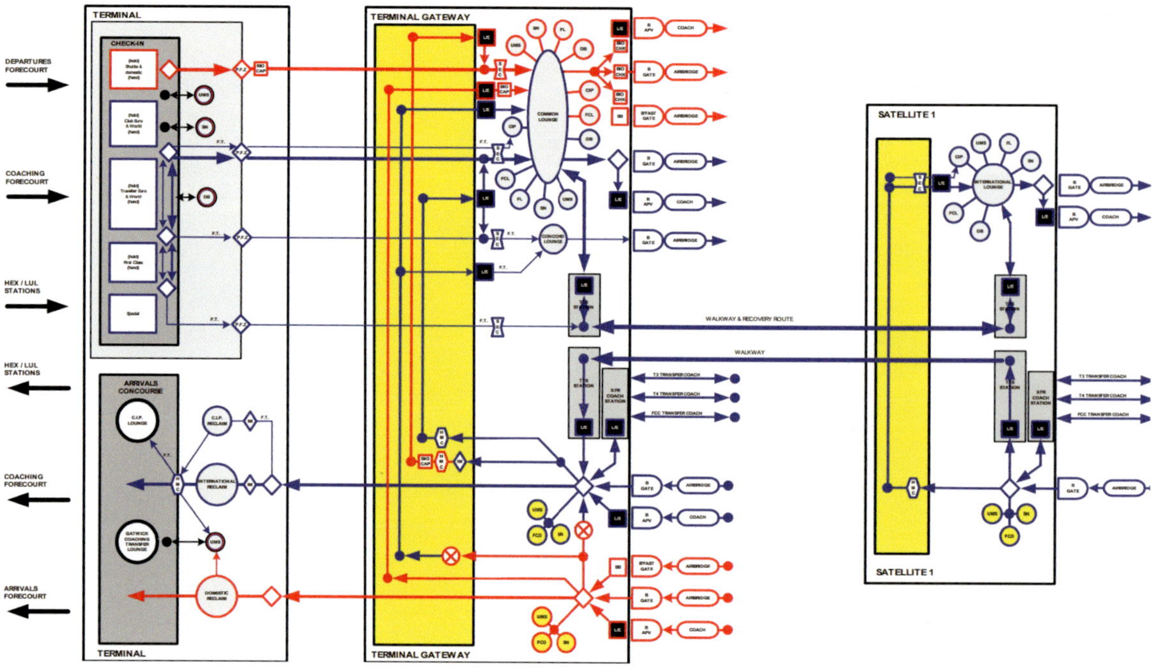

Figure 7.57 Heathrow Terminal 5 2008 – passenger flow logic showing the complexity associated with centralising transfer security.
SOURCE: YRM (REPRODUCED WITH PERMISSION FROM RMJM)

Key:

TTS – tracked transit system or automated people mover
ART – airside road tunnel linking Terminal 5 with Terminals 1, 2 and 3
HEX – Heathrow Express and Crossrail high-speed rail link to London
LUL – London Underground Ltd Piccadilly line
M25 – London orbital motorway

The resultant passenger and baggage flow logic is very complex, but has been integrated within the simple but striking form of the terminal and satellites (Figures 7.57–7.61).

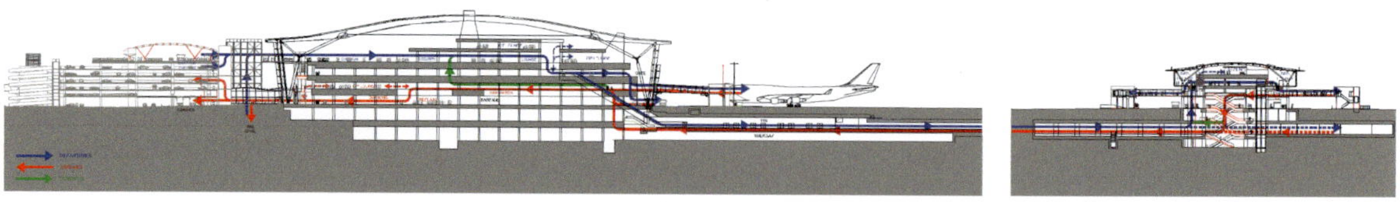

Figure 7.58 Heathrow Terminal 5 cross section showing passenger flow through terminal and satellite.
SOURCE: YRM (REPRODUCED WITH PERMISSION FROM RMJM)

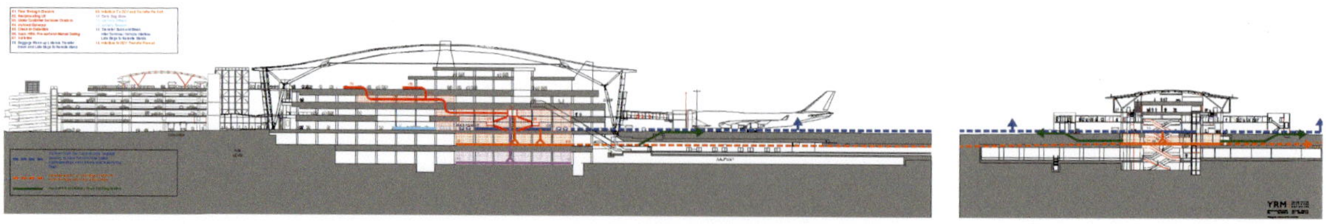

Figure 7.59 Heathrow Terminal 5 cross section showing baggage flow through terminal and satellite.
SOURCE: YRM (REPRODUCED WITH PERMISSION FROM RMJM)

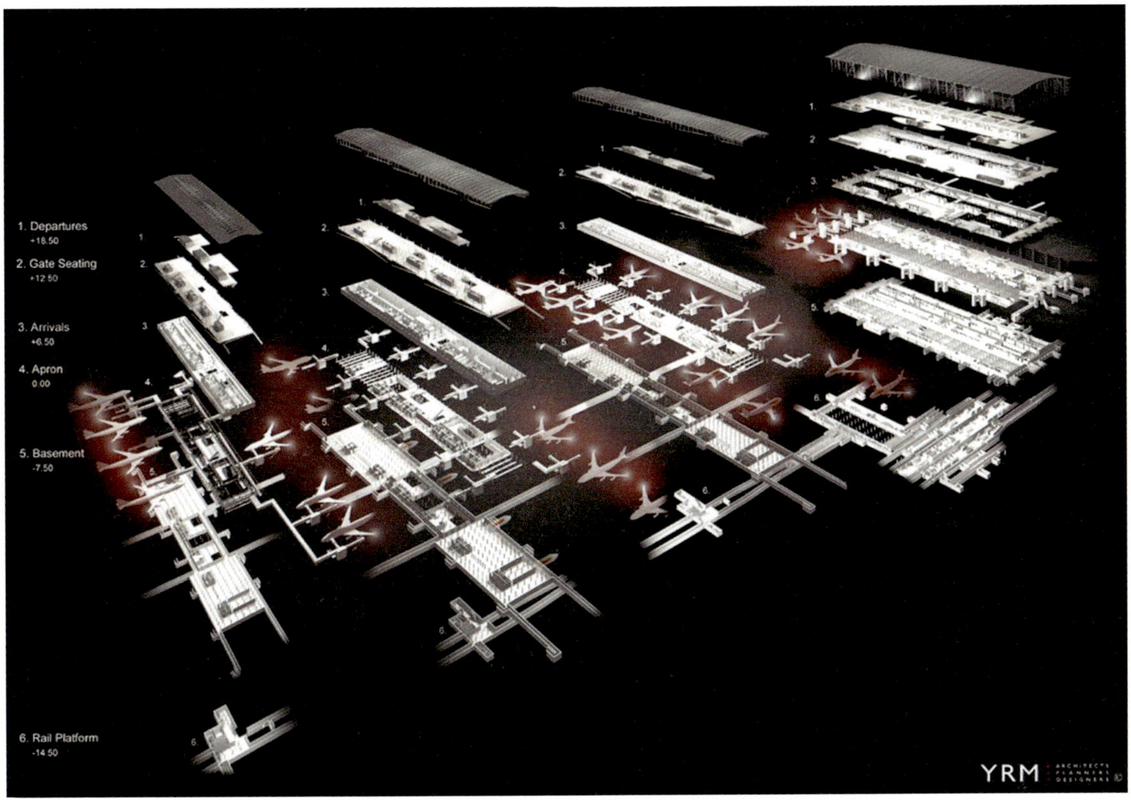

Figure 7.60 Heathrow Terminal 5 2008 – exploded axonometric of the Hub Infrastructure.
SOURCE: YRM (REPRODUCED WITH PERMISSION FROM RMJM)

Figure 7.61 Heathrow Aerial Perspective of Terminal 5 Complex 2008.
SOURCE: YRM REPRODUCED WITH PERMISSION FROM RMJM

Figure 7.62 Heathrow Ultra personal rapid transit (POD) 2010.
SOURCE: HEATHROW AIRPORT

The introduction of electric autonomous vehicles at Heathrow *'POD'* saw the beginning of a new personalised rapid transport system (PRT). An upgraded version of the system using larger minibus-sized vehicles is being considered as group rapid transit (GRT) for the 3rd runway expansion project. (Figures 7.62 and 7.63)

Figure 7.63 Ultra Personal Rapid Transit 2010.
SOURCE: ULTRA AND HEATHROW AIRPORT

Figure 7.64 Heathrow Terminal 2, Concept Design.
SOURCE: FOSTER + PARTNERS

Terminal 2 at Heathrow, originally known as Heathrow East, was designed by Foster + Partners to replace the original 1955 Frederick Gibberd Terminal 2 and Queens Building to become the new home for Star Alliance airlines and handle an estimated 20 million passengers every year (Figure 7.64). Architect Luis Vidal was later appointed by Heathrow's Spanish owner Ferrovial to take up the terminal design where the Foster+Partners master plan left off.

Heathrow Terminal 2 is unusual in two respects. First, the wavy roof incorporated northlights, a feature more usually found in factories but adopted because it allowed good daylight penetration into the interior without the accompanying heat gain, an important consideration as airports become more environmentally conscious. Second, due to the proximity of the London Underground under the ground floor, it was designed without both a deep basement and a departure baggage system. It relied instead on reusing the obsolescent baggage system in the adjacent mothballed Terminal 1. This will in turn produce additional challenges as the second phase of the terminal is designed and will result in additional costs and complications and an extended phasing programme to maintain business continuity (Figures 7.65 and 7.66).

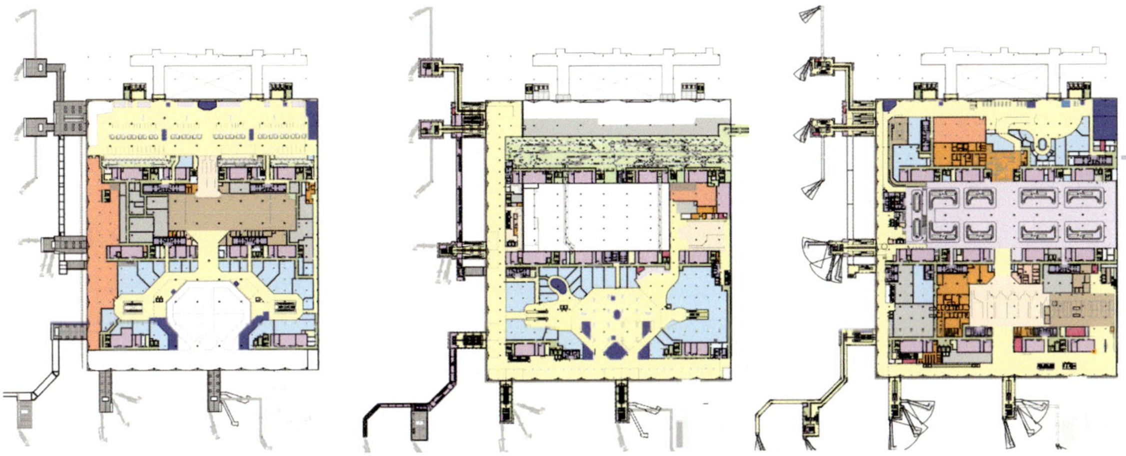

Figure 7.65 Heathrow Terminal 2, floor plans.
SOURCE: HEATHROW AIRPORT VIA ARCHITECTURE DAILY

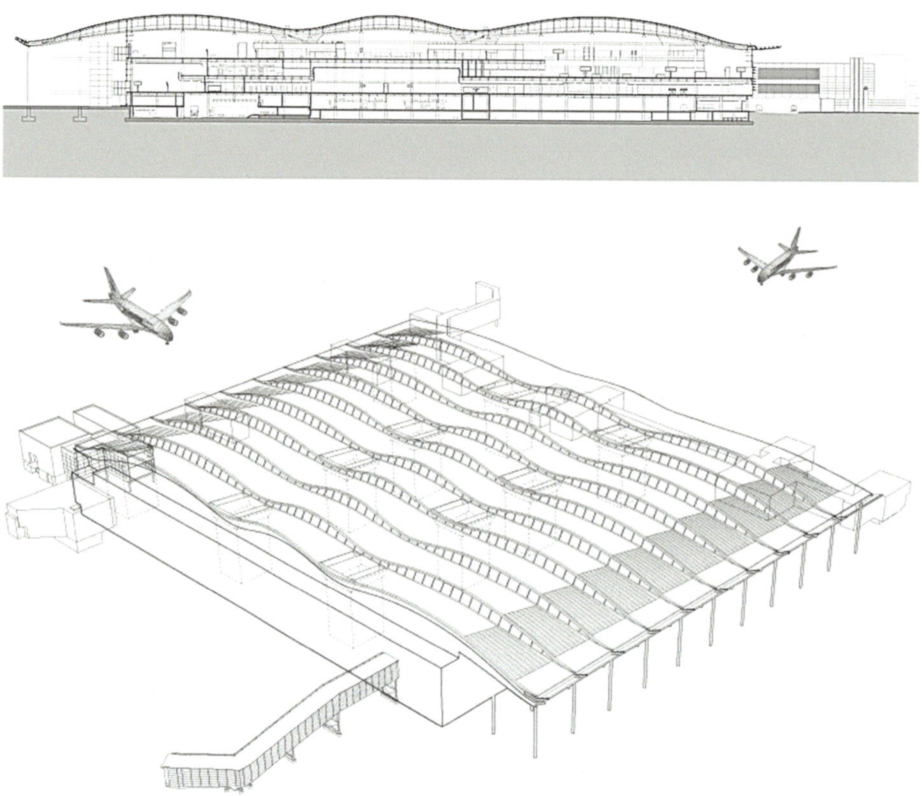

Figure 7.66 Heathrow Terminal 2, cross section.
SOURCE: HEATHROW AIRPORT VIA ARCHITECTURE DAILY

The Heathrow Expansion Project is planned to increase the capacity of Heathrow from the current 80 million passengers a year to 142 million passengers a year by building a third parallel runway, increasing terminal capacity and providing additional stands and ancillary facilities while keeping within the UK Government's environmental targets (Figure 7.67). The Heathrow Expansion Project is currently on hold, but Heathrow argues that demand for aviation will recover from COVID-19 and the additional capacity at an expanded Heathrow, the UK's only hub airport, would allow Britain as a sovereign nation to compete more effectively for trade.

Figure 7.67 Heathrow Expansion Project – Existing buildings are coloured orange while the proposed new buildings are highlighted in yellow.
SOURCE: HEATHROW AIRPORT

34. Beijing Capital (PEK) 2008

Beijing Capital Terminal 3 can be seen as an evolution on a larger scale, of the concepts that Foster developed initially at London Stansted and refined and expanded at Hong Kong Chek Lap Kok (Figures 7.68–7.70).

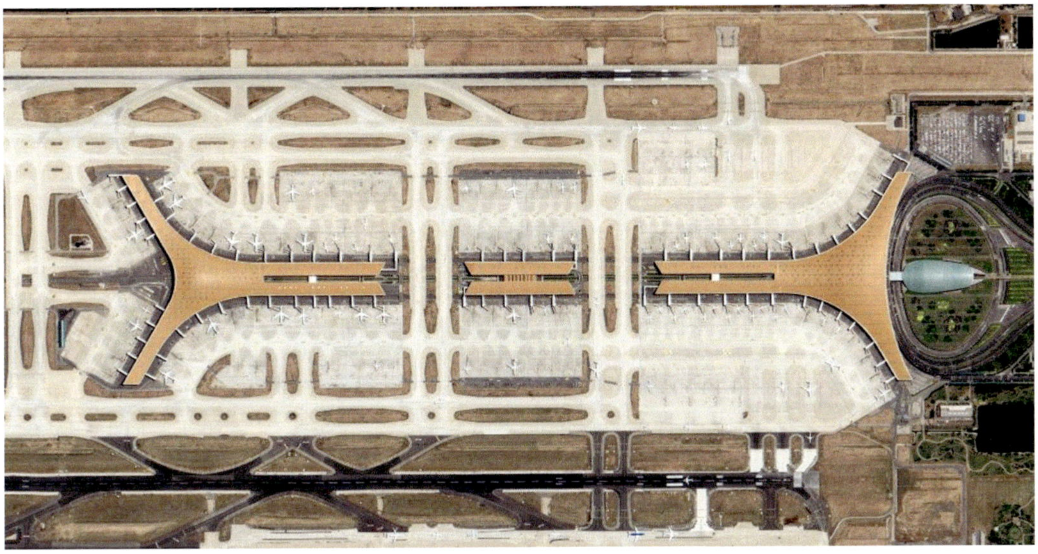

Figure 7.68 Beijing Capital Terminal 3 aerial view.
SOURCE: GOOGLE EARTH

Figure 7.69 Beijing Capital Interior.
SOURCE: FOSTER + PARTNERS, PHOTOGRAPHER: NIGEL YOUNG

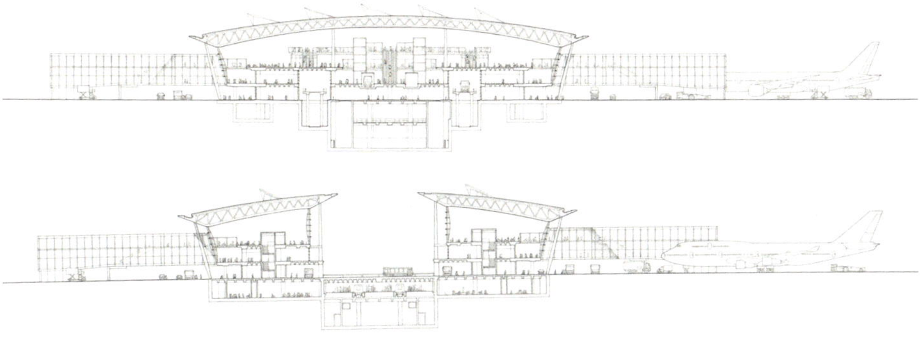

Figure 7.70 Beijing Capital sections through satellite.
SOURCE: FOSTER + PARTNERS

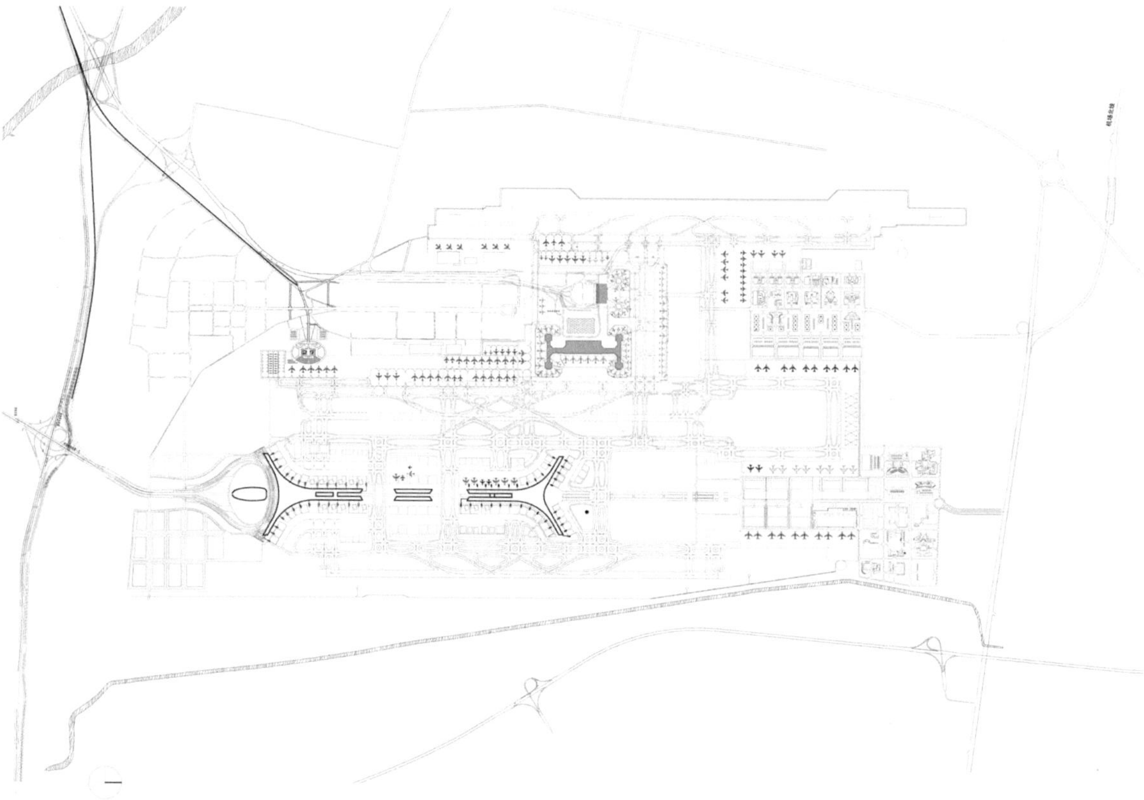

Figure 7.71 Beijing Capital master plan.
SOURCE: FOSTER + PARTNERS

The efficiency of the layout has enabled Beijing Capital Airport to be the second airport in the world to exceed 100 mppa, the great majority of whom pass through the Foster-designed terminal.

The airport has three widely spaced runways and three main terminals. Terminal 1, with 60,000 m^2 of space, opened on 1 January 1980 and replaced the smaller existing terminal, which had been in operation since 1958. The latter continues as a VIP and charter terminal. Terminal 2 opened on 1 November 1999, with a floor area of 336,000 m^2. This terminal was used to replace Terminal 1 while the latter was undergoing renovation. It can handle twenty aircraft on stands connecting directly to the terminal building. The much larger terminal 3, covering approximately 1,300,000 m^2, was designed for 50 million passengers by a consortium of Netherlands Airport Consultants (NACO), UK Architect Foster and Partners and ARUP and opened for the Olympics in 2008 (Figure 7.71).

Beijing Capital Airport Terminal 3 adopts a composite form with both a finger pier and satellites and is one of the biggest in the world. The building, which opened in 2008, resembles a giant dragon and has an aerodynamic roof (Figure 7.72 and 7.73).

The mid-field terminal, together with the satellite concept and associated transit, can most often be found in the largest of airports, because it can create the greatest possible number of jetty-served stands within reasonable walking distance and far more than any equivalent finger per concept.

Foster + Partners describe Terminal 3 as follows:

> *Located between the existing eastern runway and a planned third runway, the terminal building and Ground Transportation Centre (GTC) together enclose a floor area of 1.3 million square metres and were designed to accommodate 50*

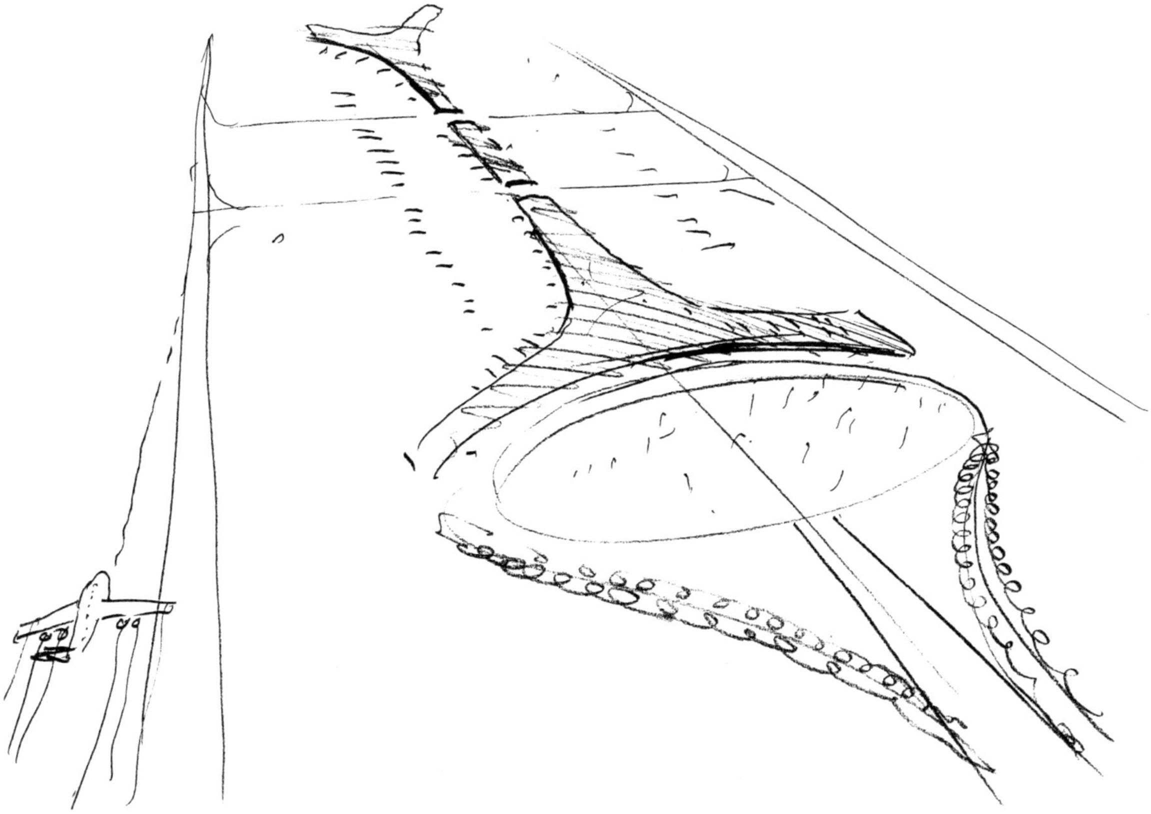

Figure 7.72 Beijing Capital, concept diagram by Norman Foster.
SOURCE: FOSTER + PARTNERS

million passengers per annum by 2020. Although conceived on an unprecedented scale, the terminal's design expands on the new airport paradigm created by Stansted and Chek Lap Kok. Designed for maximum flexibility to cope with the unpredictable nature of the aviation industry, like its predecessors, it aims to resolve the complexities of modern air travel, combining spatial clarity with high service standards.

Transport connections are fully integrated, walking distances for passengers are short, with few level changes and transfer times between flights are minimised. Like Chek Lap Kok, the terminal is open to views to the outside and planned beneath a unifying roof canopy, whose skylights are both an aid to orientation and sources of daylight – the colour cast changing from red to yellow as passengers progress through the building.

Figure 7.73 Beijing Capital, 2008.

SOURCES: TOP LEFT: MA WENXIAO/SINO; TOP RIGHT: WIKIPEDIA; BOTTOM: FOSTER + PARTNERS

35. Dubai International (DXB) 2010

Today, Dubai International boasts the world's only underground terminal. It provides direct access to the satellite. This was necessitated by the limited size of the site, as development had grown up around what had originally been a desert location, constraining expansion (Figures 7.74 and 7.75). When completed, Terminal 3 was the largest building in the world by floor space, with over 1,713,000 m^2 (18,440,000 ft^2) of space, capable of handling 43 million passengers in a year. The airport has now surpassed Heathrow as one of the world's busiest international airports, serving more than 89 million passengers a year in 2018, and is unusual with more than 80% of passengers being transfers.

Figure 7.74 Layouts of Underground Terminal Departures and Arrivals areas.
SOURCE: DUBAI AIRPORTS

Figure 7.75 Dubai International Underground check-in concourse.
SOURCE: DUBAI AIRPORTS

Figure 7.76 Dubai International Retail Mall in the satellite concourse.
SOURCE: DUBAI AIRPORTS

Paul Griffiths, Chief Executive of Dubai Airports, has said of the underground terminal:

> *Terminal 3 is unique. Firstly, it's one million square metres – twice the size of Heathrow's Terminal 5. Secondly, the whole terminal has been built underground, yet it feels like a large cathedral. Thirdly, it is the first terminal in the world that has been designed specifically for the A380.*
>
> (2020 ACI EUROPE Airport Business, 2008)

As the new Dubai World Central has been delayed, plans are being developed for Dubai International to grow even further, largely by creating additional remote stands. At present, its underground terminal is unique, but it shows an approach that maximises efficient land use and so may be applicable to constrained airports in future. The satellites are dominated by an extensive retail offer making passenger circulation and wayfinding less intuitive and more complex (Figure 7.76).

Dubai International Airport opened in 1960 with a simple terminal and runway of just 1,800 m after the late Ruler of Dubai, Sheikh Rashid bin Saeed Al Maktoum, ordered its construction.

The 1970s saw the construction of a new terminal to accompany the lengthened runway which included unusual external, spiral ramps to the four boarding gates. The expansion of the airport had a new impetus in the 1990s with the growth of Emirates Airlines which had only been founded in 1985.

The expansion of Dubai International has been driven by the growth of Emirates Airlines which has an all-wide-body aircraft fleet and operates the largest number of A380 aircraft in the world. The result is a very large transfer hub primarily serving Emirates Airlines and handling more than 80 million passengers per annum (mppa), which is anticipated to grow to over 100 mppa. Faced with a very constricted site for such a large hub, airport planners Aéroport de Paris (ADPi) proposed that the terminal should be placed under the taxiway system that surrounded the satellite concourse. The layout maximizes the number of aircraft stands provided in combination with providing an efficient taxiing system. The underground concourses feature very high ceilings and are brightly lit with artificial lighting, so the passenger has no real perception that they are indeed underground and the resultant vertical circulation is more effortless than other satellite designs (Figures 7.77–7.81).

Figure 7.77 Dubai Terminal 3 2008 Entrance Canopy.
SOURCE: GULF NEWS

Figure 7.78 Dubai Terminal 3 Underground Reclaim.
SOURCE: BUSINESS TODAY ME

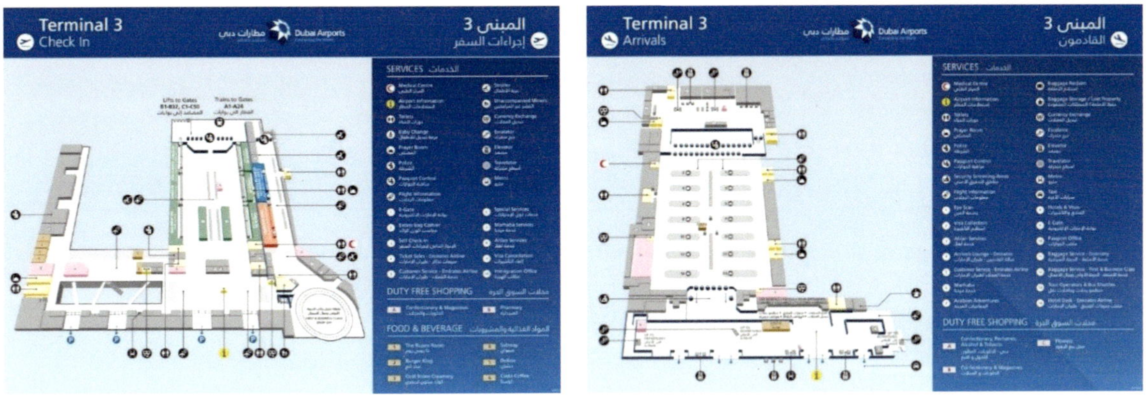

Figure 7.79 Layout of Underground Terminal 2008.
SOURCE: DUBAI INTERNATIONAL AIRPORT

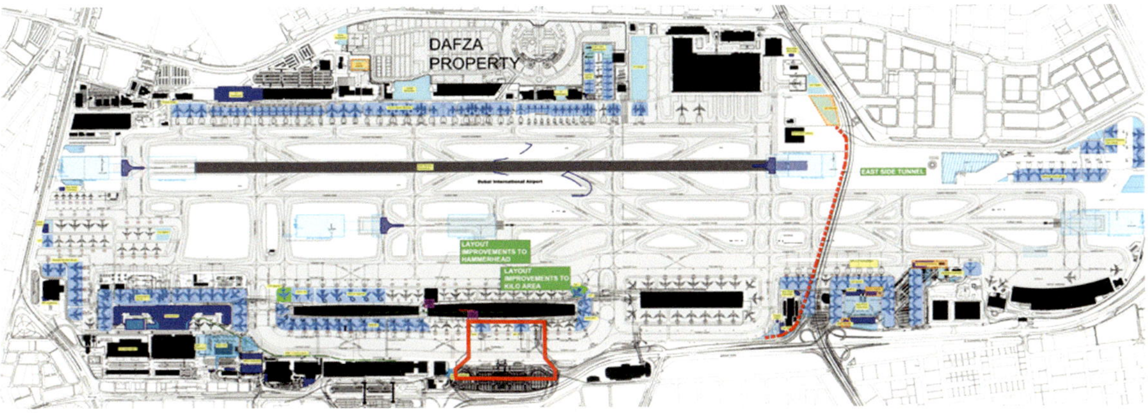

Figure 7.80 Dubai International Site Plan 2008 – Underground Terminal outlined in red.
SOURCE: ATKINS

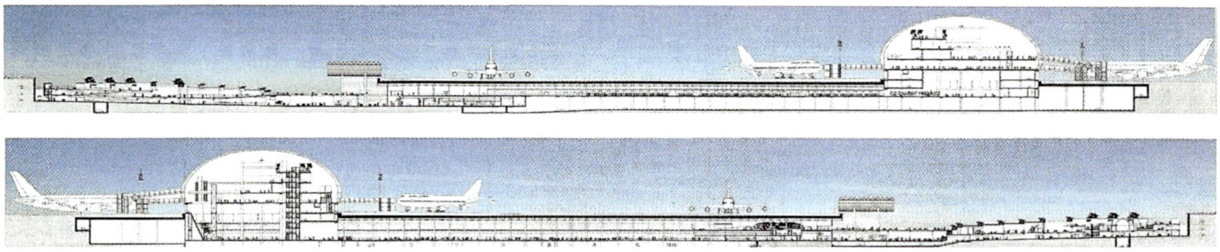

Figure 7.81 Dubai International, Terminal 3, cross section.
SOURCE: AIRPORTS INTERNATIONAL SEPTEMBER 2002

Dubai, designed by Yamaski's successor practice HOK, reuses some of the motifs of the St. Louis window pattern in a simplified linear version of the roof.

An underground terminal was briefly considered for Heathrow Terminal 5 and the new Beijing Daxing airport to ptimize the use of the available site area, but both airports have to date followed a more conventional approach. However, as airports reach capacity and run out of development space, it is anticipated that the underground terminal concept will, in the future, gain more support. Concourse A is the world's first A380 purpose-built facility and also features the world's first multi-level boarding for First and Business Class passengers directly from the respective lounges. The concourse includes one 4-star hotel and one 5-star hotel, First and Business Class Lounges and duty-free areas. Concourse B is directly connected to Terminal 3 and is dedicated exclusively to Emirates. The terminal has 10 floors (4 basement, ground floor and 5 above-ground floors).

One of the most developed concepts of specialised passenger segmentation can be found at the airport where premium business class passengers are provided with a completely separate floor level and first-class passengers are on another floor above the main departures level (Figure 7.82 and 7.83). It demonstrates that organisational evolution of the passenger terminal is one of increasing specialisation and separation of passenger flows and facilities, either horizontally or vertically, to serve smaller and more targeted segments of the market.

Figure 7.82 Dubai International (DXB) Concourse A 2013.
SOURCE: DUBAI AIRPORTS VIA PERKINSWILL.COM

Figure 7.83 Emirates First Class Lounge Dubai 2013.
SOURCE: DUBAI AIRPORTS VIA LOUNGEREVIEW.COM

36. iGA Istanbul (IST) 2019

iGA Istanbul Airport, located in the Arnavutköy district of Northwest Istanbul, was planned as a major hub to replace Ataturk Airport which has now closed. It was designed to accommodate 90 million passengers per annum in its first phase but ultimately planned to handle 150 million passengers annually, to become one of the world's busiest airports (Figures 7.84 and 7.85).

As the largest Public–Private Partnership (PPP) project of the Republic of Türkiye, iGA Istanbul Airport is being delivered under a Build–Operate–Transfer model, where iGA, a 100% private Special Purpose Vehicle (SPV) Company, is the operating company with 25 years of concession rights and responsible for the delivery of the entire project, including the financing.

The first phase of the project, completed in 42 months, started 'soft operations' on October 29, 2018. After a transition period of approximately 6 months, all existing operations at Ataturk Airport were relocated to the new iGA Istanbul Airport during 5 and 6 April 2019. Relocation of the old airport to the 45-km-away iGA Istanbul Airport as part of the Operational Readiness and Airport Transfer (ORAT) was the largest and one of the most successful transfer operations, and was completed in 33 hours. This included the relocation, set-up and operation of 47,300 tonnes of equipment and materials along with 150 aircraft of Turkish Airlines and more than 70,000 personnel.

In addition, the entire airspace of the Istanbul Terminal Manoeuvring Area (TMA) has been changed to accommodate the runways of iGA Istanbul Airport with dual parallel independent operations. The first day of operations started on April 6, 2019, with 260 flights. By July 2023, Eurocontrol records show an average daily operation of some 1,500 Air Traffic Movements (ATMs) and 240,000 passengers per day.

With its current Phase 1 set-up, ATM capacity is for up to 2,000 air traffic movements per day, facilitated by a parallel runway system (3 main and 2 auxiliary runways) that enables the arrival and departure of over 120 aircraft per hour. The terminal is capable of welcoming over 20,000 passengers during peak hours. In 2022, Istanbul handled 64,486,178 passengers, and by the end of May 2023, there was further traffic growth of 35%.

Terminal building covers 1.4 million m^2, and is one of the largest terminals yet designed. It has 14 landside entrance/exits, 566 check-in desks, 112 self-check-in kiosks, 258 passport control points, of which 30 are e-passport, 112 security check points, 114 contact and 33 bus gates, 22 information kiosks with live agents, and convenient connectivity with the city of Istanbul via 2 major highways and a dedicated metro line.

Figure 7.84 iGA Istanbul – location plan.
SOURCE: WIKIPEDIA

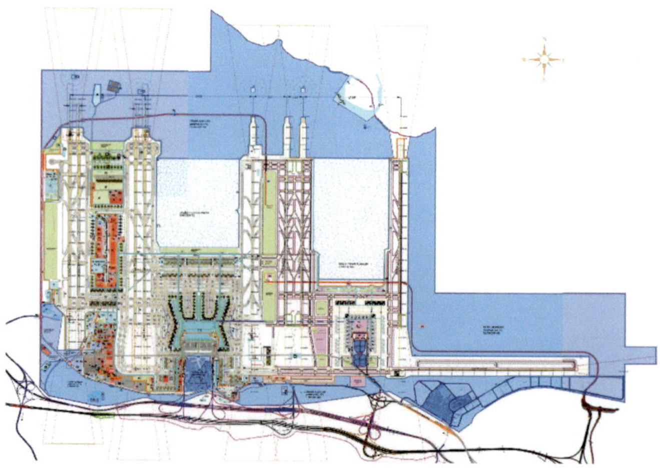

Figure 7.85 iGA Istanbul – master plan.
SOURCE: IGA ISTANBUL AIRPORT

Figure 7.86 Aerial view of iGA Istanbul.
SOURCE: WIKIPEDIA – KULTTUURINAVIGAATTORI

The architects for the terminal were a consortium comprising Grimshaw Architects, Nordic Office of Architecture and Haptic Architects. Grimshaw describe their design approach as follows (Figures 7.86–7.91):

> *The design draws on the architectural character of Istanbul – a city rich with colour, pattern and history since its Byzantine origins – and includes vaulted ceilings pierced with skylights that draw in diffuse daylight and create an open, lofty volume with clear lines of sight. Focused beams of sunlight illuminate key areas in the terminal, such as check-in, security, customs and retail areas.*

To facilitate the easy movement of luggage, iGA Istanbul Airport is equipped with an individual-carry type, destination-coded vehicle (DCV) baggage handling system that processes 21,600 bags per hour with 28 latest-technology explosives detection systems

Figure 7.87 iGA Istanbul Departures forecourt and main entrance.
SOURCE: WIKIPEDIA A. SAVIN, WIKICOMMONS

Figure 7.88 Aerial view of iGA Istanbul in operation.
SOURCE: IGA

Figure 7.89 View of iGA Istanbul at dusk with the forecourt and car parks in the foreground.
SOURCE: IGA

Figure 7.90 iGA Istanbul – main entrance.

SOURCE: WIKIPEDIA PHOTOS: ARNE MÜSELER/ARNE-MUESELER.COM/CC-BY-SA-3.0/HTTPS://CREATIVECOMMONS.ORG/LICENSES/BY-SA/3.0/DE/DEED.DE

Figure 7.91 iGA Istanbul – interior of pier.

SOURCE: WIKIPEDIA PHOTOS: ARNE MÜSELER/ARNE-MUESELER.COM/CC-BY-SA-3.0/HTTPS://CREATIVECOMMONS.ORG/LICENSES/BY-SA/3.0/DE/DEED.DE

(EDS) machines, 10,600 bag fully robotic early bag storage capacity, 48 make-up positions and 28 baggage reclaim carousels, providing an in-system time of less than 20 minutes and an almost zero baggage loss rate.

Additionally, Istanbul Airport Terminal provides extensive commercial space of over 250,000 m^2, including 142 branded retail outlets and 42 duty-free shops, a wide variety of 82 food & beverage options distributed over 150+ locations. It also has an in-terminal hotel with 451 rooms, a spa centre, a museum, 6 airline passenger lounges and many more tailor-made services for special groups, ensuring a stress-free and convenient travel experience for passengers visiting Istanbul and Türkiye or just travelling hrough iGA Istanbul Airport (Figures 7.92 and 7.93).

Figure 7.92 iGA Istanbul – Duty Free and Retail Mall.

SOURCE: WIKIMEDIA CC-BY-SA 2.0 + MATTI BLUME

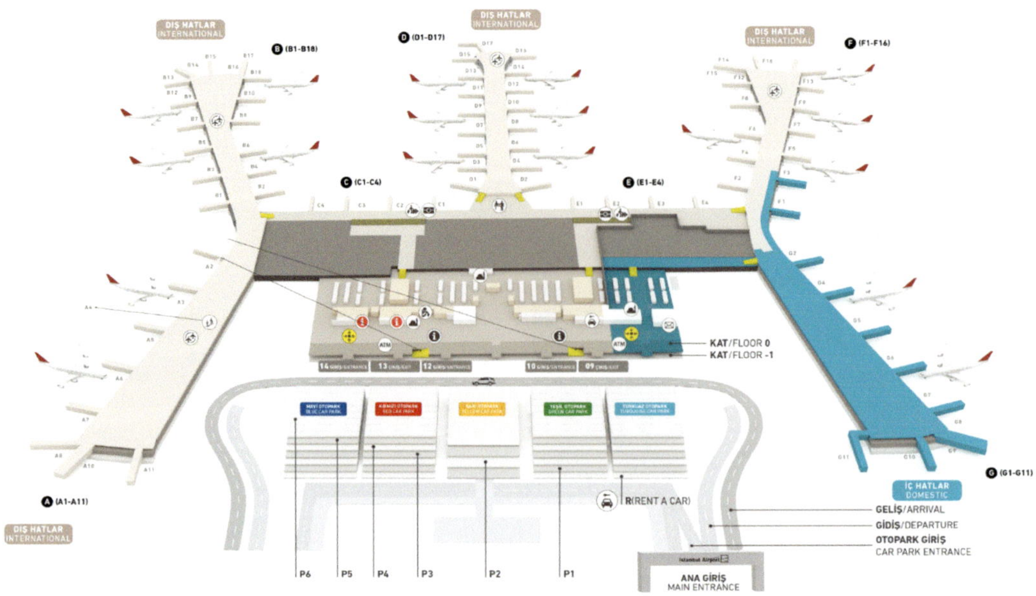

Figure 7.93 iGA Istanbul – floor plan showing the simple orthogonal terminal layout and extensive pier system that obviates the need for remote satellites and transit links in the first phase. The extensive central airside retail zone is highlighted in grey.
SOURCE: IGA ISTANBUL AIRPORT

Istanbul Airport is certified by Skytrax, well respected as an independent airport assessor, as a 5-Star Airport for facilities, comfort, cleanliness, shopping, food & beverages, staff service and security/immigration.

Istanbul Airport provides an excellent range of services and facilities for customers in a hub context, and this includes a broad range of shopping and dining options, in addition to practical elements such as showers, rest areas, workspaces, and family-friendly areas. While the terminal is large, there are regular moving walkways and buggy assistance points for PRM, elderly, and families. The airport continues to develop its cultural offering which currently includes museum exhibits, musical performances, and art projects, as well as many areas of natural landscaping to enhance the terminal decor.

Besides passenger operations, Istanbul Airport has several specially planned zones for various airport and airport-related operations:

- *Cargo area*: Located between Runways 16/34 and 17/35, direct access to runways on the airside and 2 major highways on the landside, providing a total cargo space of 1.4 million m^2 with a capacity of 5.5 million tonnes and 41 dedicated aircraft stands.
- *Aircraft maintenance, repair and overhaul (MRO) area*: Located between Runways 16/34 and 17/35, it provides an area of approx. 650,000 m^2 for MRO hangars, including dedicated aprons with 25 aircraft stands.
- *Sea port*: On the Black Sea coast side of the airport, there is a dedicated port in operation capable of handling 100,000 and 50,000 gross tonne ships at the same time for jet fuel delivery via ships.
- *Airport city development*: An area of about 40 million m^2 (4,000 hectares) of a total airport site area of 78 million m^2 (7,800 hectares) has been reserved and planned for airport city development. The airport city includes, but not limited to, hotels & accommodation zones, commercial & shopping zones, a free trade zone, park and recreational zones, residential zones, healthcare facilities, exhibition facilities, education facilities, religious facilities and supporting infrastructure.

37. Beijing Daxing (PKX) 2019

The original master plan for Beijing Daxing was developed by NACO, but the competition-winning scheme from Aeroport de Paris and Zaha Hahid proposed a large terminal with five arms as piers. The 6th arm was formed by a landside transport interchange. The new mega-airport hub was designed to handle up to 45 million passengers per year by 2021 and reach 100 million in the future with the development of a second terminal. Unlike the Atlanta 'Toast Rack', this concept prioritises short passenger walking distances over ease of aircraft movement (Figures 7.94 and 7.95).

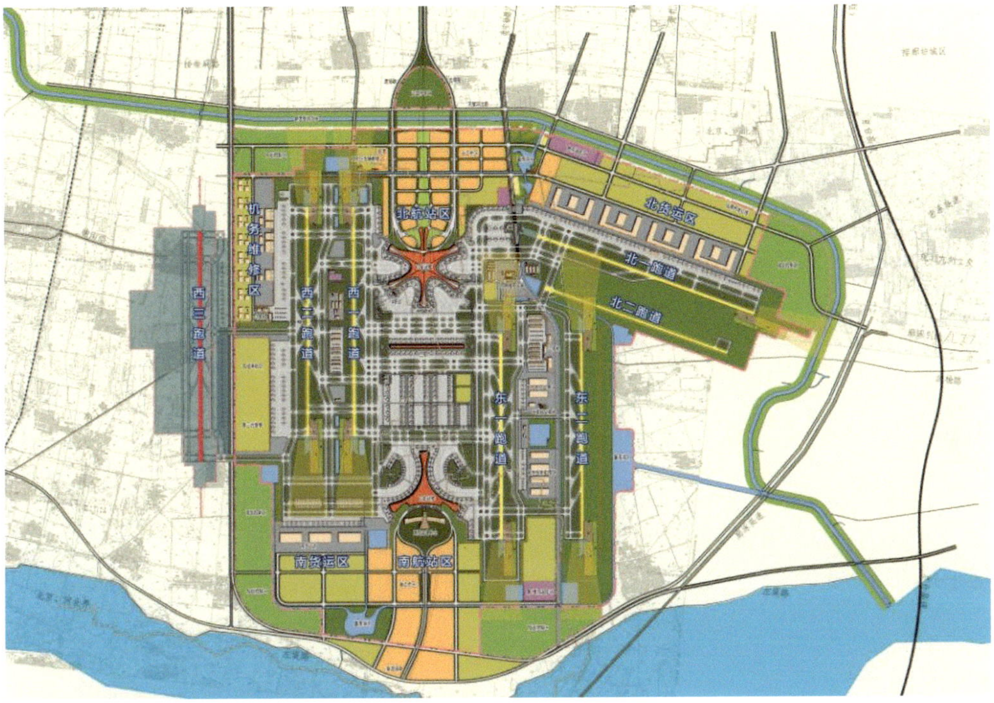

Figure 7.94 Beijing Daxing master plan.
SOURCE: PKK VIA KASKUS.CO.ID

Figure 7.95 Beijing Daxing, aerial view, nearing completion 2019.
SOURCE: WIKIPEDIA PHOTO: 王之桐

The definitive closed and curved form of the terminal, like Saarinen's earlier TWA building, may make later modification difficult, though the very large internal spans do provide scope for alteration of the layout. Only time will tell whether this concept provides sufficient flexibility to accommodate growth and change.

At the time of opening in September 2019, Daxing Airport had four runways and 79 stands, of which 60 were jetty-served. Eventually, the airport will have 7 runways in total and will be able to serve about 620.000 flights annually (Daxing-PKX, 2019). The whole airport is aligned on its north-south axis with the gates that lead to Tianamen Square and the Holy City to become the newest entry point to Beijing. The airport has been named Phoenix Gate. The terminal building design team was led by the late architect Zaha Hadid. Nicknamed The Starfish of Beijing because of its shape, it has a large central space and 6 'tentacles', resembling a giant 'octopus' (Figure 7.96). The sectional concept has international check-in at the top level and a mixed domestic departures and arrivals level on the floor below. Below that again is an international airside departure lounge and the lowest passenger level is international arrivals. The passengers will have to pass through the central area, where all the shops and leisure facilities are located and then head to their departure gate, situated along one of the arms.

ZHA's completed scheme conceptually consists of a compact radial design of six sculptural forms within the terminal's vaulted roof. The 100-m-long structural spans allow flexibility for future reconfiguration (Figures 7.97–7.103).

Figure 7.96 Roof Plan – Beijing Daxing International Airport by Zaha Hadid Architects © Zaha Hadid Architects.

A benefit of the hexagonal geometry is that the walking distance between the centre and each gate has been limited to 600 metres, the maximum recommended by IATA.

Natural light is brought into the building through a network of linear skylights, which also act as a method of wayfinding to guide passengers to and from their departure gates. All passenger amenities are located in a court at the centre of the terminal (Figure 7.104 and 7.105).

A new high-speed railway service, the Beijing–Xiong'an intercity railway, is under construction and will start from Beijing West railway station. It will connect the urban area of Beijing, Daxing District of Beijing, the new airport, Bazhou and Xiong'an. The section between the airport and Beijing will operate at speeds of 250 km/hour and the section between the airport and Xiong'an will

Figure 7.97 Level 4 – Beijing Daxing International Airport by Zaha Hadid Architects © Zaha Hadid Architects.

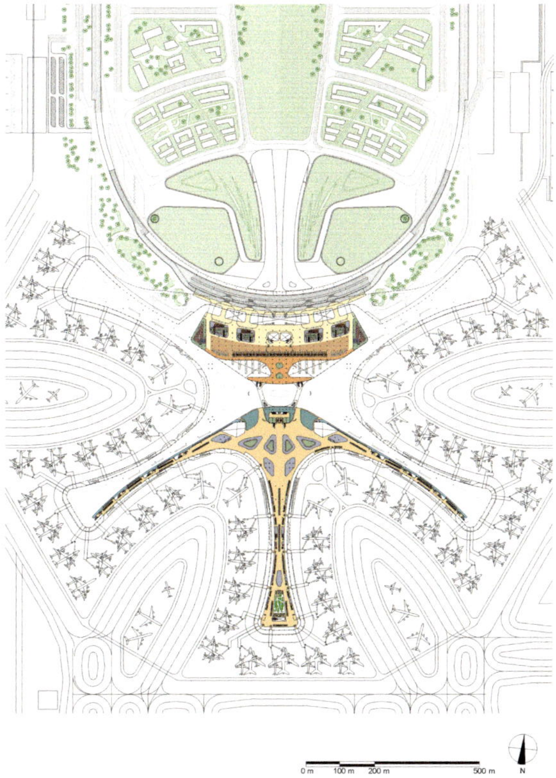

Figure 7.98 Level 3 – Beijing Daxing International Airport by Zaha Hadid Architects © Zaha Hadid Architects.

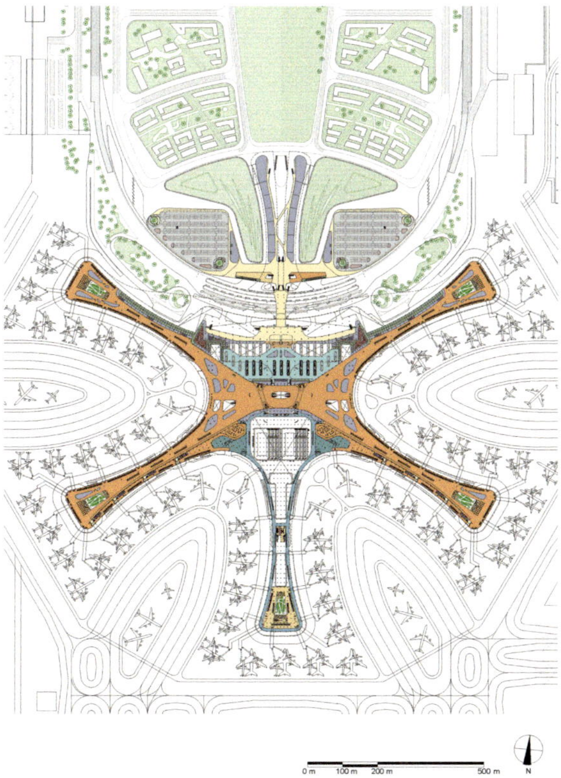

Figure 7.99 Level 2 – Beijing Daxing International Airport by Zaha Hadid Architects © Zaha Hadid Architects.

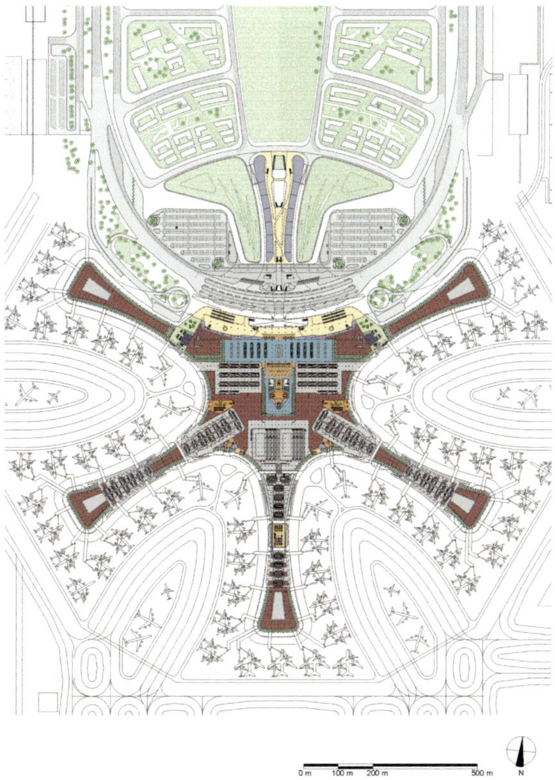

Figure 7.100 Level 1 – Beijing Daxing International Airport by Zaha Hadid Architects © Zaha Hadid Architects.

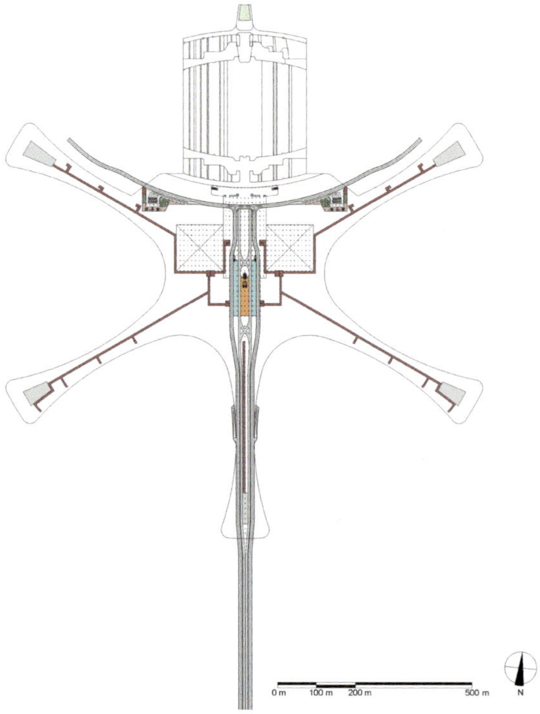

Figure 7.101 Basement Level 1 – Beijing Daxing International Airport by Zaha Hadid Architects © Zaha Hadid Architects.

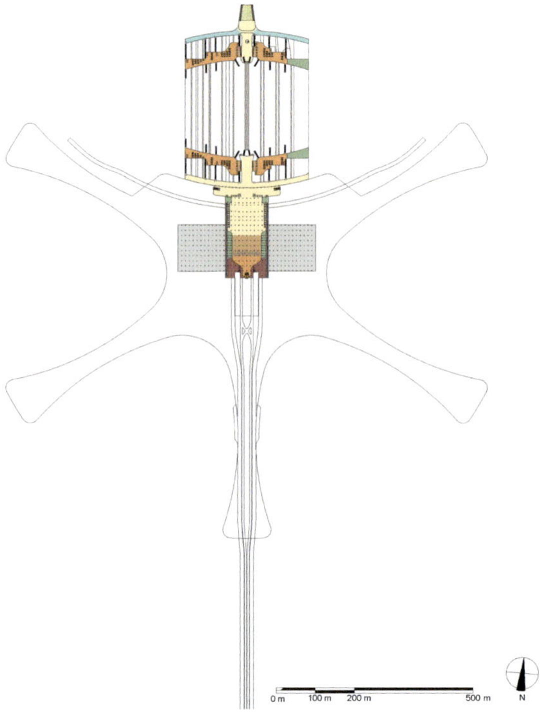

Figure 7.102 Basement Level 2 – Beijing Daxing International Airport by Zaha Hadid Architects © Zaha Hadid Architects.

Figure 7.103 Computer rendering, Beijing Daxing, International Airport by Zaha Hadid Architects, © Zaha Hadid Architects.

Figure 7.104 Beijing Daxing Roof.
SOURCE: WALLPAPERCAVE.COM

Figure 7.105 Beijing Daxing Interior.
SOURCE: ARCHITECTS' JOURNAL

Figure 7.106 Beijing Daxing location plan.
SOURCE: BKX VIA AUSTRALIAN BUSINESS TRAVELLER

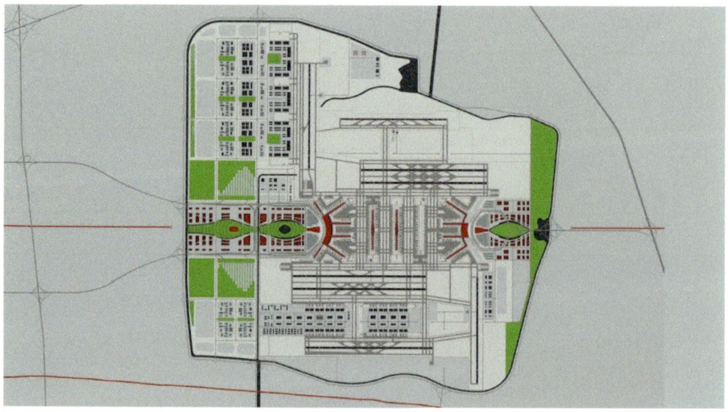

Figure 7.107 Beijing Daxing – original master plan.
SOURCE: NACO

operate at speeds of 350 km/hour, cutting the end-to-end journey time between the two cities to about 30 minutes and Beijing to the airport in about 11 minutes (Figure 7.106).

The master plan incorporates two terminals, each supported by a major airport city type of development on the approach to the terminal. The runways are widely spaced to allow ancillary facilities, such as cargo and logistics to be located between them, while peripheral areas are reserved for aircraft maintenance, repair and overhaul (MRO) and other airport supporting facilities.

Aerial photographs show there are many undeveloped sites around the airport, but with its exceptionally good connectivity, it is a natural candidate to become an aerotropolis. Some of the close-in developments that will form part of the aerotropolis are shown on the original master plan by NACO (Figure 7.107).

CHAPTER 8

FUTURE DEVELOPMENTS

There is a strong link between economic well-being, measured by GDP and the propensity to travel. Social mobility within Europe has been helped by the growth of low-cost carriers (LCCs) and has in turn stimulated significant growth, both in business and visiting friends and relatives (VFR) travel. For the future, both Boeing and Airbus forecast the greatest growth in Asia. So that is where we should expect the most innovative design in the future, while in Europe and the USA, where the market is mature, we should expect a greater focus on process improvement to eke more capacity out of existing infrastructure. Political ambitions can also make airport developments vulnerable, as can be seen in the cancellation of the new Mexico City Airport (MEX).

The Covid-19 pandemic has brought an abrupt halt to the growth of air travel and indeed has resulted in a dramatic contraction during 2020 and 2021. Whether this will result in a significant, permanent change to the growth trajectory cannot be determined at this stage. In the past where events have slowed growth, for example, the 9/11 attack of 2001, the SARs epidemic of 2003 or the global recession of 2008, growth has quickly resumed.

There is, however, one example of an event that had a permanent effect on the structure of the industry. The 9/11 attack on New York's Twin Towers has led directly to the imposition of much stricter security controls, altering the infrastructure of airports and the perception and cost of air travel. The tragedy also contributed directly to the demise of Concorde, as many of the clientele using the supersonic service were killed in the attack and demand never recovered. The demise of the British Airways Concorde service had an important commercial dimension rather than being solely a safety or re-certification issue.

As a direct result of Covid-19, many multinational companies have elected to do business through video conferencing rather than face-to-face meetings and have even suggested they will never return to using expensive air travel. Covid-19 could therefore lead to a requirement for more sophisticated passenger screening and Port Health medical facilities, a permanent reduction in business travel and the associated provision of airline lounges in airports and the further commoditisation of air travel through the growth of LCCs.

The concept of the airport as an interchange will continue to evolve as the synergy between land, sea and air transfer is realised, giving rise to new linking spaces between each of the processing elements that have been called Third Space. Poland's Centralny Port Komunikacyjny (CPK) arguably boasts the most ambitious Rail/Air Interchange to date and incorporates a major atrium as a linking space between air and rail terminals.

Retail concepts will have to change to respond to online buying through the internet, potentially making the conventional approach of walk-through duty-free shops redundant and giving way to the show rooms of the experience economy (e.g. Changi, Vancouver).

DOI: 10.4324/9781003323624-8

Health, safety and security measures will become ever more demanding and are expected to require additional space, but the focus will be on making the process as seamless and non-intrusive as is compatible with meeting ever higher standards.

Designing within environmental limits – noise, air quality and global warming – is only just beginning, and as yet, airports have not set themselves very demanding goals. Public opinion will no doubt require both airports and airlines to reduce radically their carbon footprint and overall environmental impact. This can be seen in the controversy that surrounds both Red Sea Umluj Tabuk and Western Sydney International (WSI).

However, there are a number of promising avenues to explore such as autonomous vehicle movement and electrification of ground vehicles and alternative fuels for aircraft.

It is expected that aviation will take some time to recover following the Covid-19 pandemic, and in the meantime, airports will have to concentrate on process improvements to reduce their cost base while simultaneously introducing systematic health checks that will inevitably require both additional infrastructure and new processes. While traffic volumes are reduced, it is obviously a good time to make improvements that might otherwise disrupt the business, but few airports will have the financial resilience to take advantage of the opportunity this downturn in traffic presents. Dubai World Central (DWC) is an example of an airport development that is challenged by a new economic reality.

There are few instances of airports investing in new infrastructure to cater to reduced traffic levels. However, a notable exception is Pittsburgh which was faced with a major and permanent downturn in traffic following the demise of US Airways which used it as its main hub. It has elected to rebuild its landside terminal functions and related surface access to be adjacent to the cruciform airside satellite, reducing the number of stands and the size of the complex and eliminating the underground transit system that linked the terminal and satellite, while releasing land for more commercial development.

Lastly, to answer some of the challenges identified above, digital design has much to contribute. As well as facilitating the introduction of more sophisticated processes, it will be able to offer more personalised experiences to those who embrace new technology, with the risk of alienating those on the wrong side of the digital divide and, in particular, first-time travellers, who will still need assistance.

Following the Covid-19 outbreak, the airline industry may take a different form, as businesses in particular have adapted to travel restrictions and more meetings are held electronically. It is expected that in turn airport design will have to change due to reducing demand for business-class travel and increasing pressure for routine health checks and operational cost reduction. Further automation, aided by the expansion of Artificial Intelligence, is likely to be the response, presenting new opportunities for change in the subsequent era.

So 2020 may yet be regarded as a new turning point in the evolution of airport design, which leads to retrenchment, downsizing and simplification, rather than further growth, which has to date resulted in continuing expansion and complications of airports. The latter has been the story of the first 100 years of airport design.

Five airports currently being developed have been selected to illustrate some of the challenges airports face, whether they be political, financial, or environmental.

- Mexico City (MEX)
- Dubai World Central (DWC)
- Red Sea Umluj Tabuk (RSI)
- Western Sydney International (WSI)
- Centralny Port Komunikacyjny (CPK)

38. Mexico City (MEX)

Mexico City Texcoco Airport is a partially built megastructure which was supposed to be Mexico City's New International Airport, but it was abruptly cancelled in late 2018 for political reasons (Figure 8.1). Instead, the current Mexican government plans to build a less ambitious international airport at Santa Lucía airbase north of the city. The Covid-19 pandemic may yet affect other airports in a similar way, resulting in delays, cancellations or downscaling of many projects.

The terminal was designed by Norman Foster and Mexican Architect Fernando Romero. The terminal has a monumental scale inspired by Mexican architecture and symbolism. The roof of the terminal represented an eagle with its wings open to take flight. The X was also intended to be symbolic of the country's name "México".

Mexico City would have been one of the new types of terminal architecture, a megastructure which is characterised by being so large that it cannot be understood from the ground. So instead of the main entrance being the key view, it is the aerial view that becomes the focus, a view that no one can see in reality and interior views, which only give a partial understanding of the terminal, that predominate (Figures 8.2 and 8.3).

The development of an airport at Texcoco was planned in four stages. Phase I, scheduled to open by October 2020, would have consisted of one main terminal of 8,000,000 square feet (743,000 m^2) and three independent runways, which would have a capacity for 68 million passengers annually. In its final phase (2065), it would have consisted of six runways, an additional main terminal and two satellite terminals, with capacity for 125 million passengers, making it one of the largest airports in the world (Figures 8.4 and 8.5).

The X-shape is also promoted as an efficient alternative to the Beijing Daxing concept and provides the maximum number of gates within reasonable walking distances without

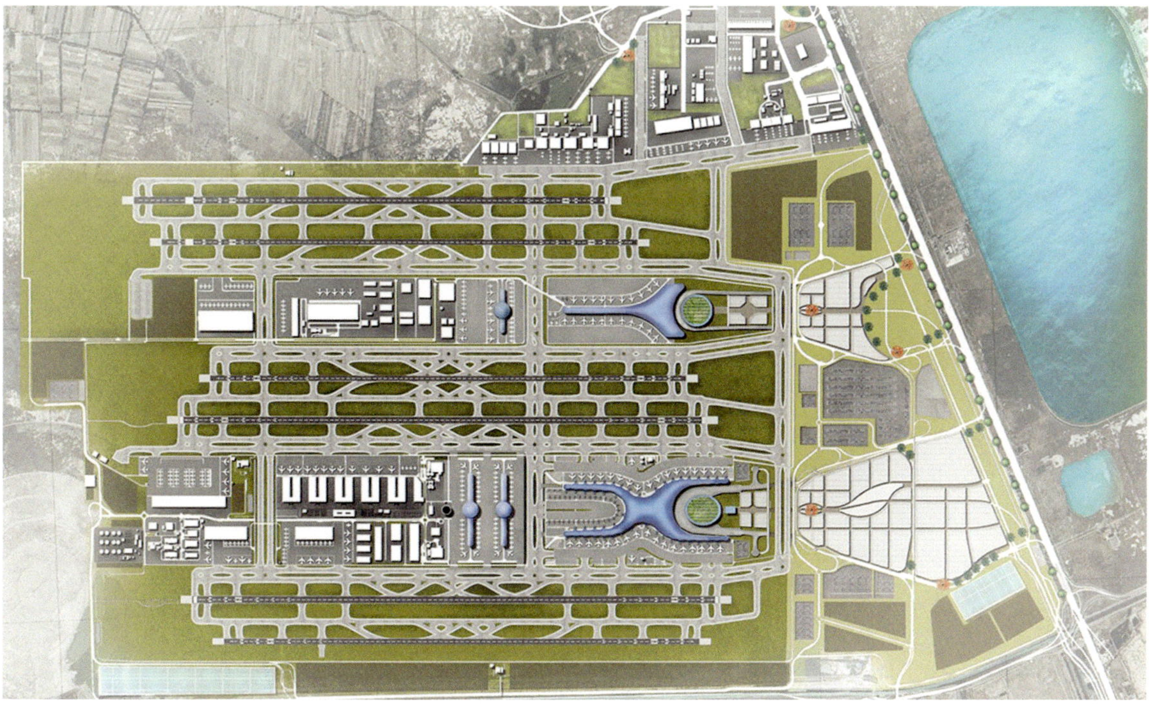

Figure 8.1 Mexico City, Master Plan with 6 runways and 2 terminals.
SOURCE: LANDRUM AND BROWN

Figure 8.2 Mexico City International, Departures Forecourt.
SOURCE: DBOX FOR FOSTER + PARTNERS

Figure 8.3 Mexico City International, Departures Lounge.
SOURCE: DBOX FOR FOSTER + PARTNERS

a transit system. The X shape is also used in the airports of Abu Dhabi, Mumbai, Nairobi, Incheon, Pittsburgh and SeaTac (Figure 8.6). Foster describes the concept as follows (Figures 8.7–8.10):

> *The maximum span internally is 170 metres. The lightweight glass and steel structure and soaring vaulted roof were designed for Mexico City's challenging soil conditions. Its unique pre-fabricated system could be constructed rapidly, without the need for scaffolding – the airport was to be a showcase for Mexican innovation, built by Mexican contractors and engineers. The entire terminal is enclosed within a continuous lightweight gridshell, embracing walls and roof in a single, flowing form, evocative of flight. The design ensures short walking distances and few level changes, it is easy to navigate.*
>
> (Foster, 2018)

> *The entire building is serviced from beneath, freeing the roof of ducts and pipes and revealing the environmental skin. This hardworking structure harnesses the power of the sun, collects rainwater, provides shading, directs daylight and enables views – all while achieving a high performance envelope that meets high thermal and acoustic standards.*

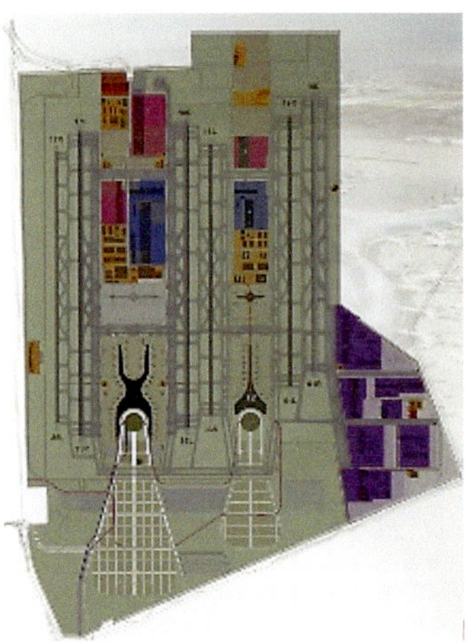

Figure 8.4 Mexico City Master Plan.
SOURCE: FOSTER + PARTNERS

Figure 8.5 Mexico City CGI aerial view.
SOURCE: FOSTER + PARTNERS

Figure 8.6 Mexico City, Winning Competition Design 2018.
SOURCE: FOSTER + PARTNERS

Figure 8.7 Mexico City International, aerial view.
SOURCE: FOSTER + PARTNERS

Figure 8.8 Mexico City International, cut-away CGI of pier.
SOURCE: FOSTER + PARTNERS

Figure 8.9 Mexico City International, alternative views of the Departures Lounge.
SOURCE: FOSTER + PARTNERS

Figure 8.10 Mexico City International, Departures Lounge.
SOURCE: FOSTER + PARTNERS

39. Dubai World Central (DWC)

Dubai World Central is planned to become the world's largest airport, with an ultimate capacity exceeding 220 million passengers a year and 16 million tonnes of cargo per annum. The airport forms the heart of a greater Aerotropolis project covering 140 km^2 in six zones: Logistics City, Commercial City, Residential City, Aviation City and Golf City (Figure 8.11).

The airport will ultimately have five parallel runways, each 4500 metres long. The main terminal building will house a fully automated baggage handling system capable of handling around 240 million bags per annum. The cargo facility will allow air-to-sea connections to Jebel Ali Port and Free Zone in less than four hours via a dedicated road.

What makes Dubai World Central different is that it is designed primarily as a transfer hub, with over 80% of passengers transferring through the satellites, and it handles only international but no domestic traffic. It is also much larger than all other terminals, with a design capacity of 220 mppa so there is no opportunity to avoid the use of a transit system. The other difference is that it eschews the curvilinear approach of the previous airports in favour of a simpler orthogonal geometry which will allow it to be constructed more easily using a modular phased approach as traffic increases (Figure 8.12).

The project is currently under review as the economy in the region slows down and load factors on Emirates Airlines have sunk to below 80%. It is now not expected to open until 2030. Dubai Airports said in 2016 it was expanding Dubai International to handle 118 million passengers a year by 2023, 18 million more than its initial cap in case the development of Al Maktoum International was delayed. Further delay or downscaling of the project as a result of the Covid-19 hiatus should be anticipated (Figures 8.13 and 8.14).

The first phase of the airport opened with cargo operations in 2010, followed by passenger services in 2013, but the major expansion featured here is currently on hold

Figure 8.11 Dubai World Central, model.
SOURCE: LESLIE JONES ARCHITECTURE

Figure 8.12 Dubai World Central, satellite.
SOURCE: DUBAI AIRPORTS VIA LESLIE JONES ARCHITECTS

Figure 8.13 Dubai World Central and Aerotropolis.
SOURCE: YOUTUBE

Figure 8.14 Dubai World Central master plan c2050.
SOURCE: DUBAI AIRPORTS

Figure 8.15 Dubai World Central, terminal building in foreground and satellites in background.
SOURCE: DUBAI AIRPORTS VIA INTERNATIONAL AIRPORT REVIEW

(Figure 8.15– 8.17). The initial phase of the airport, located in the cargo zone, currently handles 7 million passengers a year but has been expanded to 15 million passengers a year as an interim update.

Dubai World Central is the most ambitious Aerotropolis to date. It is focused on Al Maktoum Airport, which is planned to grow to 220 million passengers a year and ultimately supersede the existing Dubai International Airport as Dubai's main hub airport. It contains several zones over an area of 145 km^2 that combine to form the Aerotropolis:

Figure 8.16 Dubai World Central model.
SOURCE: DUBAI AIRPORTS – AL MAKTOUM INTERNATIONAL

Figure 8.17 Dubai World Central, Retail Mall.
SOURCE: DUBAI AIRPORTS VIA LESLIE JONES

Commercial, Golf, Business Park, Logistics, Aviation, Exhibition and Residential Districts. The exhibition area was planned to host Expo 2020; however, the exhibition was delayed due to the Covid-19 outbreak and is now open between autumn 2021 and spring 2022 (Figures 8.18–8.20).

These ambitious plans are complemented by equally far-reaching schemes for investment in infrastructure projects, including extensions of two metro lines to the airport and the proposed high-speed rail system, designed to link the various Emirates, especially Dubai and Abu Dhabi via the new airport.

Figure 8.18 Dubai World Central Aerotropolis.
SOURCE: DUBAI AIRPORTS VIA EMIRATES247.COM

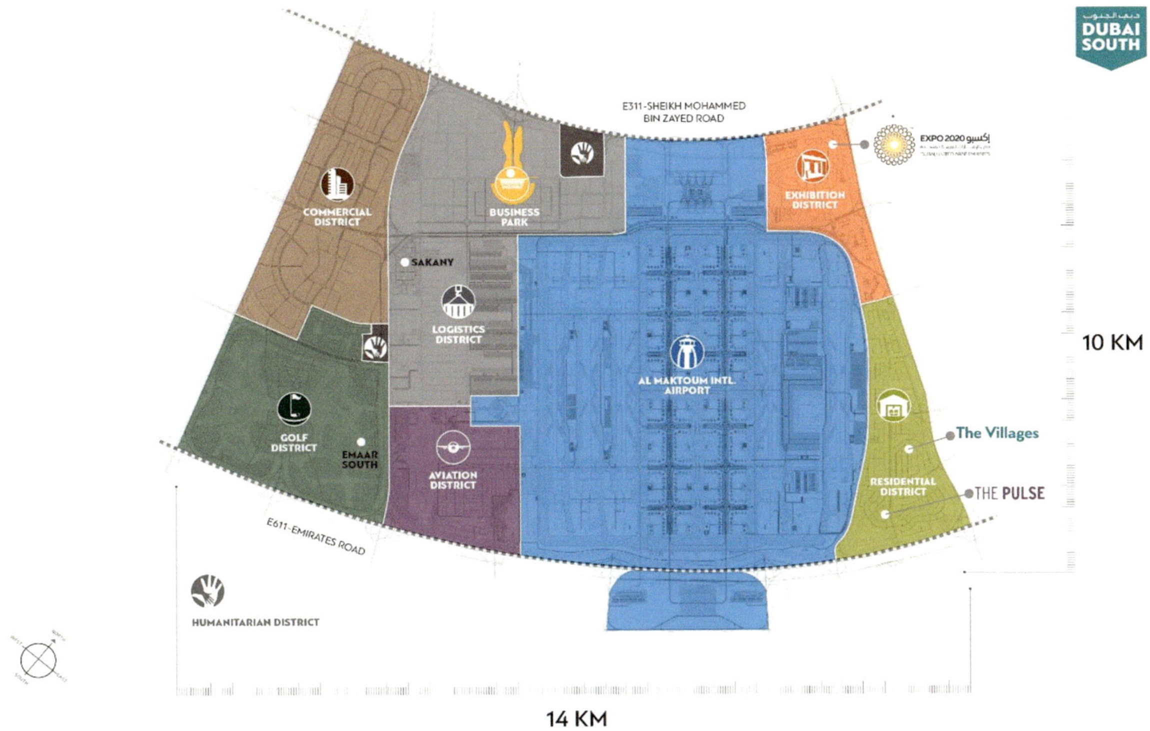

Figure 8.19 Dubai World Central master plan.
SOURCE: DUBAI AIRPORTS – DUBAI SOUTH

Meanwhile, there are currently doubts about the financing of the project that is slowing down progress. This has resulted in the existing Dubai International airport publishing plans to handle more traffic, increasing the existing capacity from 90 million passengers a year to 118 million passengers a year in 2023, primarily through process improvement rather than further investment in infrastructure.

Figure 8.20 Dubai Metro 2030.
SOURCE: WIKIPEDIA

40. Red Sea Umluj Tabuk

The design for Red Sea International is the subject of close scrutiny by environmentalists and has given rise to great controversy over its environmental credentials, overshadowing consideration of the design quality of this small-scale exclusive airport. Once complete, the Red Sea International Airport is aimed at serving the one million annual tourists and visitors expected by 2030. The new international airport will also serve for the two new resort developments which will be built nearby the airport, including Foster + Partners' Southern Dunes Hotel and a ring-shaped hotel for Ummahat AlShaykh island. Visuals, released by The Red Sea Development Company, show that the five fluid dune-shaped roofs that form the airport are organised in a radial layout (Figure 8.21).

Foster + Partners, the architects for the scheme, say (Figures 8.22 and 8.23)

> *'The airport has been inspired by the forms of the desert, the green oasis and the sea. The design of the terminal aims to bring the experience of a private aircraft terminal to every traveller by providing smaller, intimate spaces that feel luxurious and personalised. The form of the roof shells cantilevers on the landside and airside to provide shade to the passengers. An internal green oasis with an indigenously planted garden forms a green focus, creating a relaxed, resort-like atmosphere within the airport terminal. The airport will be powered by 100 percent renewable energy.'*

However, the Architects' Climate Change Network (ACAN) has stated that

> *'Anything that encourages the expansion of the aviation industry and the associated rise in the burning of fossil fuels should be viewed as off-limits for architects concerned about the rate at which our planet is warming. It is as simple as that we should be working to bring about a moratorium on the expansion of air travel.'*

Figure 8.21 Red Sea Airport aerial view. The scheme revives the cantilever roof form to provide shelter to passengers boarding aircraft rather than employing the now ubiquitous airbridge.
SOURCE: FOSTER + PARTNERS

Figure 8.22 Interior of one of the five executive-style waiting lounges under the shell-like cantilever roofs that provide shade from the desert sun to passengers boarding the aircraft.
SOURCE: FOSTER + PARTNERS VIA ARCHDAILY

In his statement, Foster called for a *'sense of proportion and serious consideration of the facts'* when discussing aviation's impact on the climate crisis, pointing out aviation emissions account for 2% of the global total. He added: *'If carbon imprints are properly taxed, then it may mean a reset for fast food and cheap flights but agriculture and aviation are not going to go away and they will both need the most sustainable buildings to serve them together with the architects who can most responsibly design them.'*

This environmental debate foreshadows what must be expected when future airport schemes are evaluated against evolving societal expectations for sustainable development.

Figure 8.23 Internal view of the atrium circular circulation area evoking the character of a green oasis.
SOURCE: FOSTER + PARTNERS VIA ARCHDAILY

41. Western Sydney International (Nancy-Bird Walton) (WSI)

Western Sydney International has been designed to serve the western suburbs of Sydney, promote growth in the area and relieve pressure on Sydney Kingsford Smith, Sydney's international airport on the east coast. The latter has served the city since 1921, making it second only to Amsterdam in longevity. It had been progressively extended from a grass field by constructing runways on land reclaimed from Botany Bay, but lack of room to expand and noise nuisance affecting adjacent communities now constrains future expansion.

Western Sydney International is being built on the traditional lands of the Aboriginal People of the great Dharug nation. It is designed to accommodate 10 million passengers per year from day one and reach around 81,000 air traffic movements a year on a single runway in 2023, operating domestic, international and freight flights 24 hours a day (Figure 8.24). As demand grows, a second runway will be added, allowing it to expand progressively, increasing the overall airspace capacity of Sydney to its ultimate design capacity of 82 million passengers per year and handling both low-cost and full-service carriers (Figures 8.25). The Qantas Group plans to operate up to 15 narrow-body aircraft – 10 Jetstar and five Qantas – from Western Sydney International (Nancy-Bird Walton) Airport within the first year, flying domestic routes such as Melbourne, Brisbane and the Gold Coast.

The design for WSI is an evolution of the initial concepts developed by international firm Zaha Hadid Architects and Australian firm Cox Architecture. David Holm, Cox director, states that:

> *The design takes inspiration from the local flora of Western Sydney and its unique natural qualities, such as the Great Australian Light. These spaces, colours and materials have been incorporated into the design to give the project an unmistakable regional identity focused on the experience of the passenger and the user.*

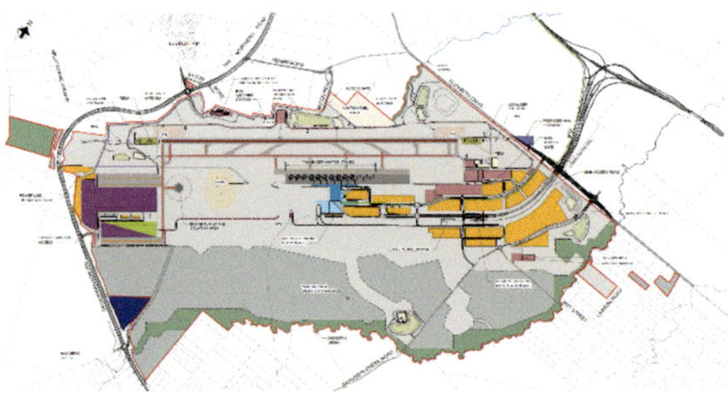

Figure 8.24 Western Sydney International phase 1 master plan for 10 mppa.
SOURCE: © WESTERN SYDNEY INTERNATIONAL AIRPORT

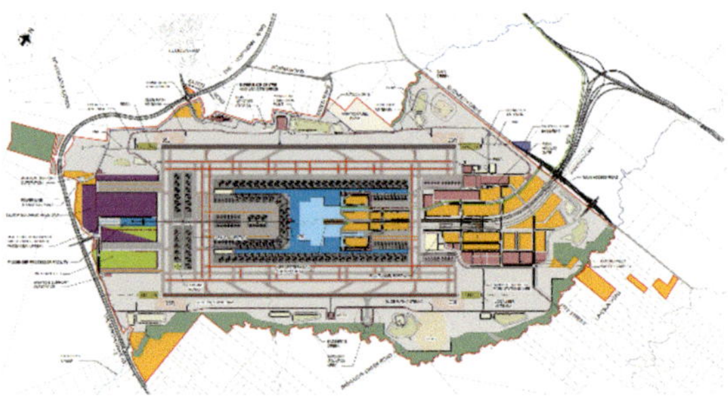

Figure 8.25 Western Sydney International Ultimate Phase master plan for 82 mppa.
SOURCE: © WESTERN SYDNEY INTERNATIONAL AIRPORT

Figure 8.26 Aerial view of WSI at 82 mppa.
SOURCE: © WESTERN SYDNEY INTERNATIONAL AIRPORT

The midfield master plan and terminal design eschew the complex curves found in many contemporary designs of global hubs and return to a simpler indeterminate orthogonal layout and neo-Miesian International Style aesthetic for the envelope which will facilitate expansion as the airport grows and changes. The design follows the current trend of avoiding the use of an airside transit system with an extended pier system that will involve long walking distances for some passengers. The airport states it will support nearly 28,000 direct and indirect jobs by 2031 and underpin the development of the Western Sydney Aerotropolis, at the centre of Western Parkland City. The airport will be linked to an extension of the metro system and regional road network. The Aerotropolis will support around 200,000 new jobs in industries such as aerospace and defence, manufacturing, freight and logistics, tourism and research (Figures 8.26 and 8.27).

Cox Architects go on to state (Figures 8.27 and 8.28):

> *The airport's terminal roof design assists in intuitive navigation for users. Inspired by the light filtration of the local eucalyptus tree and the curly forms of its bark, the design follows a set of simple geometrical and structural principles that provide a sophisticated sequence of discovery. In addition, the roof's elegant, single-ridge shape was inspired by the aerofoil profile of an airplane wing, tapered so that when seen from below, it gracefully caps the Main Terminal Building and piers.*

Figure 8.27 Artist's impression of the developed concept for Western Sydney International.
SOURCE: © WESTERN SYDNEY INTERNATIONAL AIRPORT

Construction is well under way, and opening is planned for 2026. However, the developed details of the proposed flight paths were released in June 2023 based on the following principles: minimising noise in residential areas, minimising the impacts of air operations on natural and visually sensitive areas and avoiding aircraft converging over a single merge point in a single residential area (Figures 8.29–8.31).

However, this has provoked alarm from local conservation and community groups and shows that noise nuisance, along with climate change, continues to be one of the most intractable environmental issues facing the expansion of air travel.

Figure 8.28 WSI Developed Interior Concept.
SOURCE: © WESTERN SYDNEY INTERNATIONAL AIRPORT

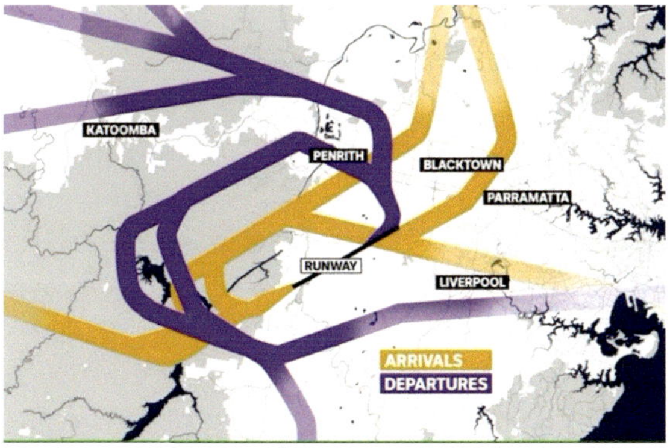

Figure 8.29 Projected flight preference Routes R 05.

SOURCE: DEPARTMENT OF INFRASTRUCTURE, TRANSPORT REGIONAL DEVELOPMENT, COMMUNICATIONS AND THE ARTS

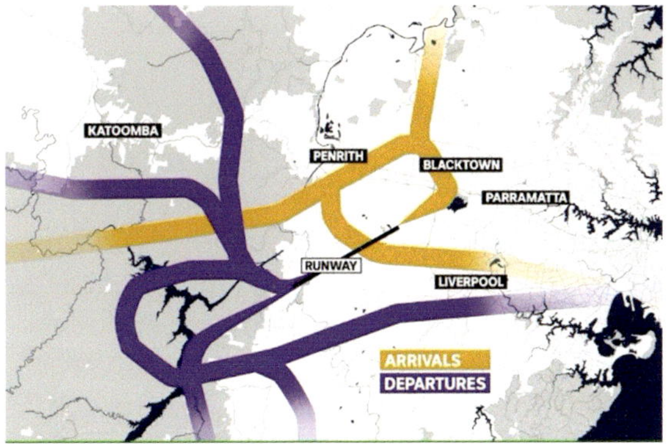

Figure 8.30 Projected flight preference Routes R 23.

SOURCE: DEPARTMENT OF INFRASTRUCTURE, TRANSPORT, REGIONAL DEVELOPMENT, COMMUNICATIONS AND THE ARTS

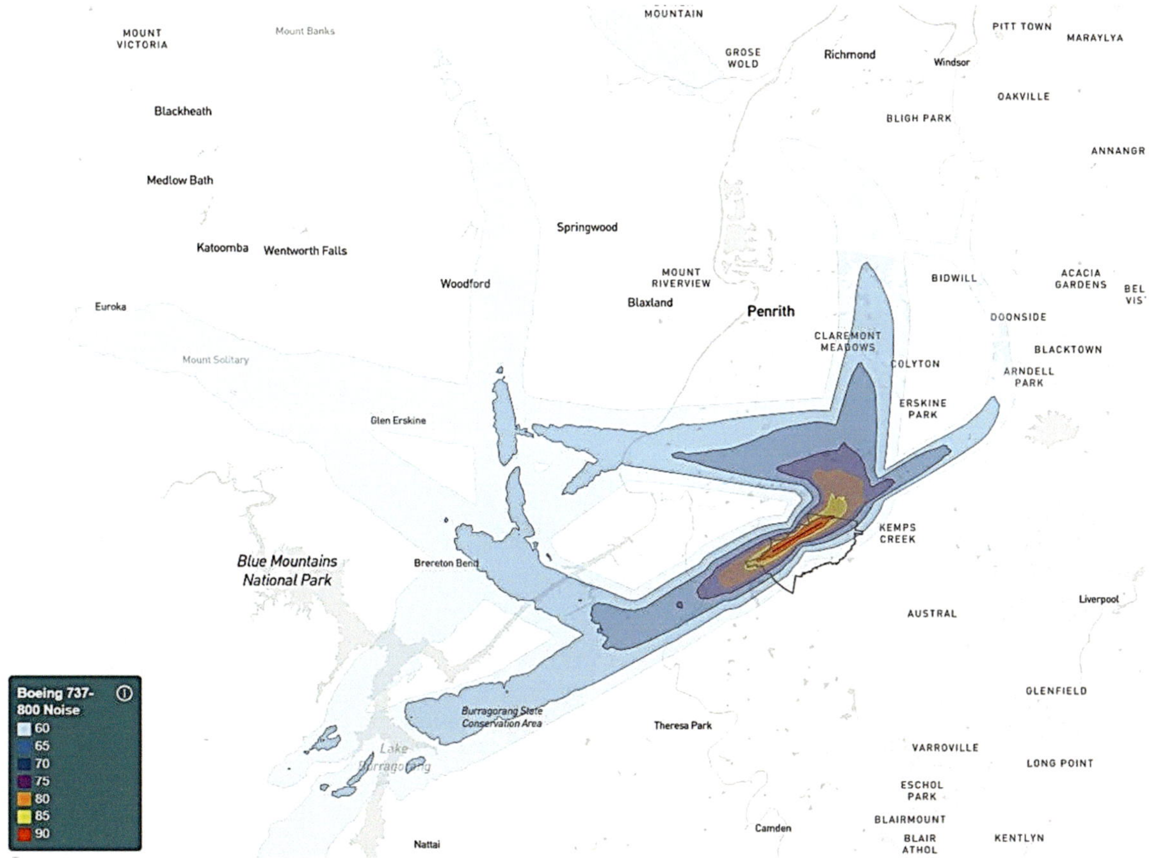

Figure 8.31 Projected noise contours for B737-800.

SOURCE: DEPARTMENT OF INFRASTRUCTURE, TRANSPORT, REGIONAL DEVELOPMENT, COMMUNICATIONS AND THE ARTS

42. Centralny Port Komunikacyjny (CPK)

The new global hub is designed to serve up to 40 million passengers initially, with a projected opening date of 2028, and then expand seamlessly to meet a 65 million passenger target in 2060 and become the main gateway to Poland. CPK airport is to be built on land around 25 miles southwest of Warsaw and linked to both Warsaw and Łódź by a new high-speed rail network and road system to become a major air, rail and road interchange (Figure 8.32). Combining the greenfield international airport with a major public transport interchange (railway and bus station) and road connectivity, the CPK will be the main gateway to Poland.

The plan assumes that the airport will, at its opening, have two parallel runways, infrastructure for approximately 330,000 aircraft operations (take-offs and landings) per year and a terminal suitable to serve 40 million passengers per year.

CPK is planned as a dual-use infrastructure, i.e., as a civil airport that can be used by NATO support and rapid response forces in the event of a threat of armed conflict. This is also why the CPK Master Plan assumes that military infrastructure is envisaged in the northern part of CPK. The latter's planning and development are the responsibility of the Ministry of Defence in cooperation with CPK.

The master plan follows the current trend of handing a large number of aircraft through multiple piers attached directly to the terminal, also found at Beijing Daxing (DAX) and iGA Istanbul (IST), thus avoiding the provision of an airside people mover, at least in the initial phase. In this respect, it bears comparison with the original master plan for Chicago O'Hare (1948) (Figures 8.33–8.35).

The Foster + Partners and Buro Happold consortium won the design competition for master architect. Foster + Partners said that the project:

> *will act as a symbolic gateway to Poland, a 21st century transport interchange which brings together air, rail and road, while reflecting the country's national identity and providing exceptional passenger experience.*

Figure 8.32 Centralny Port Komunikacyjny (CPK) location plan of the airport between Warsaw and Łódź on the planned new Polish high-speed rail network.

SOURCE: CENTRALNY PORT KOMUNIKACYJNY (CPK)

(a) (b)

Figure 8.33 Centralny Port Komunikacyjny (CPK) master plans (a) initial phase 1.1 (two runways) and (b) ultimate phase 2.2 (three runways).
SOURCE: CENTRALNY PORT KOMUNIKACYJNY (CPK)

Figure 8.34 Centralny Port Komunikacyjny (CPK) winning competition design aerial view of landside approach and railway interchange. Note the landing pads in the foreground for air taxis.
SOURCE: FOSTER + PARTNERS

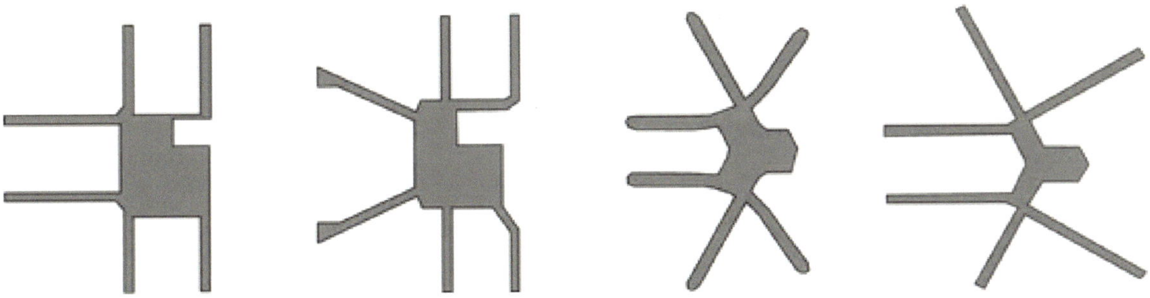

Figure 8.35 Centralny Port Komunikacyjny (CPK) studies of alternative configurations for the terminal and multiple piers showing the evolution of the design.
SOURCE: CENTRALNY PORT KOMUNIKACYJNY (CPK) MASTER PLAN REPORT 2023

Figure 8.36 Centralny Port Komunikacyjny (CPK) developed aerial view of Phase 1 Terminal and Rail Interchange in the foreground. Note the unusual diagonal forecourt that increases the length of interface with the terminal and interchange.
SOURCE: CENTRALNY PORT KOMUNIKACYJNY (CPK) – VISUALISATION FOSTER + PARTNERS

Grant Booker, head of studio, Foster + Partners explained (Figure 8.36):

> *Our design for CPK is centred on the passenger. The ambition is to create an accessible, calm and legible building that enhances the experience of travel. The terminal's minimal level changes and generous structural spans create clear lines of sight – from landside to airside – making it extremely easy to navigate. The vaulted roof fills the space with light and leads passengers through the building, guiding them towards their destinations. Bringing together three modes of transport in one place – and integrating a new railway station that is part of a national high-speed rail system – we believe that CPK will completely transform the way people travel across Poland and become a powerful new gateway to Europe and the rest of the world.*

The centrepiece of the multimodal interchange is an atrium, seamlessly connecting the airport terminal, railway station, and other modes of transport under a singular, unifying canopy. The atrium links the terminal and the railway station – an accessible space where all passengers meet, regardless of which mode of transport they are using (Figures 8.37–8.39).

CPK railway station will accommodate the needs of CPK's operations and serve as a national transport hub, integrating rail connections from major cities in Poland. It will consist of six underground platforms (12 tracks) serving regional and long-distance trains. The platforms will be covered, and it will be possible to reach the station on the north

Figure 8.37 Centralny Port Komunikacyjny (CPK) – External view of the forecourt and main entrance to the terminal.
SOURCE: CENTRALNY PORT KOMUNIKACYJNY (CPK) – VISUALISATION FOSTER + PARTNERS

Figure 8.38 Centralny Port Komunikacyjny (CPK) – Interior view of the main entrance to the terminal.
SOURCE: CENTRALNY PORT KOMUNIKACYJNY (CPK) – VISUALISATION FOSTER + PARTNERS

Figure 8.39 Centralny Port Komunikacyjny (CPK) atrium at the main entrance linking the terminal and rail interchange.
SOURCE: CENTRALNY PORT KOMUNIKACYJNY (CPK) – VISUALISATION FOSTER + PARTNERS

side of the terminal, via the atrium, in comfort and without risk of being exposed to the elements (Figure 8.40). The ambitious railway interchange will facilitate connections within Poland, enabling transfers between Warsaw and the largest Polish cities in less than 2.5 hours. The Airport City, to be created within the CPK's region, will include exhibition, congress and conference facilities as well as offices. The investment should provide employment for over 150,000 people.

Figure 8.40 Centralny Port Komunikacyjny (CPK) railway interchange platform level.
SOURCE: CENTRALNY PORT KOMUNIKACYJNY (CPK) – VISUALISATION FOSTER + PARTNERS

CHAPTER 9

CONCLUSION

These case studies show why, when, where and how new airport master planning and terminal concepts have arisen. The drive for change has mainly been from the increasing scale of operation and the particular spatial constraints of individual sites. Perfectly workable examples of each of the terminal concepts can be found today, for example:

- The pioneering hangar/depot processor first seen at Berlin Johannisthal can still be found at the Virgin Galactic Spaceport and numerous executive terminals with boarding inside the hangar, where passenger volumes are small.
- The simple terminal that was the norm in the pioneering period is still being built at regional airports that focus on the low-cost carrier market such as Bristol International, Bordeaux Billy and Copenhagen Go.
- The linear terminal first seen at Le Bourget continues to be relevant, particularly at single-runway airports with large-scale examples at Kansai, Medina and Western Sydney International.
- The unit terminal as seen at Dallas Fort Worth has primarily been built in the USA, though ADPI developed this concept at Charles de Gaulle T2 and a number of international airports. It has now largely fallen out of favour as safety, security and commercial imperatives lead to centralisation.
- New finger pier concepts, first seen at Chicago O'Hare, can be found at a number of the largest new international airports, such as Hong Kong, Singapore, Bangkok, Seoul Terminal 2, Istanbul and Poland's Centralny Port Komunikacyjny.
- Small circular satellites, seen in prototype at Gatwick, have largely fallen out of fashion, but a recent example is Tel Aviv Ben Gurion.
- The larger linear satellite developed at Atlanta with underground transit links has been repeated at Denver and Heathrow and other locations where site area is at a premium.
- The mid-field hub has become the norm for most international hub airports with two or more runways, following the example of Washington Dulles.
- While several underground terminals have been proposed only Terminal 3 in Dubai has been constructed to address the extreme shortage of land.

There has been a progressive separation or specialisation of functions such as hangars, control towers and administration as they have been separated from terminal buildings. Conversely, centralised terminals now dominate over decentralised ones to minimise the staffing costs of meeting security requirements and respond to retail pressures to maximise footfall.

Dry boarding has been a preoccupation of designers since the pioneering days with drive-through and cantilever roof concepts giving way to apron-drive airbridges once the

DOI: 10.4324/9781003323624-9

technology had evolved sufficiently. The only exception is that low-cost carriers avoid the use of airbridges in favour of steps to the front and rear aircraft doors to speed passenger boarding and alighting during aircraft turnaround while also reducing operational costs.

Commercial design initially focused on spectator events and catering as a means of subsidising operating costs. More recently, the trend towards privatisation has been accompanied by a radical growth in retail, which can now deliver enough revenue to cover all day-to-day airport operating costs. In the future, it is expected that the growth of internet shopping will lead to a refocusing retail towards experiences rather than direct sales. Commercial demands have often allowed natural passenger circulation to be distorted by extended walking distances but the rise of internet shopping may in the future provide a remedy as maximising commercial footfall may no longer be a priority. There will also be an increasing focus on leisure experiences, such as the Changi Jewel, as transfer hubs try to find new ways to compete.

There have been a variety of responses to the need to find an appropriate airport style, including Neo-Classical, Art Deco, International Style, High Tech, Contextual Design, Expressionism and, most recently, Mega-Structuralism focusing on Aerial and Interior Architecture. No one style has prevailed, but architects have generally tried to reflect the modern era and capture the spirit of flight and the excitement of travel, most recently with the floating, long-span steel roof.

Some planning concepts and architectural styles have proven to be better able to withstand the challenges of growth and change. The finger pier and the indeterminate architecture, facilitated by the International Style, are the most amenable to incremental expansion by virtue of the use of orthogonal layouts and modular repetitive interchangeable components. This is perhaps why there has been a resurgence of this approach in the most recent European terminal designs at Berlin, Frankfurt, Schiphol and Marseilles.

Long-span roof structures have also offered opportunities for internal reorganisation, at least at the upper level, which demonstrates one of the key benefits of the original concept for Stansted with departures and arrivals traffic in a side-by-side configuration at the main passenger level.

The case studies show that individual airports have successfully employed many different concepts as they have developed over time. All the early terminals followed the simple terminal formula without planning for growth. The first airport to take a long-term view was the second terminal in Berlin, but it did this by grossly oversizing the facilities compared to the anticipated traffic. However, it did last for the 60 years for which it was planned. It finally closed in response to criticism of its unacceptable environmental impact on the nearby city centre.

In the post-war period, a variety of concepts developed to cope with increasing size, including the linear and unit terminal, finger pier and circular and linear satellite. The 1947 masterplan for Chicago O'Hare with its multiple-finger piers has proved adaptable to larger aircraft and increased volumes of traffic and has influenced subsequent designs at St. Louis Lambert, London Gatwick, Amsterdam Schiphol and Frankfurt Main.

The pier concept was revived in a larger form and supported by a transit at Hong Kong International, which has been until recently the largest terminal under a single roof. The multiple pier concept has enjoyed a revival in the 21st century as the benefits of being able to access a large number of stands from the main terminal without resorting to a transit system have been better appreciated, as can be seen in Istanbul.

The linear unit terminal, which offers short walking distances and close parking, was realised in its most developed form at Dallas Fort Worth. It was supported by a uniquely elaborate transit system to facilitate easy transfers and allowed it to become one of the major US hubs. In the later development of the international terminal, the unit terminal concept was abandoned in favour of the now universally adopted centralised approach, which minimises staffing costs for security control and maximises retail footfall. Linear terminals are most often planned where developments are proposed parallel with the runway, such as Heathrow Terminal 4, the main terminal building Madrid Barajas T4, Western Sydney,and Kansai, but unless, like Kansai, they employ a transit system, they are inherently limited to a relatively small scale of say 10–20 mppa before walking distances become unacceptable.

The airports with smaller satellites – New York (TWA), Los Angeles and Paris Charles de Gaulle – have all abandoned them in later developments. The TWA building has been abandoned, and a linear terminal has been built around it for the low-cost airline Jet Blue. The small satellites at Los Angeles have been joined back to the terminal to form finger piers, and later developments at Paris Charles de Gaulle have adopted a linear terminal development supported by large linear satellites. The airport is planning a large finger pier concept for the proposed Terminal 4 at Charles de Gaulle.

It is interesting that Atlanta Hartsfield has progressively morphed from a simple terminal to a linear terminal, then a finger pier and finally to a mid-field terminal with linear satellites accessed by an underground transit. It is currently the busiest airport in the world. In many ways, it epitomises the evolution of airport design in its response to the challenges of growth and change. Its influence can be seen in Denver and the progressive redevelopment of Heathrow and Washington Dulles. It is recognised as the most efficient format in terms of land use and aircraft movement, but it requires an expensive underground transit system. Stansted also employed the transit as an architectural device to separate the satellites from the terminal when walking distances would have been perfectly well satisfied by piers; indeed in later phases, the pier concept has been reintroduced. Stansted, however, did pioneer the floating steel frame roof which has been employed in most subsequent major airports as the ultimate symbol of flight.

New York Idlewild (JFK) shows what can happen when masterplan concepts break down in the face of commercial expediency that allows a free for all with separate terminal designs that compete in an uncoordinated way. Most of their terminals have become obsolete, especially the most architecturally striking ones, the TWA terminal and Pan American World Port, because they were too small to adapt to increasing demand and overdesigned to meet the very specific airline needs of the day. It is perhaps significant that the Airlines that commissioned such bespoke designs – Pan American and TWA – no longer exist and their terminals have become obsolete.

By contrast, the indeterminate planning of the midfield terminal at Washington Dulles has stood the test of time and shown itself to be adaptable and flexible and able to respond to growth and change while still delivering the highest quality architectural design.

Dubai International is unique in being the only underground terminal in the world, but its importance lies in that, for each passenger handled, the airport occupies the smallest site area of any major international airport, so we may well see a revival of this concept as airports become increasingly congested.

The current generation of new airports has seen the re-emergence of large multiple-finger pier concepts at Bangkok Suvarnabhumi, Seoul Incheon Terminal 2, Istanbul, Beijing Daxing and Poland's Centralny Port Komunikacyjny with the objective of avoiding the use of a transit system, at least in the early years of development. The largest planned airport, to date, Dubai Word Central will have four very large multiple cruciform satellites, each handling 50–60 million passengers a year and, like Stansted, will have no passengers boarding aircraft directly from the terminal. The airport is unusual in having a very high transfer rate of over 80%; perhaps, it is not surprising that it prioritises ease of transfer over direct departures and arrivals.

Airport terminals' prime function is to process passengers and their baggage, and with each generation, the processing requirements have become onerous, as customs, immigration, security and now health checks have evolved and become more complex. While many hope to disguise this with more sophisticated IT that seeks to make the journey more effortless, it also involves providing more area at the key processing points. The introduction of security has already made a large number of terminals obsolescent and future public health checks to combat the spread of pandemics may also require more space at key locations that will generate new capacity constraints and planning solutions.

The environment is belatedly emerging as a key business management topic for airports around the world. In response to societal expectations, environmental factors are increasing in importance in order to meet society's expectations for sustainable development. However, airports have been slow to respond to challenges of environmental sustainability, with only Japan changing its airport development policy to mitigate noise nuisance by building their newer airports on artificial islands. Beyond that, airports tend to employ changes to operating practices, rather than changes to design to provide some level of mitigation. More sophisticated environmental mitigation is expected at every new airport development, including the electrification of vehicles, more sustainable energy sources, a reduction in embedded carbon and strategies to deliver net zero operational carbon. However, noise, air quality and traffic impact along with global warming remain for the time being the most serious concerns of local communities.

In conclusion, airport masterplan and terminal design concepts have largely evolved in response to increasing traffic volumes and those that have proved most durable are those that have been planned for growth and change from the outset. The airports that survive best are arguably those which combine careful site selection far enough away from city centres but with good connectivity to the city centre, with the ambition to create a landmark of architectural merit that celebrates the spirit of flight as well as a planned approach to accommodating growth and change.

We do need to remember that for all our best intentions to foresee the future, the level of investment in airports is such that longevity is a prerequisite. We have to live with the legacy of airport designs for many generations. In the words of Winston Churchill speaking in the House of Commons on 28 October 1944:

> ***We shape our buildings; thereafter they shape us*.**

His words are particularly apposite as airports increasingly have to focus on expansion, and have to contend with legacy issues, as there are unlikely to be many totally new airports in the developed world.

APPENDIX

AIRPORT MASTER PLANS THROUGH THE AGES

Airport master plans, USA 1919 and 1922

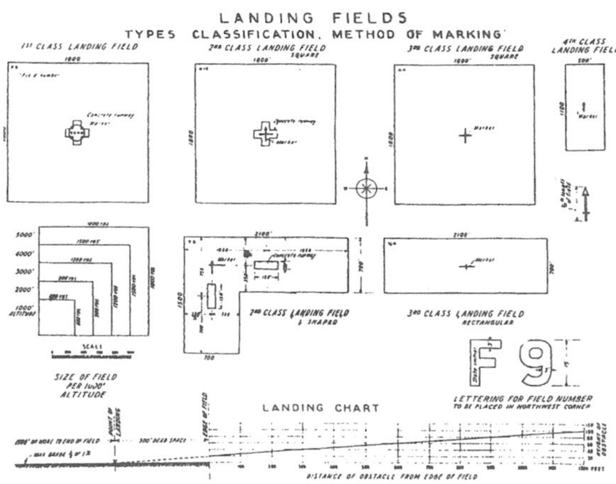

Figure A.1 Plans for Many landing Fields 1919.
SOURCE: AIRCRAFT JOURNAL NO 4

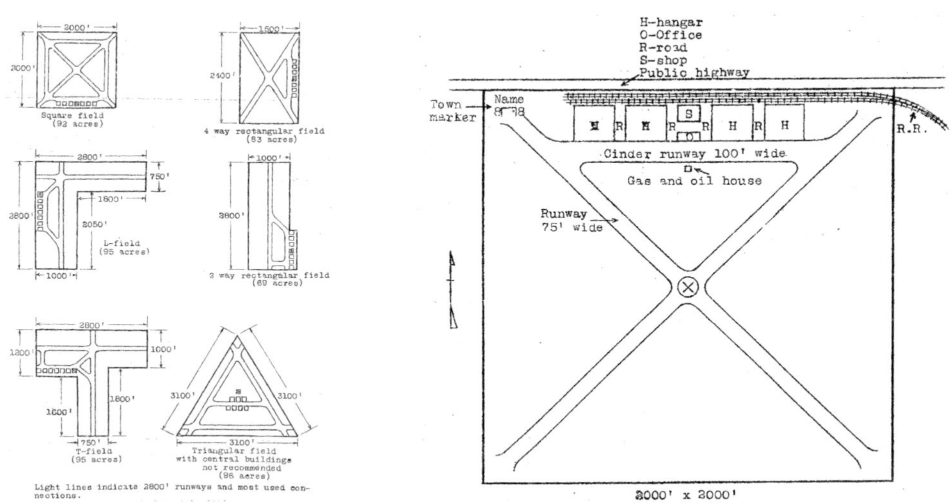

Figure A.2 How to lay out and build an Airplane landing Field 1922.
SOURCE: NATIONAL ADVISORY COMMITTEE FOR AERONAUTICS, TECHNICAL MEMORANDUM 164

The above design guides are the earliest masterplans for civil airports, based on the First World War experience.

Airport master plans – USA 1930 and Europe 1932

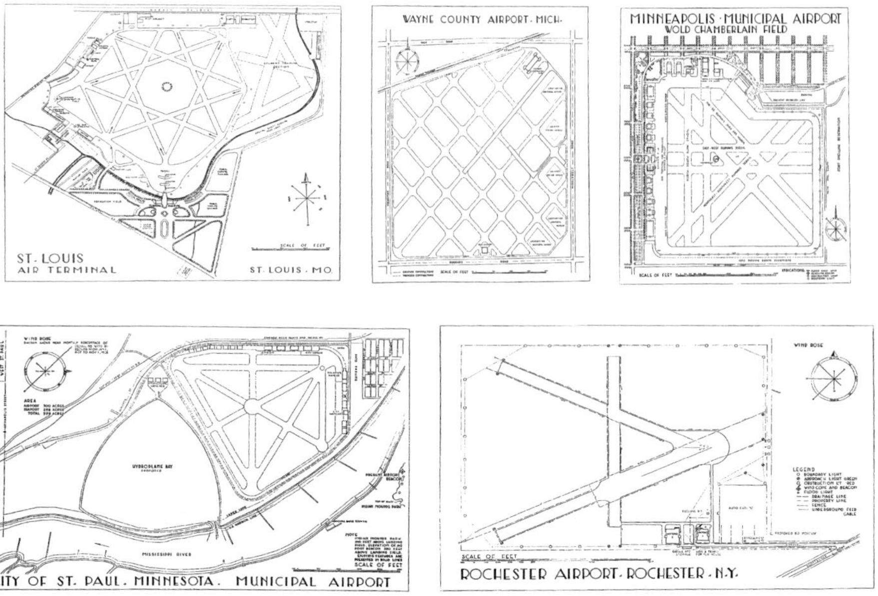

Figure A.3 American Airport master plans 1930.
SOURCE: AIRPORTS THEIR LOCATION, ADMINISTRATION AND LEGAL BASIS

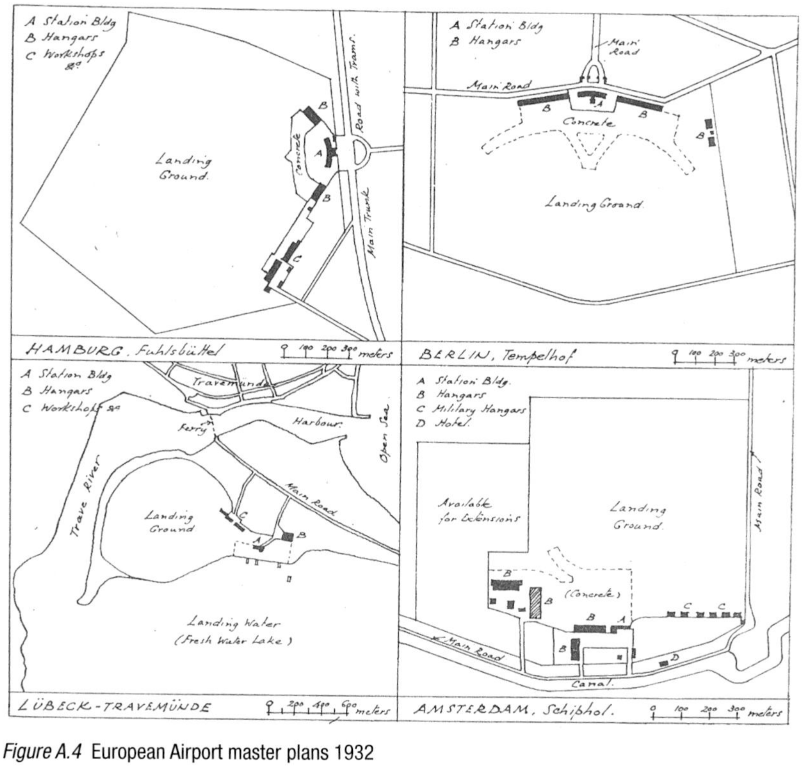

Figure A.4 European Airport master plans 1932
SOURCE: AVIATION AND THE AERODROME, H. ANGLEY LEWIS-DALE

The American Airport master plans of 1930, above, show that hard-surfaced runways were already the norm, while the European Airport master plans of 1932 show that airports were persevering with all-direction grass fields. Some commentators have suggested that this is a reflection of generally better ground conditions in Europe, while others suggest it was due to different views on safety at the time.

Airport master plans – USA 1939

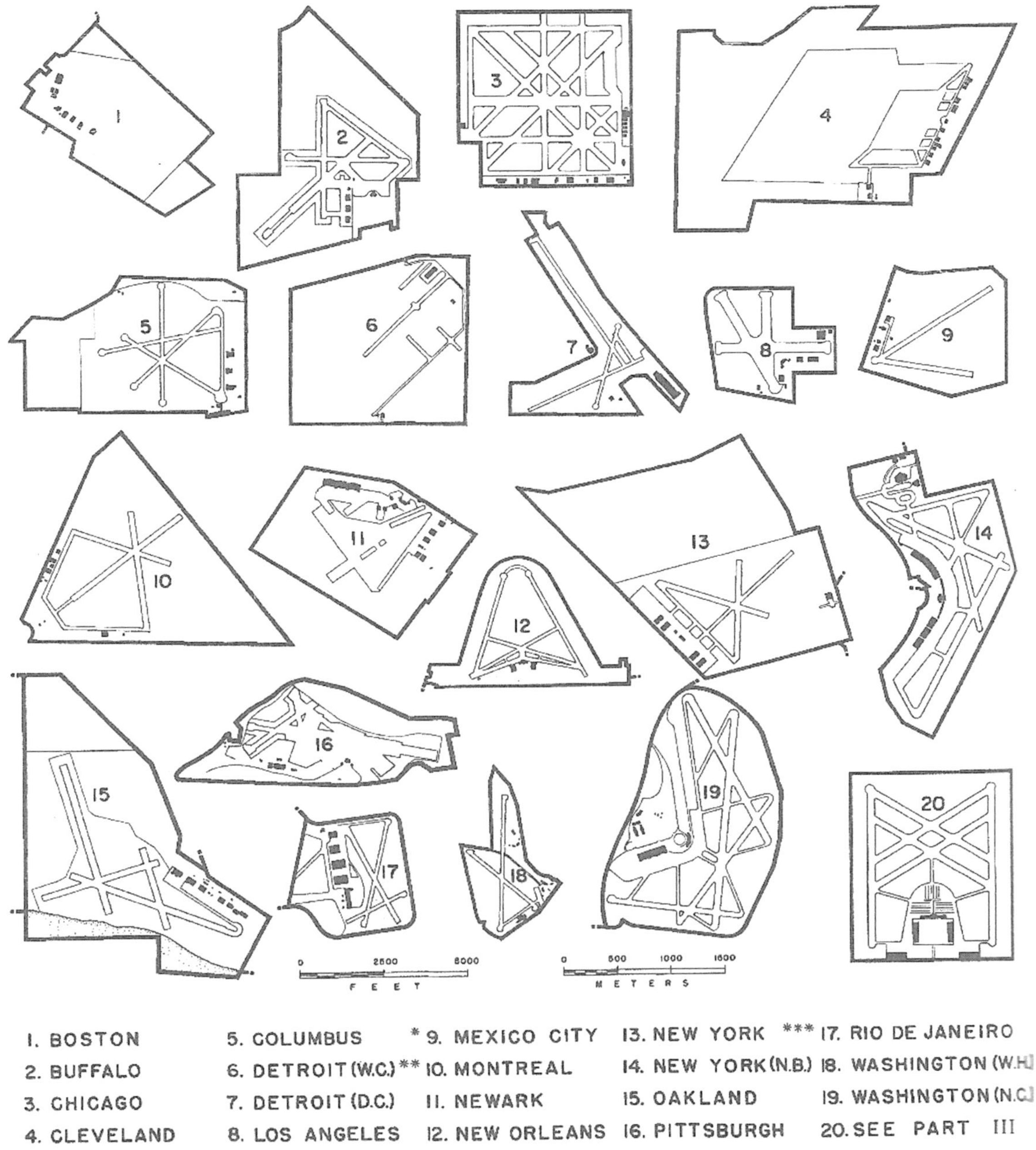

Figure A.5 Comparative American Airport master plans 1940.

SOURCE: AIRPORTS, SOME ELEMENTS OF DESIGN AND FUTURE DEVELOPMENT

The airports' layouts above have all been drawn to the same scale and were featured in John Walter Wood's 1940 book. It is notable that the vast majority of American airports had hard-surfaced runways orientated to suit the main anticipated wind directions. A number followed a new trend, with terminals that projected into the airfield in the gap between runways, as seen at New Orleans Sushan (12), New York, La Guardia (14) and Washington, National (19).

Airport master plans – Europe 1939

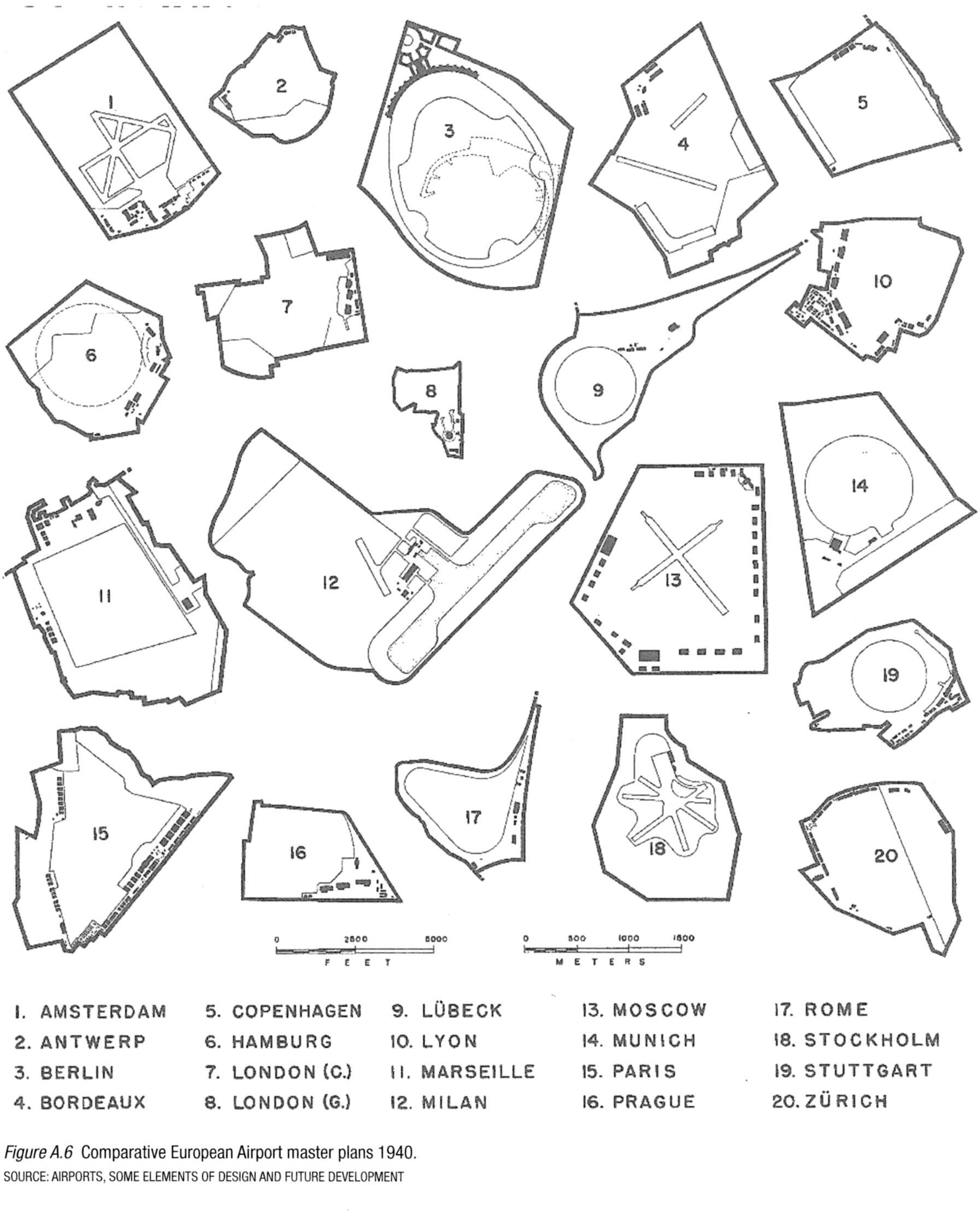

Figure A.6 Comparative European Airport master plans 1940.
SOURCE: AIRPORTS, SOME ELEMENTS OF DESIGN AND FUTURE DEVELOPMENT

Like the preceding examples, these layouts have all been drawn to the same scale and allow direct comparison with their American counterparts. The majority still maintains the all-direction grass field in preference to hard-surfaced runways but these are about to become obsolete due to the ever-increasing weight and performance of the newly introduced aircraft. The first European airport to be designed with concrete runways was Stockholm Bromma in 1936 to respond to soft ground conditions. It was followed by the conversion of Amsterdam Schiphol from an all-direction grass field to concrete runways in 1939.

Post-war airport master plans 1940s

Figure A.7 Post WW2 Airport master plans.
SOURCES: AIRPORT PLANNING, CHARLES FROESCH & WALTER PROKOSCH, 1946 BRITISH AIRWAYS (LHR)

Master plans of the 1940s included New York Idlewild (JFK), Chicago O'Hare using tangential runway forms analogous to a Catherine Wheel or in the case of Heathrow the 'Star of David'. This generation of masterplans all shows a move to locate the terminal area centrally between multi-directional perimeter runways.

Airport master plans 1954–1956

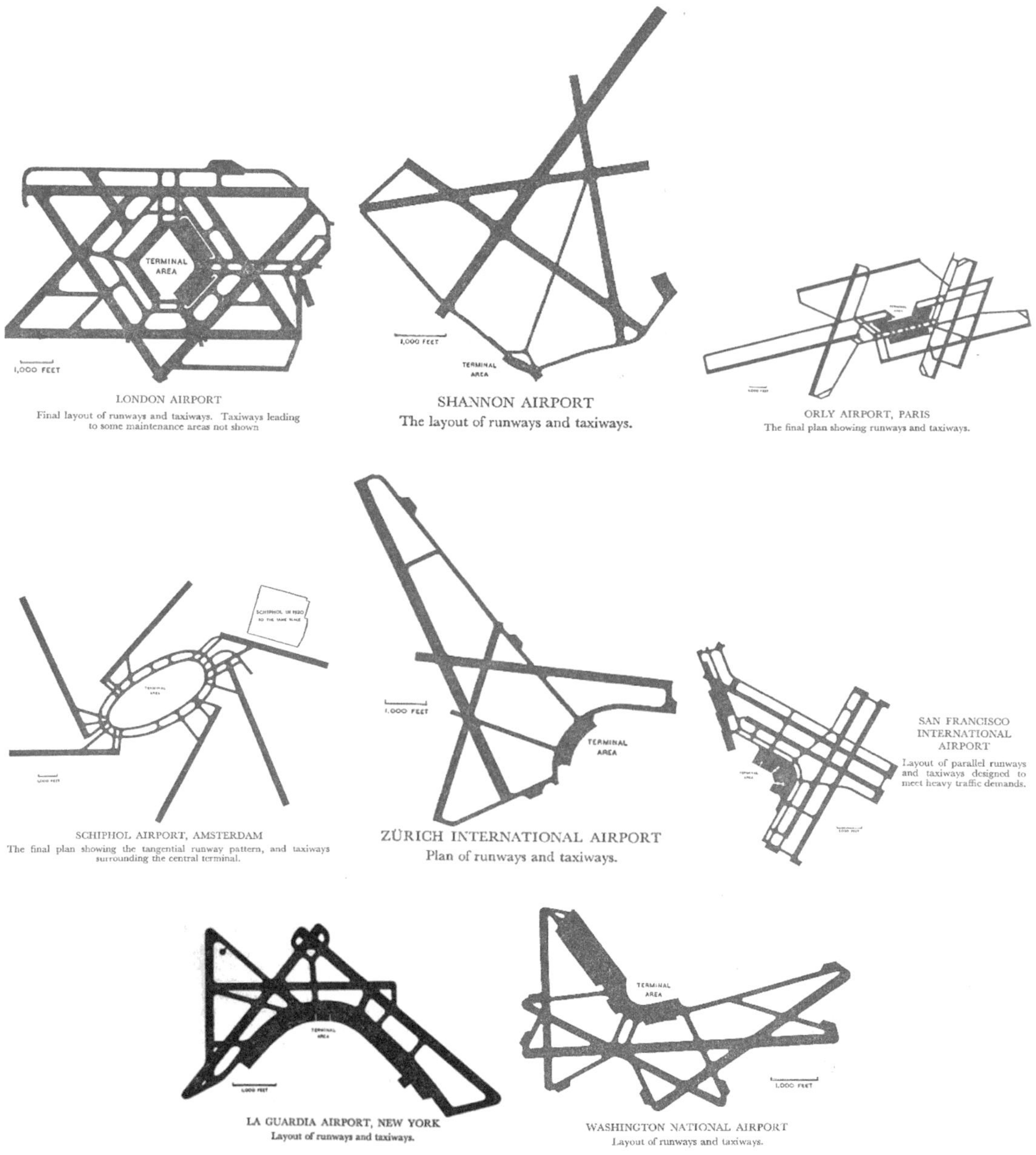

Figure A.8 Airport master plans 1954–56.
SOURCE: FAMOUS AIRPORTS OF THE WORLD, JOHN STROUD

Airport master plans of the 1950s followed the principles set down in the war-time precedents, but with some simplifications in runway layout as aircraft performance in cross winds improved. For example, the New York Idlewild (JFK) plan was progressively simplified and the number of runways reduced as the 1950s advanced. The San Francisco plan of 1954 and the Orly plan of 1955 only showed runways orientated in two directions, and single runway layouts were developed for London Gatwick (1958) and the expansion of San Diego Airport which was originally opened in 1928 as the Lindberg Field. They are, respectively, now the busiest single runway airports in the UK and the USA.

Airport master plans 1970s

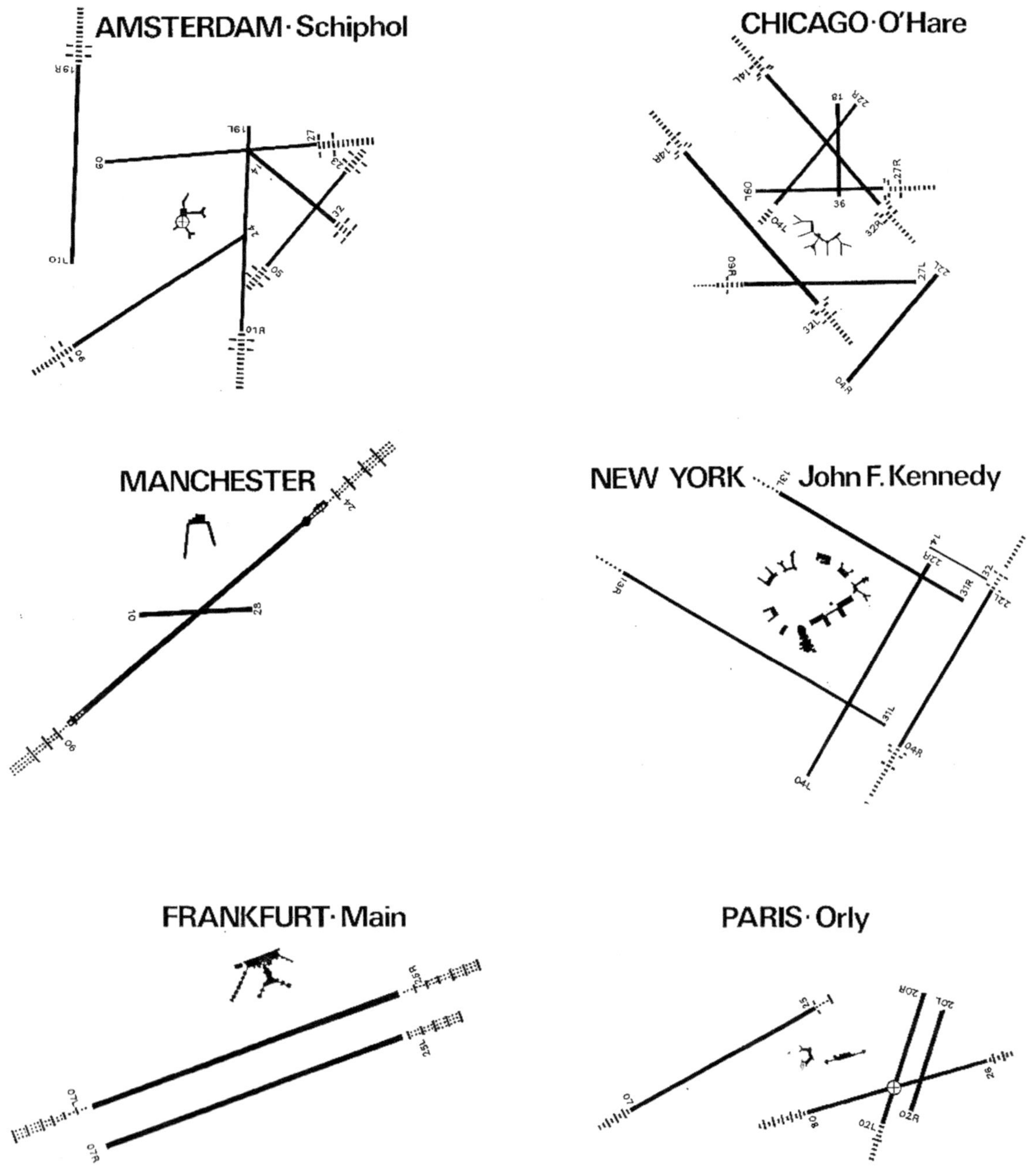

Figure A.9 Airport Master Plans 1973.

SOURCE: THE WORLD'S AIRPORTS, JOHN STROUD

AIRPORT MASTERPLANS OF THE 1970S SHOWED FURTHER SIMPLIFICATION OF THE RUNWAY LAYOUTS AS AIRCRAFT PERFORMANCE IN CROSS-WIND CONDITIONS IMPROVED.

Airport master plans 1980

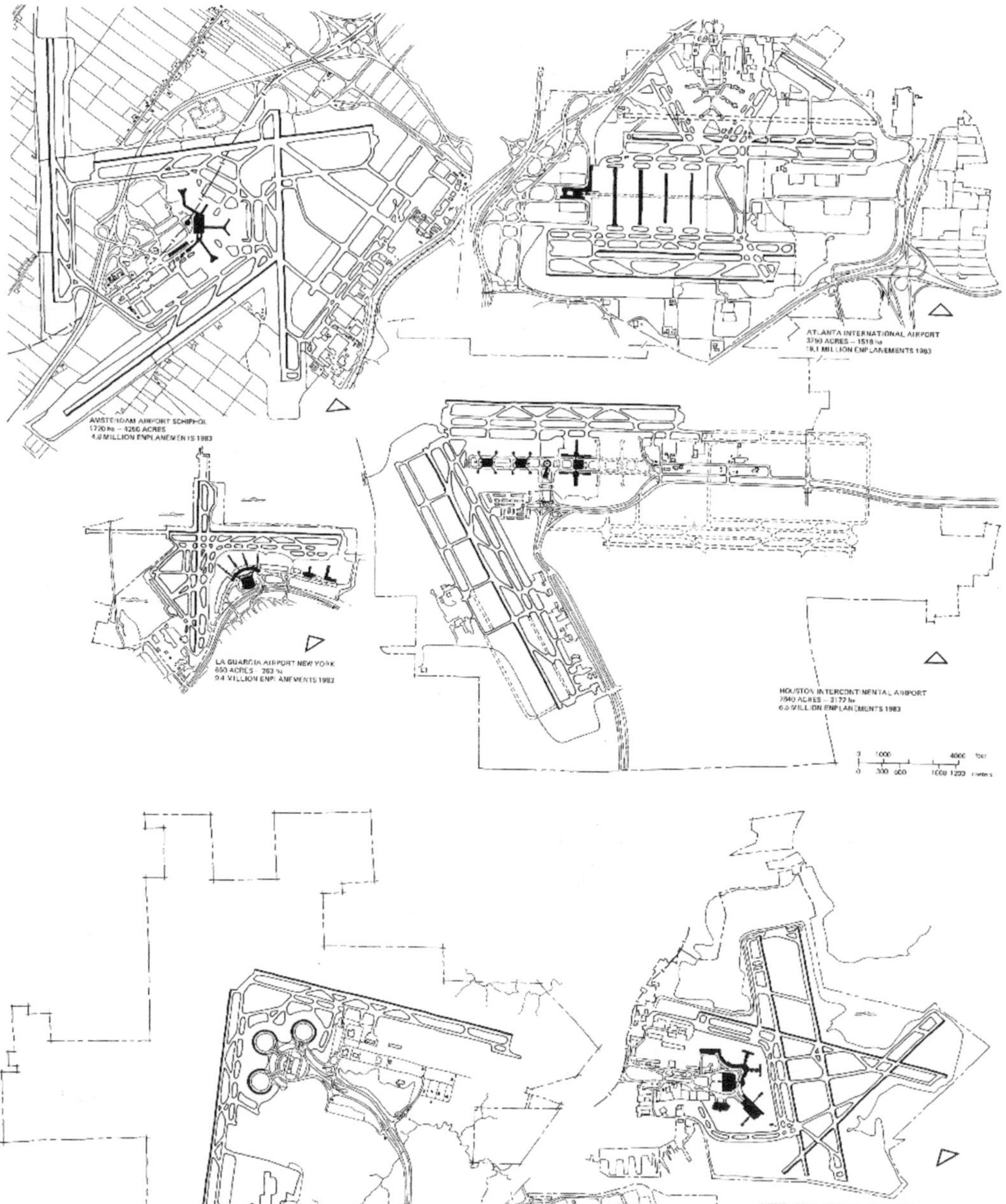

Figure A.10 Airport Master Plans – 1980.

SOURCE: THE AIRPORT PASSENGER TERMINAL, WALTER HART

1. AMSTERDAM SCHIPHOL 2. ATLANTA HARTSFIELD 3. NEW YORK LA GUARDIA 4. HOUSTON INTERCONTINENTAL. 5. KANSAS CITY 6. BOSTON LOGAN 7. PHOENIX SKY HARBOUR 8. SAN FRANCISCO

Airport master plans 2016 & 2030+

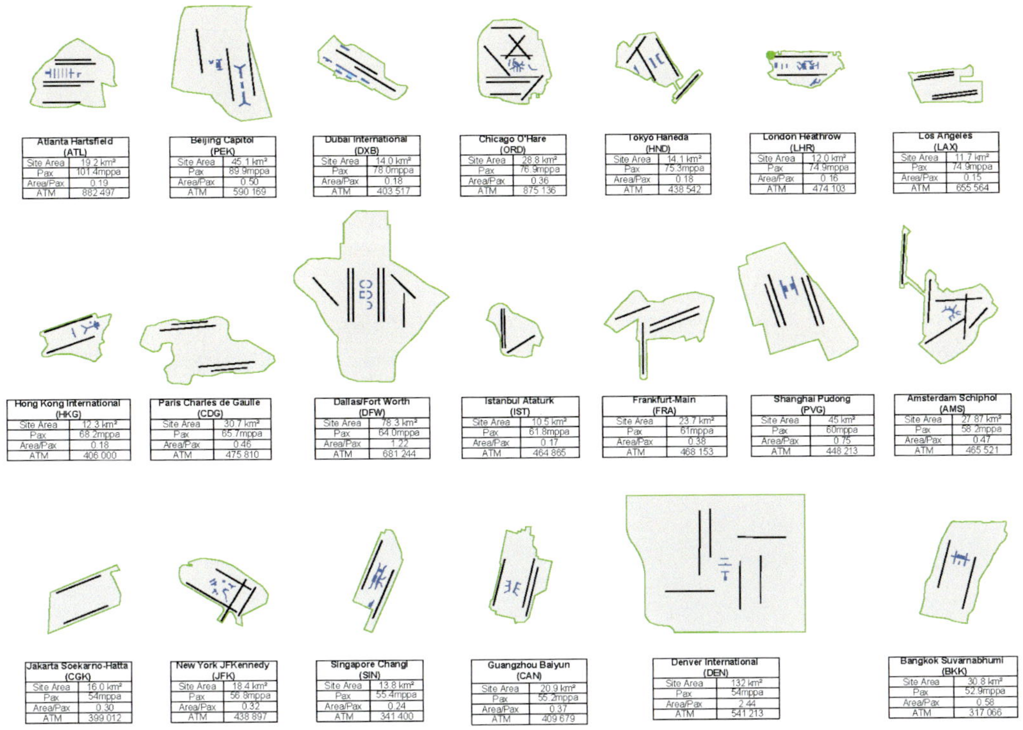

Figure A.11 Top Twenty Busiest Airports master plans 2016.

SOURCE: ATKINS LTD

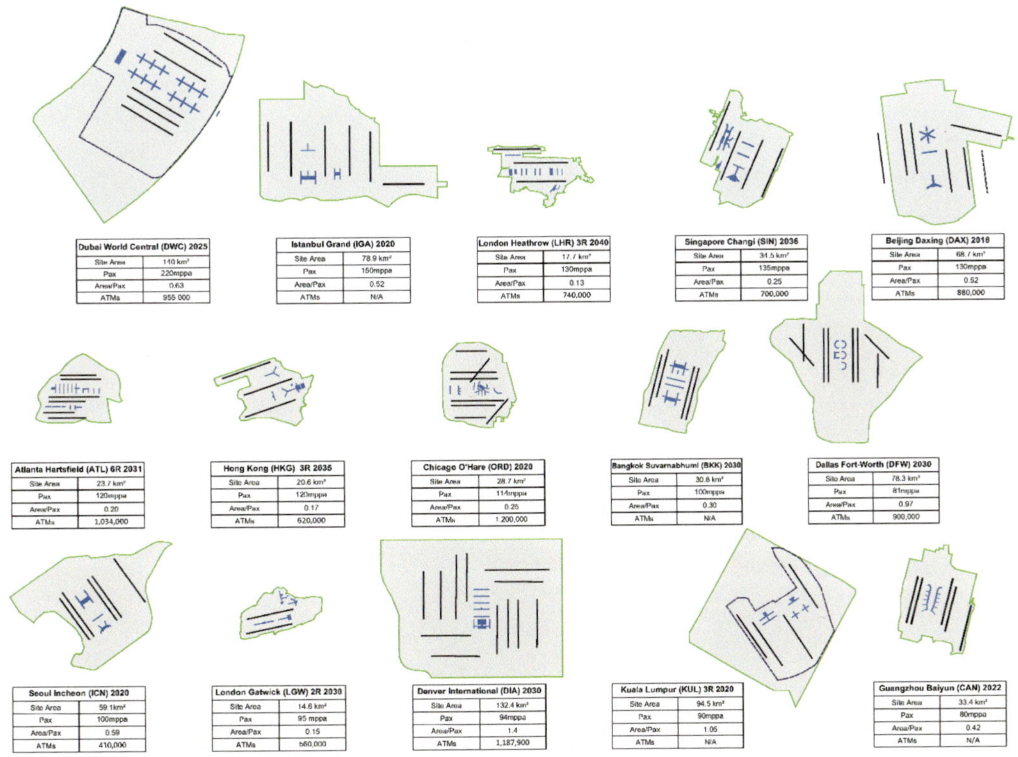

Figure A.12 Future Airport master plans showing additional runways and terminals.

SOURCE: ATKINS LTD.

BIBLIOGRAPHY

2020 ACI EUROPE Airport Business, 2008. T3: Dubai's latest infrastructural marvel. *Airport Business.*

Abramson, D. M., 2017. *Obsolescence*. Chicago & London: The University of Chicago Press.

Agache, D.-A., 1912. *Canberra Master Plan Competition, Entry No 4*, Canberra: National Archive of Australia.

Ahlbrecht, D. B.-R., 2017. *Der Flughafen Tempelhof in alten Bildern*. Erfurt: Sutton Verlag GmbH.

Ahrend, G., 1988. *Great Airports Worldwide*. New York: Air Cargo News Books.

Aircraft Engineering and Aerospace Technology, 1936. The Low Flying Nuisance: The Anti-noise League's Complaint and the Air Ministry's reply. *Aircraft Engineering and Aerospace Technology*, 8(5), pp. 119–120, https://doi.org/10.1108/eb030040

Airport Technology, E., 2017. *Airport Technology*. [Online] [Accessed 28.02.2020 February 2020].

Airport Technology, E., 2020. *https://www.airport-technology.com/*. [Online] [Accessed 10 February 2020].

Airports Council International, 2003. *Airports in the First 100 Years of Flights*. Twickenham: Insight Media.

Airports, D., 2017. *Dubai International Airport (DWC)*. [Online] Available at: http://www.dubaiairports.ae/corporate/about-us/dwc-dubai-world-central [Accessed 20 April 2017].

Airports, D., 2017. *Dubai International Airport (DXB)*. [Online] Available at: www.dubaiairports.ae/corporate/about-us/dubai-international-(dxb) [Accessed 8 August 2017].

Airways, 2014. *airwaysmag.com*. [Online] [Accessed 7 December 2019].

AJ, 1965. Airport Building at Gatwick, Sussex, Yorke, Rosenberg & Mardall. *AJ Information Library*, Issue 3rd November, p. 1019.

Allen, R., 1983. *Major Airports of the World*. Shepperton, Surrey: Ian Allen Ltd.

Allward, M. & M, R., 1962. *London's Airports (Heathrow and Gatwick)*. 4th ed. Hampton Court: Ian Allen.

Anderson, I., 2014. *Heathrow, From Tents to Terminal 5*. Gloucestershire: Amberley Publishing.

Anglo Japanese Airport Consortium SDN BHD, 1992. *KL International Airport Masterplan Volume B*, Kuala Lumpur: Government of Malaysia.

Architects' Journal, 1958. Gatwick Airport. *Architects' Journal*, Issue 14th August, p. 236.

Ashford, N. M. C. A., 1999. *Airport Finance*. 2nd ed. Woodhouse Eaves, Leics: The Loughborough Airport Consultancy.

Ashford, N. J., Mumayiz, S. A. & Wright, P. H., 2011. *Airport Engineering, Planning, Design, and Development of 21st Century Airports*. 4th ed. Hoboken, New Jersey: John Wiley and Sons, Inc.

Assendorf, C., Bosma, K., Eisenbrand, J. & Entwistle, J., 2004. *Airworld: Design and Architecture for Air Travel*, Vitra Design Museum.

AT Profile 1 and Building Design, 2002. Transportation Centre, Incheon International Airport. *Architecture Today*, Issue 11th January.

Atkins, 2013. *Medina Airport*. [Online] [Accessed 30 January 2017].

Atkins, 2014. *TfL.gov.uk - A New Airport for London*.

BAA Group Planning A, 1997. *Planning Standards*, London: BAA.

BAA Group Planning B, 1997. *Planning Guidelines*, London: BAA.

BAA Group Planning C, 1997. *Calculations*, London: BAA.

Bangkok, A., 2020. *bangkokairportonline.com*. [Online] [Accessed 10 march 2020].

Banham, R., 1960. *Theory and Design in the First Machine Age*. London: The Architectural Press.

Banham, R., 1962. The Obsolescent Airport. *Architectural Review*, pp. 252–3.

Bates, J. C., 2011. *Beyond Expectation Seoul Incheon 10th Anniversary Report*, Twickenham: Insight Media Ltd.

Becker, H.-J. & H, R., 2000. *100 jahre Luftschiffe*. Stuttgart: Motorbuch Verlag.

Beckman, M. T., 2014. *Airports in Architectural Competitions 1920-1940*. Delft, Technical University of Delft.
Bednarek, J. R. D., 2001. *America's Airports, Airport Development 1918–1947*. College Station: Texas A & M University Press.
Belogolovsky, V., 2017. Interview with Paul Andreu. *Architecture Daily*, Issue 07.03.2017.
Benthem, J., 2013. *Megastructure Schiphol, City Centre Led to Airport City*, naiolo publishers.
Berkers, M. et al., 2013. *Megastructure Schiphol, Design in Spectacular Simplicity*. Amsterdam: Naioio Publishers.
Biernacki, W., 1950. Thinking Ahead. *Flight*, Issue 04 May.
Bingham, N., 2012. *100 Years of Architectural Drawing 1900–2000*. London: Laurence King Publishing Ltd.
Binney, M., 1999. *Airport Builders*. Academy Editions ed. Chichester: John Wiley & Sons.
Black, A., 1922. *How to Lay Out and Build an Airplane Landing Field*. Washington, D.C.: National Advisory Committee for Aeronautics, NCA Technical Memorandum 164.
Black, A., 1929. *Civil Airports and Airways*. New York: Simmons-Boardman Publishing Company.
Black, A., 1930. *American Airport Designs*. Chicago: Taylor, Rogers & Bliss, Inc for the Lehigh Portland Cement Company.
Blacklock, M., 2005. *Recapturing the Dream, A Design History of New York's JFK Airport*. London: Mark Blacklock.
Blankenship, E. G., 1974. *The Airport, Architecture - Urban Integration - Ecological Problems*. New York: Praeger Publishers.
Bloom, D. B., 2015. *The Metropolitan Airport, JFK International and Modern New York*. Philidelphia: University of Pennsylvania Press.
Blow, C., 2005. *Transport Terminals and Modal Interchanges, Planning and Design*. Oxford: Architectural Press.
Blow, C. J., 1991. *Airport Terminals*. 2nd ed. Oxford: Architectural Press, Reed Educational and professional Publishing Ltd.
Bosco Thomas, L. & Director, A.D., 2015. *Airport Sustainability Report Appendix 10*, New York: PANYNJ, Port Authority of New York and New Jersey.
Botton, A. d., 2002. *The Art of Travel*. London: Hamish Hamilton, Penguin Books.
Botton, A. d., 2009. *A Week at the Airport: A Heathrow Diary*. London: Profile Books Ltd.
Bouverie, D. P., 1937. *Ramsgate Municipal Airport, RIBA*.
Bowdler, R. et al., 2000. *Berlin Tempelhof, Speke, Le Bourget*. Paris: Caisse nationale des monuments et des sites, Editions du patrimone.
Bowen, D. J. & R. D, J.-P., 2020. The Rise of Air Transportation. In: D. J. Rodrigue, ed. *The Geography of Transport Systems*. New York: Routledge.
Braden, B. & H, P., 1989. *A Dream Takes Flight, Hartsfield Atlanta International Airport and Aviation in Atlanta*. Athens, Georgia: University of Georgia Press.
Bradley, A., 2012. *A Comparison of the Costs of Robotic, Semi-Automated and Manual Build Airport Baggage Systems*, Cranfield: Cranfield University, Department of Air Transport, School of Engineering.
Branigan, M., 2011. *A History of Chicago's O'Hare Airport*. Charleston: The history Press.
Brodherson, D. P. P., 1993. *What can't go up can't come down, the history of American airport policy, planning and design, PhD Thesis*. Ithaca, New York: Cornell University.
Budd, L. C., 2007. *Producing Airspace: The Contested Geographies of Nottingham East Midlands Airport*, Loughborough: Loughborough University.
Bullock, F., 1998. *LAX, Los Angeles International Airport*. Shrewsbury: Airlife Publishing Ltd.
Burke, R. H., 1948. *Chicago Orchard (Douglas) Airport, Master Plan*, Chicago: City of Chicago.
Butcher, L., 2011. *Aviation: Security SN/BT/1246*, London: House of Commons Library.
Butcher, L., Ares, E., Midgley, H. & Smith, L., 2017. *Heathrow Expansion, Briefing Paper, No. CBP 12.06.2017*, London: House of Commons, UK.
Cairns, S. & J, J. M., 2014. *Buildings Must Die, a Perverse View of Architecture*. Cambridge, Massachusetts: The MIT Press.
Calder, S., 2002. *No Frills, The Truth Behind the Low Cost Revolution in the Skies*. London: Virgin Books Ltd.

Calver, G., 2008. *Terminal 5, Transforming Heathrow*. London: BAA.
Caro, R. A., 2015. *The Power Broker, Robert Moses and the Fall of New York*. London: The Bodley Head, A penguin Random House Company.
Cauter, G., 1980. *International Airport*. London: Octopus Books.
CDA, C. D. o. A., 2019. *flychicago.com*. [Online] [Accessed 10 December 2019].
Centre for Cities, R., 2017. The Centre for Cities 2017 Report. *Crawley Observer.*
Chandos, J., 1956. *London Airport*. London: HMSO, for the Ministry of Transport and Civil Aviation.
Chaplin, J. C., 2001. *Safety Regulations, The First 100 Years*. London: Journal of Aeronautical History, Paper, Royal Aeronautical Society.
Clancy, J., 2008. First class all the way. *The Guardian*, 11 january.
Clarke, D., n.d. *Aircraft and Airports*. London: Cavendish House.
Cluett, D., B, N. J. & L, B., 1977. *The First Croydon Airport 1915–1928*. 2nd, Croydon Airport Society, 2001 ed. Sutton, Surrey: London Borough of Sutton.
Cluett, D., N, J. & L, B., 1980. *Croydon Airport 1928–1939, The Great Days*. Sutton, Surrey: London Borough of Sutton, Libraries and Arts Services.
Cohen, A. P. & Gensler, D. H. D., 2012. *Design for Aviation*, Gensler Publications.
Corbusier, L., 1923, English Edition 1927, first paperback edition 1970. *Towards a New Architecture*. 1946, Architectural Press Reprint 1970 ed. London: The Architectural Press.
Corbusier, L., 1935. *Aircraft*. London: The Studio Ltd.
Corbusier, L., 1970 reprint. *Towards a New Architecture*. London: Architectural Press.
Cortzen, J., 2000. *til vejrs, Kobenhavns Lufthavn 1 75 AR*. Copenhagen: Borsens Forlag A/S.
Cuadra, M., 2002. *World Airports: Vision and Reality: Culture and Technique: Past and Present*. Hamburg: Julius Verlag GmbH.
D'Silva, C. J., 2006. *Constructing Suvarnabhumi, A Photographic Study During Construction of the New Bangkok International Airport*. Suthep, Chiang Mai 50200, Thailand: Silkworm Books.
Davies, M. P. D. T., 2008. T5: The Concept. *International Airport Review*, 12(2), p. 12.
Dawson, R. T. p., 1962. *Phaidon News.* [Online] [Accessed 28 February 2020].
Daxing-PKX, 2019. *Daxing-PKX-Airport.com*. [Online] [Accessed 1 October 2019].
De Neufville, R., 1976. *Airport Systems Planning*. London: The Macmillan Press.
Delius, P. & Slaski, J., 2005. *Airline Design*. Dusseldorf: teNeus Verlag GmbH.
Department of Trade, UK, 1974. *Maplin, Review of Airport Project*, London: Her Majesty's Stationery Office.
Dierikx, M., 1999. *Schiphol, Amsterdam's Airport Below the Sea*. Shrewsbury: Airlife Publishing Ltd.
Dittrich, E., 2012. *Der Flughafen Tempelhof*. Berlin: Lukas Verlag.
Doganis, R., 1992. *The Airport Business*. London & New York: Routledge.
Doherty, S., 2008. *Heathrow's Terminal 5, History in the Making*. Chichester: John Wiley & Sons.
Douglas, D. G., 1996. *The Invention of Airports: A Political, Economic, and Technological History of Airports*. Philadelphia: University of Pennsylvania.
DXB, 2018. *dubaidxbairport.com*. [Online] [Accessed 10 July 2019].
Eames, J., 2000. *Sydney Airport, 80 Years as the Gateway to Australia*. Edgecliff, NSW: Focus Publishing Pty.
Edwards, B., 1998. *The Modern Terminal, New Approaches to Airport Architecture*. London: E & FN Spon.
Eggebeen, J., 2007. *Airport Age: Architecture and Modernity in America*. New York: The City university of New York, Proquest information and Learning Company.
Eiselin, S., Frommberg, L. & Gutzmer, A., 2016. *The Art of the Airport, The World's Most Beautiful Terminals*. London: Frances Lincoln Limited, Quarto Publishing Group.
Elke, D., 2005. *Der Flughafen Tempelhof*, Lukas Verlag.
Elliott, D. M. & J, N., 2010. An introduction to airport APM systems. *Journal of Advanced Transportation*, 33(1), pp. 35–50.
Emmerson, A., 2013. *The Marvels of Flying*. London: Capital Transport Publishing Ltd.
Engineers, I. o. C., 1967. *Airports for the Future*. London: Institution of Civil Engineers.
Falconer, J., 1990. *Heathrow*. Shepperton: Ian Allen Ltd.
Fayazbakhsh, M., 1996. *Environmental Impacts of Airports*. Salford: Department of Environmental resources, University of Salford.

Federal Aviation Administration, 1962. *Comments by the Late Eero Saarinen, World-Renowned Architect, about Dulles International Terminal Building before his death in September 1961*, Washington, D.C.: Federal Aviation Agency.
Federal Aviation Administration, Department of Transportation, 1962. *Dulles International Airport: Handbook of Information*, Washington, DC: FAA.
Fentress, C., 2013. *Now Boarding, Frentress Airports + The Architecture of Flight*. Denver: Denver Art Museum in association with Scala Publishers.
Fentress, C. W., 2006. *10 Airports, Fentress Bradburn Architects*. New York: Edizioni Press Inc.
Flight, 1919. Editorial. *Flight.*
Flight, 1955. Tower of London & L.A.P.'s New Tower. *Flight* (), p. 565
Flight Global, 1956. B.E.A. Decade: The story of the Corporation's ten eventful years. *Flight*, 08(03), p. 193.
Flight, E., 1909. La Grand Semaine d'Aviation. *Flight*, September.
Flight, E., 1929. RIBA Competition for a new airport for 1943. *Flight*, January.
Flight, Editor, 1952. Another International Airport, Friendship Airport. *Flight* (7th March), p. 266.
Flughafen, Frankfurt/Main AG, 1996. *60 Jahre Flughafen Frankfurt*. Frankfurt: Flughafen Frankfurt/Main AG.
flystl, 2019. *flystl.* [Online] [Accessed 12 May 2019].
Forgione, M., 2014. *Los Angeles Times, Encounter, Travel.* [Online] [Accessed 28 February 2020].
Foster, N., 1996. *Reinventing the Airport*. London: Foster + Partners.
Foster, N., 2008. *fosterandpartners.com*. [Online] [Accessed 28 February 2020].
Foster, N., 2018. *https://www.fosterandpartners.com*. [Online] [Accessed 4 October 2019].
Foster, +. p., 2020. *fosterandpartners.com*. [Online] [Accessed 1 May 2020].
Foster, +. P., 2010. *Airports & Infrastructure*. London: Foster + Partners.
Freeman, P., 2017. *Abandoned and little known Airfields: Michigan, North East Detroit.* [Online] Available at: www.airfields-freeman.com/MI/Airfields_MI_Detroit_NW.htm [Accessed 20 June 2017].
Froesch, C. & P, W., 1946. *Airport Planning*. New York: John Wiley & Sons, Inc.
Fuller, G. & H, R., 2004. *Aviopolis, A Book about Airports*. London: Black Dog publishing Limited.
Geddes, N. B., 1932. *Horizons*. New York: Little, Brown and Company.
Gensler Publications, 2012. *Design for Aviation*. San Francisco: ORO Editions.
Gerhardt, P., 1932. Chicago Municipal Airport. *The Architectural Record,* p. 117.
Gesar, A., 1994. *New York's Airports: John F. Kennedy, Newark, LaGuardia*. Shrewsbury: Airlife Publishing Ltd.
GOMACO, 2008. *GOMACO World 36.1.* [Online] [Accessed 10th September 2017].
Gordon, A., 2008. *Naked Airport, A Cultural History of the World's Most Revolutionary Structure*. 2008 ed. Chicago and London: University of Chicago Press.
Grant, R. G., 2002. *Flight, 100 Years of Aviation*. 2004 ed. London: Dorling Kindersley Limited.
Greif, M., 1979. *The Airport Book, From Landing Field to Modern Terminal*. New York: Mayflower Books, Inc.
Grief, M., 1978. *The New Industrial Landscape, The Story of the Austin Company*. Clinton, New Jersey: The Main Street Press.
Grief, M., 1979. *The Airport Book*. New York: Mayflower Books Inc.
Groot, M., 2020. *AirportHistory.org*. [Online] [Accessed 10 April 2020].
Gropius, W., 1935. *The New Architecture and the Bauhaus*. London: Faber and Faber.
Guller, M. & G, M., 2000. *From Airport to Airport City*. Barcelona: European Airports Regions Conference ARC.
Gunston, B., 1991. *The World's Major Passenger Airliners*. Moorebank NSW, Australia: Mallard Press.
Hackelberger, C., 2004. *Flughafen Munchen*. Munich: Kock + Partner Terminal 2, Birkhauser.
Hamlin, T. F., 1940. Airports as Architecture. *Pencil Points,* Issue October, p. 639.
Handleman, P., 1998. *Chicago O'Hare, The World's Busiest Airport*. Shrewsbury: Airlife England.
Hanks, S. S., 1929. *International Airports*. New York: The Ronalds Press Company.
Harbour, I., 2020. *https://www.rsh-p.com/projects/t4-madrid-barajas-airport*. [Online] [Accessed 11 January 2020].

Harrison, I., 2002. *Ports and Airports*. London: HarperCollinsPublishing.

Harrison, M. L. I., 1997. *Airports*. Reigate: BUCHair (U.K.) Ltd.

Hart, W., 1985. *The Airport Passenger Terminal*. New York: John Wiley & Sons Inc.

Heathrow Airport Limited (BAA Heathrow), 2008. *Terminal 5, Transforming Heathrow*. London: Heathrow Airport Limited.

Hecker, M., 2005. *Historic Airports, Proceedings of the International L'Europe de l'air Conferences on Aviation Architecture*. London: English Heritage.

Heeb, C., 2007. *A Multifaceted Monument- the Complex Heritage of Tempelhof Central Airport*, Berlin: Brandenburg University of Technology Cottbus, Faculty of Architecture, Civil Engineering and Urban Planning.

Heino, A. F., 1945. Designing the large terminal. *Architectural Record* (OK), pp. 80–81.

Henderson, D. P., 2020. *sunshineskies.com*. [Online] [Accessed 1 March 2020].

Henket, H.-J., 2013. *Megastructure Schiphol*, naiolo publishers.

Herberta, 2., 2016. *Looking for these air traffic control towers*. [Online] Available at: https://www.reddit.com/r/architecture/47lq0p/looking_for_these_control_towers/ [Accessed 12 November 2017].

Hirsh, M., 2016. *Airport Urbanism, Infrastructure and Mobility in Asia2*. Minnesota: University of Minnesota.

Hoffman, L. J., 2015. *Pan American World Airways, Images of Modern America*. Charleston, South Carolina: Arcadia Publishing.

Hong Kong International Airport, 2015. *HKIA Master Plan 2030*, Hong Kong International Airport.

Hooks, M., 1997. *Croydon Airport, Images of England*. Stroud, Gloucestershire: Tempus Publishing.

Hooks, M., 1999. *Lufthansa, Images of Aviation*. Stroud, Gloucestershire: Tempus Publishing Ltd.

Hooks, M., 2002. *Croydon Airport, The Peaceful Years*. Croydon: Tempus Publishing for Croydon Airport Society.

Horonjeff, R. e. a., 1958. *Exit Taxiway Location and Design*, University of California: Institute of Transportation and Traffic Engineering.

Horonjeff, R., McKelvey, F. X., Sproule, W. J. & Young, Y. S. B., 2010. *Planning & Design of Airports*. 5th ed. New York: McGraw-Hill.

Hubbard, H. V., 1930. *Airports Their Location, Administration and Legal Basis*. Cambridge: Harvard University Press.

Hunt, D., 1958. Focus on Turn-round. *Flight* (7th November), pp. 165–167.

Hunt, D., 1961. How Idlewild was planned for the jet age. *Architectural Record* (September), pp. 152–189.

Hunt, D. J., 1961. How Idlewild was planned for the jet age. *Architectural Record* (September), pp. 152–189.

Hurst, D., 2012. London Estuary Airport Expansions, The Maplin Story. *UK Progressive*.

IATA, June 2016. *Improving operations delivers green benefits*. [Online] Available at: http://airlines.iata.org/analysis/improving-operations-delivers-green-benefits [Accessed 12 May 2017].

ICAO, 2004. *ICAO Circular 303-AN/176*, Montreal: ICAO.

ICAO, October 2016. *ICAO Symposium on Aviation and Climate Change,*. Montreal, ICAO.

Jahn, H., 1991. *Airports*. Basel: Birkhauser Verlag.

Janic, M., 2013. *Airport Analysis, Planning and Design*. New York: Nova Science Publishers, Inc.

Jarvis, P., 2014. *British Airways, An Illustrated History*. Stroud, Gloucestershire: Amberley Publishing.

Jersey, P. A. N. Y. N. J, 2020. *anewlga.com*. [Online] [Accessed 20 April 2020].

Jodidio, P., 1994. *Kansai International Airport*, Renzo Piano Building Workshop.

Jodidio, P., 2004. *Paul Andreu, Architect*. Basel: Birkhauser.

Jones, G., 2000. *Gatwick Airport*. Shepperton, Surrey: Ian Allen Publishing.

Jones, W., 2006. *New Transport Architecture*. London: Octopus Publishing Group Ltd.

Joos, J. S. K. & Helmut, W., 2001. *AirPorts*. In: Helmut W. Joos, ed. Berlin: Ernst Wasmuth Verlag.

Kahn, A., 1937. *Speech by Albert Kahn to the Adcraft Club, "Thirty Minutes of American Architecture and Architects." 11/22/1937, Box 1, Albert Kahn Collection, Bentley Historical Library, University of Michigan*. Ann Arbor, Michigan: University of Michigan.

Karsardis, J. D. & L, G., 2011. *Aerotropolis*. New York: Farrar, Straus and Giroux.

Keisel, K. M., 2012. *Dayton Aviation, The Wright Brothers to McCook Field, Images of Aviation*. Charleston, South Carolina: Arcadia Publishing.
Kent, D. E., 2012. *Midway Airport, Images of Aviation*. Charleston: Arcadia Publishing.
Kidder Smith, G., 1961. *The New Architecture of Europe*. Paperback 1962 ed. Harmondsworth, Middlesex: Pelican Books Ltd. copyright The Wold Publishing Company.
Kingfisher Press, 2015. *The Wright Brothers Photographs*. New York: Kingfisher Press, LLC.
King, J. & T, G., 1980. *Golden Gatwick, 50 Years of Aviation*. London: British Airports Authority/ Royal Aeronautical Society.
Koch, +. P., 2004. *Munich Airport International*. Basel: Birkhauser.
LA, D., 1960. *Los Angeles, Department of Aviation, Annual Report*, Los Angeles, Flight Path Learning Center: LA, DOA.
Labro, G., 1938. Le Bourget Airport. *L'Architecture*, January.
Larkham, P., 2013. *The London Regional Reconstruction Committee: Architects, Exhibitions, and Post -War Visions for Re-Planning - Working Paper Series no. 16*. Birmingham: Birmingham School of the Built Environment, Birmingham City university- Centre for Environment and Society Research.
LCY, 2017. *London City Airport Surface Access Strategy 2017–2025*, London: London City airport.
Lethaby, W. R., 1912. *An Introduction to the History and the Theory of the Art of Building*. London: Macmillan.
Lewis-Dale, H. A., 1932. *Aviation and the Aerodrome, A Treatise on the Problems of Aviation in Relation to He Design and Construction of Aerodromes*. London: Charles Griffin and Company Ltd.
Lindsey, G.M, Interview by Patt Morrison, Los Angeles Times, 24 Dec 2013
Lord, A. & H, M., 2000. *Airports*. 5th ed. Reigate: BUCHair (UK) Ltd.
Luckman, C., 1988. *Twice in a Lifetime; From Soaps to Skyscrapers*. 1st edition (June 1 1988) ed. New York: W W Norton & Co Inc.
Luckman, C., 2014. *iconeye.com*. [Online] Available at: https://www.iconeye.com/architecture/los-angeles-international-airport-jet-age-modernism [Accessed 6 October 2018].
Lynch, C., 2003. *Chicago's Midway Airport, The First Seventy Five Years*. Chicago: Lake Claremont Press.
Macnair, J., 2007. Terminal Condition (Heathrow Terminal 5). *A J*, Issue 06.12.07, pp. 26–33.
Mason, G., 2018. Mind the Gap. *Airport Focus International,* Issue May/June, pp. 12–17.
McCallum, I., 1959. *Architecture USA*. London: The Architectural Press.
McKie, D., 1973. *A Sadly Mismanaged Affair, a Political History of the Third London Airport*. London: Croom Helm.
McMullen, A., 1940. The Development of Airports, New York Municipal Airport (LaGuardia). *Pencil Points* (October), pp. 615–632.
Meyrick, D. J. H., 1983. *Civil Aviation and Land Use: The Case of London's Terminal Airports 1919–1946*, London: Birkbirk College, University of London.
Ministry for Land, Transport and Maritime Affairs, South Korea & Airports Council International, 2012. *Incheon Airport, Beyond Expectation, 10th Anniversary Special Report*, Twickenham: Insight Media Ltd.
Morshed, A., 2002. *Aviator's (Re)Vision of the World: An Aesthetics of Ascension in Norman Bel Geddes's Futurama*, Massachusetts: Massachusetts Institute of Technology.
Motum, J., 1989. *The Putnam Aeronautical Review*. London: Conway Maratime Press Ltd.
Moyle, T., 2015. *Art Deco Airports, Dream Designs of the 1920s & 1930s*. London: New Holland Publishers Pty Ltd.
Museum Der Stadt, Konigsberg, n.d. [Online] Available at: http://museumkoenigsberg.tumblr.com [Accessed 24 10 2017].
National Research Council (US) Environmental Studies Board, 1971. *Jamaica Bay and Kennedy Airport, a Multidisciplinary Environmental Study'*. New York: Washington, National Academy of Sciences.
Negus, G. & S, T., 1984. *Aviation in Birmingham*. Leicester: Midlands Counties Publications.
Neufville, R. d., 1976. *Airport Systems Planning*. London and Basingstoke: The Macmillan Press Ltd.

New Orleans Lakefront Airport, 2017. *lakefrontairport.com*. [Online] Available at: http://www.lakefrontairport.com/lakefront-airport-history.html [Accessed 11 November 2017].

Niemietz, K., 2013. Depoliticising Airport Expansion: Market-Oriented Responses to the Global and Local Externalities of Aviation. *Institute of Economic Affairs (IEA)*, IEA Discussion Paper (No 51).

NJ, P. A. N. &., 2020. *anewjfk.com*. [Online] [Accessed 4 April 2020].

Pageot, H., 2013. *An Airport Did You Say?* 3rd ed. London: Amazon.

Paris, C. G., 2020. *parisaeroport.fr*. [Online] [Accessed 12 March 2020].

Partners, F. +., 2017. *Foster + Partners*. [Online] Available at: http://www.fosterandpartners.com/projects/spaceport-america [Accessed 10th Oct 2017 March 2017].

Pascoe, D., 2001. *Airspaces*. London: Reakton Books Ltd.

Pearman, H., 2004. *Airports, A Century of Architecture*. London: Laurence King Publishing Ltd.

Peck, M. C., 2005. *Washington Dulles International Airport*. Charleston, South Carolina: Arcadia Publishing.

Pelli, C., 2000. *National Airport Terminal*. Gloucester, Massachusetts: Rockport Publishers.

Pickens, B. L., 1956. Proud Architecture and the\spirit of St Louis. *Architectural Record* (April), p. 196.

Pilon, N., 2006. *Citizens' Views on Air Traffic Growth*, Eurocontrol.

Pirath, C. (., 1938 (translated from original German 1937). *Aerodromes, Their Location, Operation and Design*. London (Berlin): Sir issac Pitman Sons (Julius Springer).

Port Authority of New York and New Jersey PANYNJ, 2019. *Draft Environmental Assessment, JFK International Airport Redevelopment Programme 20.12.2019*, New York: PANYNJ.

Powell, K., 1992. *Stansted, Norman Foster and the Architecture of Flight*. London: A Blueprint monograph, Fourth Estate and Wordsearch Ltd.

Powell, K., 2006. Madrid Airport. *Architects' Journal,* Issue 27th April.

Powers, A., 1992. *In the Line of Development:* FRS Yorke, E Rosenberg and CS Mardall *to* YRM 1930–1992. London: RIBA Heinz Gallery.

Priest, J., 2015. *A Boy from Surrey, Impressions of a Lifetime*. Falmer: easyBook publishing.co.uk.

Przychowski, H. D. R. W., 2011. *Die Berliner Flughafen*. Munich: Geramond Verlag.

Rawson, R. R., 2013. *An Evaluation of the Contribution of the UK Airport Master Planning System to a More Sustainable Development*, Centre for Aviation, transport and Environmental school of Science and the Environment, Manchester Metropolitan University: Manchester.

Renzo Piano, B. W., 2020. *http://www.rpbw.com/project/kansai-international-airport-terminal*. [Online] [Accessed 10 January 2020].

Rogers, L., 1994. Lyons Made (Aerogare Lyons). *RIBA Journal,* Issue August, pp. 32–36.

Ronald, V., 2012. *forgottenairfields.com/*. [Online] [Accessed 20 Nov 2018].

Root, A. L., 1905. *Gleanings in Bee Culture*.

Russo, C. J., 2015. *The Art of the Airport Tower*. Washington, D.C.: Smithsonian Books.

Sabar, R., 2009. *An Evaluation of the Provision of Terminal Facilities for the Design of Low Cost Airport Terminals*, Cranfield: Cranfield University.

Sant'Elia, A., 1914. *Futurist Manifesto*, Milan: s.n.

Santala, S., 2015. *Laboratory for a New Architecture - The Airport Terminal, Eero Saarinen and the Historiography of Modern Architecture*, Helsinki: University of Helsinki.

Schonwetter, C., 2005. *Airport Design*. Koln: raif daab gmbh.

Scientific Institute for Air Transport, Technical College, Stuttgart, 1938. *Aerodromes, Their Location Operation and Design (Translated from the German)*. London: Sir Isaac Pitman and Sons Ltd.

Scott, F., 2008. *On Altering Architecture*. Abingdon: Routledge.

Selig, N. C., 2013. *Lost Airports of Chicago*. Charleston: The History Press.

SFO, 1960. *SFO Museum, Aviation Museum and Library Collection*. [Online] Available at: https://www.flysfo.com/museum/aviation-museum-library/collection/15062 [Accessed 8 June 2017].

Sherwood, P., 2001. *Heathrow, 2000 Years of History*. Stroud, Gloucestershire: Sutton Publishing Limited.

Sims, C. A., 1932. *British Aviation Illustrated*. London: A. & C. Black Ltd.

Spender, S., 1933. *The Landscape near an Aerodrome*. [Online] [Accessed 21st June 2018].

Stevenson, G. M., 1972. *The Politics of Noise*. Belmont, California: Duxbury Press, Wadsworth Publishing Company, Inc.

Stewart, R., 2006. Interchange Now, Manmade Modular Megastructures. *Architectural Design*.

Stewart, R., 2016. The Smart Airport. *International Airport Review,* Issue 20th October.
Stoller, E., 1999. *The TWA Terminal*. New York: Princeton Architectural Press.
Stratford, A., 1974. *Airports and the Environment*. 2nd ed. London: The Macmillan Press Ltd.
Stroud, J., 1956. *Famous Airports of the World*. London: Frederick Muller Ltd.
Stroud, J., 1973. *The World's Airports*. London: The Bodley Head, Putnam World Aeronautical Library.
Stroud, J., 1980. *Airports of the World*. London: Putnam & Company.
Stuart, J., 1940. The Washington National Airport. *Pencil Points*, pp. 603–613.
Szurovy, G., 2003. *The American Airport*. St, Paul: MBI Publishing Company.
Talbert, A. E., 1953. *Famous Airports of the World*. New York: Random House, Inc.
TeNeues, 2005. *Airline Design*. Dussledorf, New York, West Byfleet, Paris and Madrid: TeNeues Publishing Corporation.
Texas, S. H. A., 2019. *tshaonline.org*. [Online] [Accessed 1 February 2020].
The Architect & Building News, Editor, 1939. Birmingham Airport, Architects Norman & Dawbarn. *The Architect & Building News* (18th August), pp. 188–192.
The Architect and Building News, 1939. Birmingham Airport, Architects Norman & Dawbarn. *The Architect and Building News* (18 August), pp. 188–192.
The Beauty of Transport, E., 2017. *thebeautyoftransport.com*. [Online] [Accessed 20th June 2019].
The Environmental Protection Agency, Hong Kong Special Administration Area, 2005. *Hong Kong's Air Quality Objectives*. [Online] Available at: http://www.epd.gov.hk/epd/english/environmentinhk/air/air_quality_objectives/air_quality_objectives.html [Accessed 11 May 2017].
The Government of Hong Kong, Special Administration Region of the People's Republic of China, 2017. *Methodology Adopted for Assessing Aircraft Noise Impact*. [Online] Available at: http://www.cad.gov.hk/english/ac_noise.html [Accessed 10 May 2017].
The Institution of Civil Engineers, 1970. *World Airports, The Way Ahead, Proceedings of the Conference, 1969*. London: The Institution of Civil Engineers.
The Institution of Civil Engineers, 1973. *Airports for the 80s, Proceedings of the Fourth World Airports Conference, 1973*. London: Thomas Telford.
The Institution of Civil Engineers, 1976. *Airports, The Challenging Future, Proceedings of the 5th World Airports Conference, 1976*. London: Thomas Telford.
The Institution of Civil Engineers, 1980. *Airports for the Community, Proceedings of 1987 Conference*. London: Thomas Telford.
The Institution of Civil Engineers, 1988. *Airports for People, Proceedings of the 8th World Airports conference, 1987*. London: Thomas telford.
The Institution of Civil Engineers, 1991. *Airports and Automation, Proceedings of the 9th World Airports Conference*. London: Thomas Telford.
The Institution of Civil Engineers, 1994. *Airport capacity and development funding, Procceedings of the 10th World airports Conference*. London: The Institution of Civil Engineers.
The Photographers Gallery, 1997. *Airport*. London: The Photographers Gallery.
The Science Museum UK, 2015. *Tim Boon Blog*. [Online] Available at: https://blog.sciencemuseum.org.uk/music-noise-and-silence/ [Accessed 13 Dec 2017].
Thomas, L., 1927. *European Skyways, The Story of a Tour of Europe by Airplane*. Boston and New York: Houghton Mifflin Company.
Thomas, P. E., 1937. *The RIBA Exhibition of Airports and Airways*. London: RIBA.
Thomas-Emberson, S., 2007. *Airport Interiors, Design for Business*. Chichester: John Wiley & Sons Ltd.
Treibel, W., 1992. *Die Deutschel Luftfart - Geschichte der Deutchen Verkehrsflughafen*. Bonn: Bernard & Graefefe Verlag.
U.S. Department of Commerce, Civil Aeronautics Administration, 1952. *Airport Planning*. Washington, D.C.: U.S. Department of Commerce.
Uffelen, C., 2012. *Airport Architecture*. 1st ed. Berlin: Braun Publishing AG.
U.S. Army Air Service, 1923. *Boston Public Library*. [Online] [Accessed 20th July 2017].
Veale, S. E., 1945. *To-morrow's Airliners, Airways and Airports*. London: The Pilot Press Ltd.
Vidal, L., 2013. Terminal 2A Heathrow. *Architecture Daily,* Issue 07 August.
Vines, M., 1991. *Birmingham International Airport*. Shrewsbury: Airlife Publishing Ltd.

Vitra Design Museum, 2006. *Airworld.* 2nd ed. Weil am Rein: Vitra Design Stiftung gGmbH.
Voigt, W., 1999. *Flughafen Hamburg 1929 - 1999*. Zurich: Verlag Niggli AG.
Von Gerkan, M. &. P., 1995. Aerial Drama. *The Architectural Review,* Issue February, pp. 56–59.
Voogt, A. d., 2007. *Heliodrome Architecture*. Rotterdam: 010 Publishers.
Vries, M. d., 2011. *Brilliant on the Skyline, General Aviation Terminal Schiphol*. Amsterdam: Sun Architecture Publishers.
Waugh, J. C., 2004. *Kansai International Airport, Airport in the Sea*. New York: Children's Press.
Welch, A., 2016. *e-architect.* [Online] Available at: https://www.e-architect.co.uk/turkey/new-istanbul-airport-terminal [Accessed 10 September 2017].
Welsh, J., 1994. On a wing and a layer. *RIBA Journal,* Issue July, pp. 22–29.
Wheat, G. S. E., 1920. *Municipal Landing Fields and Airports*. New York and London: G.P. Puttnam's Sons.
Wood, J. W. & L, C. L., 1940. *Airports, Some Elements of Design and Future Development*. New York: Coward- McCann Inc.
Woodley, C., 2014. *Gatwick Airport, The First Fifty Years*. Stroud, Gloucestershire: The History Press.
Wragg, D., 2009. *The World's Top 500 Airports*. Sparkford, Yeovil: Haynes Publishing.
Wright, A. J., 1979. *British Airports*. 5th 1996 ed. Shepperton, Surrey: Ian Allen Publishing.
Wright, A. J., 1992. *World Airports*. Shepperton, Surrey: Ian Allen Ltd.
Wright, F. L., 1935. Broadacre City: A New Community Plan. *Architectural Record, 77,* pp. 249–253.
Wright, F. L., 1958. *The Living City, Broadacre City*. Bramhall House: Bramhall House, New York.
Wright, H. M., 1929. Sky Harbors of the Future. *Popular mechanics,* Issue December, pp. 970–975.
Wright. W.,1908 Letter to Octave Chanute- US Library of Congress
Wright, W., 1919. The Future of Civil Flying. *Aviation and Aeronautical Engineering,* 1(January), p. 676.
Zoeller, H., 2019. *http://hugojunkers.bplaced.net/*. [Online] [Accessed 30 Nov 2019].
Zukowsky, J., 1996. *Building for Air Travel*. Munich & New York: The Art institute of Chicago and Prestel Verlag.

INDEX

Note: Page references in *italics* denote figures and in **bold** tables.

9/11 attacks 301

A380 aircraft 183, 247, 286
ACT Engineering consultants (Thailand) 255
aerodrome *3*, 21, 27, 44
Aeronautics Branch of the Department of Commerce 39
Aéroport de Paris (ADPi) 119, 286
Aerotropolis 14, 161, 232, 239, 244, 247–248, 251–252, 300, 307, *308*, 309, 315
Agache, Donat-Alfred 3, *4*
Airbus A 350 245
Airbus A 380 221, 245
Aircraft (Corbusier) 5, 18
Air France *34*, 184–185
Airline Deregulation Act of 1978 83
Airlink 110, *110*
Airport Cities 14, 78, 167, 172, 190, 239, 294, 322
airport commercial design 112, 324
airport design 1; and art movements 18; evolutionary phase 1940–1959 77; post WW2 *78*
airport master plans 327–335; 1954–1956 332, *332*; 1970s 333, *333*; 1980 334, *334*; 2016 and 2030+ 335, *335*; Airplane landing Field 1922 *327*; American *328*; Europe 1932 328; Europe 1939 330, *330*; European *328*; post-war 331, *331*; USA 1919 and 1922 327; USA 1930 328; USA 1939 329, *329*
Airports (Wood) 9
air traffic movements (ATMs) 290, 314
Alaska Airlines 190, 245
Al Maktoum, Sheikh Rashid bin Saeed 286
American Airlines 78, 83–84, 121
American City Magazine 30
Amsterdam Schiphol Airport (AMS) 80, 156–172; growth 1920–2005 *156*; growth in size and traffic volumes **157**
Andreu, Paul 179
apron drive airbridge 77, 118, 323
Architects' Climate Change Network (ACAN) 312
The Architects' Collaborative (TAC) 134
art movements, and airport designs 18
Arup 218, 265, 271, 282
Atkins 213, 265
Atlanta Hartsfield (ATL) 192–203; aerial view *192*, *197*, *201*; aerial view of Mid-Field Terminal and Satellites 199, *199*; airfield layout in early 1980 *198*; Candler Field *193–195*; Eastern Terminal and Satellites *201*; End-Around Taxiway *203*; Existing Master Plan 2015 *202*; Jet Age terminal with second parallel runway *196*; linear satellite concept 264; Long Range Development Plan to 20-Year Time Frame *192*; Master Plan Evolution *200*; 'Midnight Merry-Go-Round' – 'Hubbing' 194, *195*; Retail Atrium 2016 *203*; Terminal Layout in early 1980 *197*; underground automated people mover maintenance area and station *202*
automated people mover (APM) 119, 198, 199
Aviolanda Aircraft Construction Company 168

B757 aircrafts 231
BAA 112, 117, 232, 265
Baltimore Washington International Thurgood Marshall Airport 89
Bangkok Suvarnabhumi (BKK) 253–258, 325; 2030/2040 master plan *253*; aerial view *254*; aircraft maintenance facilities 254; airside retail in magenta *257*; alternative location for second terminal in yellow *258*; Bangkok 2006 airside retail in magenta *257*; capacity of 254; cross section through terminal *256*; design 253; master plan 2030/2040 *253*; phase 1 airport layout *254*; second terminal in blue *258*; sectional model of terminal *256*; sectional perspective through pier *257*; terminal check-in and departures level 4 *255*; terminal gate layout *253*; terminal lower boarding and arrivals level 2 *255*; upper level of pier *256*
Banham, R. 20
Beddington and Waddon Aerodromes 27
Beijing Capital (PEK) 281–284, *284*; concept diagram *283*; interior *281*; master plan *282*; mid-field terminal 282; sections through satellite *281*; Terminal 1 282; Terminal 3 281, *281*, 282
Beijing Daxing (PKX) 295–300, 325; aerial view *295*; interior *299*; location plan *300*; maintenance, repair and overhaul (MRO) of aircraft 300; master plan 295, *295*, *300*; roof *299*; roof plan *296*; 'toast rack' concept 295; by Zaha Hadid Architects © Zaha Hadid Architects *297*, *298*
Benoist, Thomas 15
Berlage, Hendrick 162
Berlin Flugplatz Johannisthal 12, 14, *14*, 21, *21*, 323
Berlin Tempelhof (THF) airport 18–20, 54–66, *54–67*, 118; hangars 54; John Walter Wood on 62; Lowell Thomas on 56; Norman Foster on 62; second terminal 58–59; site plan 58, *58*; Stedman Hanks on 56–57; underground station 54
Biernacki, W.O. 95
Black, Archibald 30
BOAC 118
Boeing 737 aircraft 231
Boeing 747-400 245
Boeing 747 aircraft 119, 128, 175
Boeing 777–800 aircrafts 245
Boeing 900 aircrafts 245
Booker, Grant 320
British Airports Authority 190
British Airways 95, 245, 259–260, 265, 268; advertisement *48*; Concorde service *275*, 301; self-service flow-through concept 270

British Airways Helicopters (BAH) 110
British Caledonian Airways 110
British European Airways (BEA) 95
Brooks, James 74
Buildings for Air Travel (Voight) 23
Build–Operate–Transfer model 290
Burke, Ralph 80
Buro Happold 318

Canberra Aerostatic Station 3, 15, *15*
canyon concept *267*, 267–268
Cathay Pacific 235, 239
Centralny Port Komunikacyjny (CPK) 301, 318–322; aerial view *319*, *320*; alternative configurations for terminal and multiple piers *319*; atrium linking terminal and rail interchange *322*; dual-use infrastructure 318; forecourt and main entrance *321*; interior view of main entrance *321*; location plan *318*; master plans 318, *319*; Ministry of Defence in cooperation with 318; railway interchange platform level *322*; railway station 320
Chanute, Octave 11
Chapman Taylor 265
Chicago O'Hare (ORD) 78–87, *80–87*, 118; airbridges 81–82; development plan *86*; finger pier concept 81–82; master plan *80*; radial runway pattern 80
Civil Aeronautical Authority 8
commercial flying 18, *77*
commercial revenue 190, 246
Concorde 119, 301
Corbusier, Le 4–5, 18, 246
Covid-19 pandemic 280, 301–303, 307, 310
Cox Architecture 314–315
Croydon Aerodrome plan 1920–1928 *28*
Croydon Airport, London 18, *19*, 27–38, *27–38*

Dallas Fort Worth (DFW) 119, 173–177; aerial perspective *174*; cross section through original design terminal complex *175*; Dallas Fort Worth Regional Plan *174*; End–Around Taxiways *177*; future aerial view of airport in 2001 *175*; International Terminal *173*; master plan *174*; opened in 173
Davies, Mike 265, 267–269
Dearborn Ford Airport 18, 25–26
De Havilland Comet 118
Delano & Aldrich 69, 121
Dellaert, Jan 160, 164
Delta Airways 198
Denver International (DEN) 228–231; airside underground transit linking satellites *230*; architect's sketch of tent roof concept *228*; Boeing 737 taxiing *231*; central landside atrium in terminal *230*; DCV baggage system 229; Fentress Architects 228; location of 228; master plan 2030 *228*; night-time view of *229*; pedestrian bridge over taxiway 231; roofscape of Jeppesen Terminal *229*; roofscape with Rocky Mountains *228*; 'toast rack' configuration 229
Deutsche Luft Reederei 21
Disney 149, 155
Disneyland 119
Douglas Aircraft Company 80
Douglas DC3 82
Douglas DC4 77, 81
Dower, John 31
dry boarding 43, 323
DSSR 265
Dubai International (DXB) 285–289, 325; Concourse A *288*; Emirates first class lounge *289*; layout of underground terminal *285*, *287*; retail mall in satellite concourse *286*; site plan *287*; Terminal 3 285; Terminal 3, cross section *288*; Terminal 3 entrance canopy *287*; Terminal 3 underground reclaim *287*; underground check-in concourse *286*
Dubai World Central (DWC) 245–246, 302, 307–311; and aerotropolis *308*, *310*; Covid-19 impact 310; Dubai Metro *311*; master plan *308*, *311*; model *307*, *309*; retail mall *310*; satellite *308*; terminal building in foreground and satellites in background *309*
Duintjer, Marius 166
Dulles Airport Access Road (DAAR) 139–140, *140*
Dulles Toll Road (DTR) 140

Eames, Charles 119, 140
Eastern Airlines *77*, *194*
economy fares, and Pan American Airways 78
Edmunds, James R. 88
Engler, Klaus 54
Engler, Paul 54
environmental limits, and airport design 302
environmental protest movement 118
European Skyways (Thomas) 29
explosives detection systems (EDS) 291–293

Farman Goliath Aircraft 18, 51
Farrell, Terry 248
Federal Aviation Agency (FAA) 140
Fentress Architects 228
finger pier concept *77–82*
First World War 2, 14, 18, 21, 27, 30
'Flight' (Brooks) 74
Flight Magazine 26
Flughafen Königsberg-Devau (KGD) 18, 22–24, *22–24*
Ford, Henry 3, 26
Ford Airport Dearborn 19, 25–26, *25–26*
Ford Motor Company 3, 26
Foster, Norman 8, 62, 103, 119, 215, 217, 219, 232, *234*, 281–282, *283*, 303, 304, 313
Foster+ Partners 232, 279, 282, 312, 318, 320
Foundation Manifesto 2
Frankfurt Air show 12
Frankfurt Internationale Luftschiffahrt Augglellung (ILA) *13*
Froesch, Charles 128
Frontier Airlines 229

Gatwick Airport, London *see* London Gatwick (LGW)
Gatwick Airport Limited 117
Gatwick Beehive 43, *43*, *45*; *see also* London Gatwick (LGW)
Gibberd, Frederick 95
Gimpo airport 247–248
Gleanings in Bee Culture 11
Global Alliances 244–246
Gloucester Road Tube Station 98
Göering, Hermann 59
Gogerty, H. L. 77
Greater Los Angeles Airport: master plan 1944 for *148*

'Great Hall' concept 269–280, *270*
Griffiths, Paul 286
Grimshaw Architects 291
Gropius, Walter 5, 134, 246

Hanks, Stedman 19, 56–57
Haptic Architects 291
Harbour, I. 260
Harrison, William H. *6*
Heathrow Europa Terminal *see* London Heathrow (LHR)
Hellmuth, Obata, Kassabaum and Brodsky (HOK) 175, 265
Hellmuth, Yamasaki and Leinweber 102
Henket, Hubert-Jan 172
Hoar, Marlow and Lovett 43
Holm, David 314
Hong Kong, Chek Lap Kok (HKG) 232–239; aerial view of SkyCity and SkyPier ferry terminal *238*; air, road, rail and sea interchange *236*; bridge link over taxiway *237*; Cathay Pacific Lounge 2018 at *235*; Chek Lap Kok Island *237*; departures and arrivals concourses *235*; departures level *233*; design 232–233; ferry routes from PRD *237*; gate-lounge concept *234*; HKIA SkyPier 2003 passenger flow layout *237*; Lantau Island *237*; with later phase short haul satellite *237*; Long Section through terminal and interchange *234*; master plan 2030 232, *232*, *239*; roof design *234*; sea–air interchange *232*; section through pier *234*; site plan *233*; Sky City land *232*; SkyPier 232, *232*, *236*; with SkyPlaza *236*; Tung Chung Aerotropolis *238*; Tung Chung Island *237*
Hong Kong Airport Express 235
Hopp, Hans 22
'Hub and Spoke' concept 77, 198
Huffman Prairie 9–12

iGA Istanbul (IST) 290–294; aerial view of *291*, *292*; airport-related operations 294; ATM capacity 290; departures forecourt and main entrance *291*; design 291; destination-coded vehicle (DCV) baggage handling system 291; duty free and retail mall *293*; explosives detection systems (EDS) 291–293; floor plan *294*; interior of pier *293*; location plan *290*; main entrance *293*; master plan *290*; Public–Private Partnership (PPP) project 290; Skytrax 294; view at dusk with forecourt and car parks *292*
Incheon Free Economic Zone (IFEZ) 251, 252
International Style Architecture 156
Istanbul Terminal Manoeuvring Area (TMA) 290

Jackaman, Morris 43
Jahn, Helmut 84, *85*
Japanese Airport Authority 190
Jeddah, King Abdulaziz (JED) 1981 209–214; aerial CAD image of new mid-field terminal *213*; aerial views of new mid-field terminal *214*; Hajj Terminal *211–212*; Hajj Terminal, plan, and section *209*; initial master plan *210*; site plan of the new mid-field terminal for 30 mppa 2019 *213*
JetBlue Airways 134

Kahn, Albert 26
Kasarda, John 236
Kidder Smith, G.E. 108
King Abdul Aziz International Airport in Jeddah 210
Kings Cross Underground Station fire 269
Kipling, Rudyard 5
KLM 156, 161–162, 167
Koolhaas, Rem 161, *161*
Korean War 88
Kosina, Heinrich 54
Kuala Lumpur (KUL) 240–243; Jungle Boardwalk *241*, *242*; master plan 2017 240, *240*; master plan 2050 *243*; model of KLIA T2 *242*; satellite *241*; satellite transit station *241*; Terminal 1 241; Terminal 2 241
Kurokawa, Kisho 240

Labro, Georges 51–52
La Guardia, Fiorello 68
Latham, Hubert 12
LAX Theme Building 118
'layer cake concept' 166
LeHigh Airport Competition 5, *6*, 43, *77*, 121
Leipzig airport restaurant 18, *19*
Lindbergh, Charles 49, 100, 193
Lindsey, Gina Marie 155
linear satellite concept 95, *95*
London, Stansted (STN) 215–219; cross section concept *217*; site plan *216*; stage 1 expansion of terminal *215*
London Gatwick (LGW) 19, 43–48, *43–48*, 79, 106–117; circular satellite 109–110, *110*; Gatwick North 112; major extension 108–109; master plan with second runway and third terminal *117*; North Terminal 111, *112*
London Heathrow (LHR) 79, 92–99, *92–99*, 264–284; aerial perspective *277*; airside lounge 274, *274*; architectural team 265; axonometric of Hub Infrastructure *277*; British Airways Concorde *275*; canyon concept *267*, 267–268; competition design source *270*; concept design *279*; cross section *276*, *280*; design 265; expansion of *98*, 98–99; exploded view *273*; floor plans *272*, *279*; food and beverage *274*; 'Great Hall' concept 269–280, *270*; group rapid transit (GRT) 278; Heathrow Expansion Programme *264*, 264–284, *280*; linear satellite concept 95, *95*; 'Magic Carpet' concept 265–267, *266*; master plan 264, *264*; motorway under terminal *264*; multi-level scheme 271; passenger flow logic *276*; personal rapid transit (POD) *278*; Piccadilly Underground 98; section and site plan *275*; 'Skywalker' 271, *271*; T5 landscape photos *269*; T5 landscaping concept *269*; Terminal 2 at 279; Terminal 5 sections *264*; 'Tied-Arch' concept *268*, 268–269; 'toast rack' model *264*, 265
London Hounslow Heath 49
London Makes a Mistake (Black) 30
long-span roof structures 324
Los Angeles International Airport (LAX) 118, 147–155; aerial view *147*, *149*, *154*; aerial view of circular terminal and satellites 1952 *150*; aerial view of early master plan *151*; bird's-eye view of airfield *150*; Circular Terminal 1952 *149*; satellite with parallel parking and front and rear door boarding *152*; terminal and satellite conceptual plans *153*; Terminal and Satellite Plan *150*; Theme Building *147*, 147–148, 155; Theme Restaurant 1962 *155*; welcome signs at main entrance to *155*

Los Angeles World Airports 155
low-cost carriers (LCCs) 190, 244–246, 301
low-energy displacement ventilation system 263
Luckman, Charles 151–152

Madrid Barajas (MAD) 259–263; compare with Heathrow Terminal 5 260; isometric plans of terminal and satellite *261*; low-energy displacement ventilation system 263; 'Magic Carpet' roof concept *259*; master plan *259, 260*; model of satellite *262*; South elevation and section through terminal *260*; views of terminal and satellite *262*
'Magic Carpet' concept 265–267, *266*
Mahlberg, Paul 54
Malaysia Airlines 245
Marinetti, Filippo Tomaso 2
Mendelson, Eric 2, *3*
Mexico City Airport (MEX) 301, 303–306; aerial view *304, 305*; alternative views of departures lounge *306*; Covid-19 pandemic impact 303; cut-away CGI of pier *305*; departures forecourt *304*; departures lounge *304, 306*; development of 303; master plan *303, 304*; winning competition design *305*; X-shape 303–304
minimum connection times (MCTs) 244
Moses, Robert 69
Motorflugplatz Johannisthal-Andershof 21
Mott MacDonald 232, 265
Murphy Jahn Architecture 255
Musée de l'Air et de l'Espace 52

National Register of Historic Places 134
Netherlands Airport Consultants (NACO) 282, 295, 300
The New Architecture and the Bauhaus (Gropius) 5
New Deal 68
New Orleans Sushan (Lakefront) (NEW) 20, 39, *39–42*
New York Idlewild (JFK) 77–78, 118, 120–137, 325; aerial view of Airport City 1962 *132*; aerial view of redeveloped airport *137*; American Airways Terminal *'Worldport'* at *125*; artists impression of The International Arrivals Building *124*; early centralised terminal concept *123*; image of future Terminal 2 *137*; International Arrivals Building (IAB) 118, *125*; master plan *122*, 124, *124, 133*; Master Plan Evolution 1947–1962 *122*; Pan American's Worldport at 118; Pan American Terminal *125, 126–128*; temporary terminal, 1947, prior to opening *123*; Terminal City concept 120, *120*; terminals **133**; TWA Terminal *120*
New York LaGuardia (LGA) 68–76, *68–76*; Art Deco style Landplane Terminal *71*; *'Flight'* 74; rebuilding 76
New York Municipal Airport *see* New York LaGuardia (LGA)
Nordic Office of Architecture 291

Official Secrets Act 92
OMA 161
One World Alliance 235, 245, 259, 260
Operational Readiness and Airport Transfer (ORAT) 290
Osaka Kansai (KIX) 220–227; 1994 aerial view of terminal *225*; 1994 model of roof structure *226*; 1994 showing curved and tapering pier and transit system *226*; 1994 to Kobe Sea Transfer *227*; aerial view after expansion *220*; aerial view of original artificial island *221*; landside canyon *221*; Sectional Elevation and Floor Plans *223*

Pan American Airways 78, 120
Paris Charles de Gaulle (CDG) 178–188; aerial view *180, 183*; aerial views of TGV Station 1994 *186*; Future Terminal 4 *178*; master plan *178*; photomontage of proposed replacement satellite *183*; proposed T4 *188*; Rail/Air Interchange *186–187*; Terminal 1, aerial view *181–182*; Terminal 1 plans *180*; Terminal 2 *184–185*; Terminal 2A, 2B, 2C, 2D and circular satellite *184*
Paris Le Bourget (LBG) 18, *19*, 49–52, *49–53*
Parkways 69–70
Parseval Company 14
Pascall + Watson 265
Pearl River Delta (PRD) 235–236
Pereira, Robert 149, 151, 155
Peterkins, John B. 75
Pheil, Abram C. 15
Piano, Renzo 222, *222*
Popular Science 121
Port Authority of New York and New Jersey (PANYNJ) 134
Powers, Alan 115
Prokosch, Walter 128
prototype interchanges 12–15, *13–17*

Rail/Air Interchange 2
Red Sea Development Company 312
Red Sea Umluj Tabuk 312–313; aerial view *312*; design for 312; interior of waiting lounges *313*; internal view of *313*
Reich Air Ministry 59
'Reinventing the Airport' (Foster) 232
Rheims Air show 12
Richard Rogers Partnership 265
Rio de Janeiro, Santos Dumont 18, *19*
Rocky Mountains 228, *228*
Rogers, Richard 220, 259, 260
Rogers, Stirk, Harbour 260, 265
Romero, Fernando 303
Roosenburg, Dirk 161–162
Root, Amos I. 11
Royal Aeronautical Society 19
Royal Flying Corps 27
Royal Geographical Society 5
Royal Saudi Air Force 210
Ryanair 245

Saarinen, Eero 8, 79, 103, 119, 128, 134, 139, 140, 142, 218, 296
Safdie, Moshe 149
Sagebiel, Ernst 59, *60*
San Francisco (SFO) 79, 90, *90–91*
Sant'Elia, Antonio 2
Schiphol Airport: 1967 on opening of new airport and terminal *158*; aerial photograph of *167*; aerial view 1928 *159*; airside view of first terminal 1928 *162*; Amsterdam Schiphol 1929 *163*; city master plan *172*; Concourse F with 'overwing' airbridges 2007 *169*; Control Tower1928 *162–163*; Flower Shop *171*; Future Master Plan *158*; layout plan

of first terminal 1928 *162*; Library *170*; location of Schiphol Airport with grass airfield 1930 *158*; master plan *158*, *160*; as military airfield *157*; with new asphalt runways 1938 *159*; official opening in 1967 *160*; Railway Station *167–168*; Rijksmuseum *170–171*; Schiphol Plaza *167–168*; second control tower erected in 1950 *164*; second terminal and new control tower *164*, *165*; Terminal A, landside view *172*; Terminal A layout plan *171*; wartime airfield layout with concrete runways *158*
Scott, Brownrigg & Turner 99, *99*
seaplanes 70, *77*
Second World War 18, 39, 77, 80, 88, 92, 164
Seoul Incheon (ICN) 247–252, 325; aerial view of T1 *247*; aerial view of T2 phase 1 with T1 *248*; and Aerotropolis, Yeongjong, Cheongna and Songdo *251*; CGI of second terminal *250*; design 247; location plan *251*; master plan 247, 252, *252*; rail network *249*; rail transfer interchange 248, *248*; to Seoul Gimpo AREX rail link *249*; Terminal 1 and 2 layouts *249*; Terminal 2 interior *250*
Simms Road Station 12
Singapore, Changi (SIN) 204–208; Airport Hotel *207*; Atrium Cinema *206*; Butterfly Garden *206*; Cactus Garden *206*; Check-in Concourse *207*; Future Master Plan *204*; The Jewel 2019 *207–208*; Koi Carp Pond *206*; Orchid Garden *204*, *206*; roof top swimming pool *207*; Terminal 1 *205*
Skidmore, Owings and Merrill 124
SkyCity 239
SkyPier 232, *232*, 235, 236
Sky Team Alliance 156
Skytrax 294
'Skywalker' 271, *271*
social mobility 301
SOM 209–210, 218
Southwest Airlines 190, 229
Special Purpose Vehicle (SPV) Company 290
Speer, Albert 59, *60*, 63
SPT Airlines' Benoist Type XIV *17*
Star Alliance 244
Stevens & Wilkinson, Smith Hinchman & Grylls (SmithGroupJJR) 199
St Louis Lambert (STL) 79–80, 100–105, *100–105*
Stout Metal Airplane Company 3
St. Petersburg Tampa Airboat Lines 15
Super Jumbo 220, 245
Suvarnabhumi Airport 253; *see also* Bangkok Suvarnabhumi (BKK)

TAMS consultants (USA) 255
Thomas, Lowell 29, 56
'Tied-Arch' concept *268*, 268–269
Tokyo Narita 190
Towards a New Architecture (Paris 1923) (Corbusier) 4–5
TPS 265
Trans World Airlines 128
tricycle undercarriages, aircrafts with 77, 81
TWA Flight Center 120, 128, *129–132*

United Airlines 77, 83, 229
United States Army Signal Corps 11
US Air Force 63

Vidal, Luis 279
Virgin Galactic Spaceport 21
visiting friends and relatives (VFR) travel 301
von Gerkan, Meinhard 219

Walter Prokosch 128
Washington Dulles (IAD) 8, 118, 119, 138–146; aerial view *146*; after 1996 Extension *145*; Control Tower and Control Tower Restaurant Tulip Collection *146*; cut-away perspective of terminal *142*; interim update with later satellites served by mobile lounges *138*; lower-level arrivals and upper-level departures plans of terminal *143*; model of terminal and forecourts *141*; section through terminal and control tower *142*; two-level forecourt and pedestrian underpass *140*; view of check-in concourse after extension *145*; view of check-in concourse at opening *144*; view of 'Planemate' mobile lounge loading aircraft *144*
Washington Friendship (Baltimore) (BWI) 79–80, 88–89, *88–89*
Washington National Airport 89, 139
Waterloo Air Terminal 95
Western Sydney Aerotropolis 315
Western Sydney International (Nancy-Bird Walton) (WSI) 314–317; aerial view of *315*; artist's impression of developed concept for *316*; construction 316; master plan *314*, 315; projected flight preference routes *317*; projected noise contours for B737-800 *317*; terminal design 315
West London Air Terminal 78, 95
Wood, John Walter 9, 62, 329
Works Progress Administration (WPA) 68
Wright, Orville 9
Wright, Wilbur 2, 11
Wright brothers 1, 9–12
Wright Field, Huffman Prairie, Dayton, Ohio 9–12, *11*
Wright Flyer III 11
Wright-Patterson Air Force Base 11
WSP 265

Yamasaki, Minoru 102
Yorke, Rosenberg & Mardall (YRM) 106, *108*, 111, *112*, 115, 260, 265, 268–271

Zaha Hadid Architects 296, *296–298*, 314
Zigarre airship 14
Zimmerman, A. C. 6